Real Analysis for the Undergraduate

Matthew A. Pons

Real Analysis for the Undergraduate

With an Invitation to Functional Analysis

with Illustrations by Robert F. Allen

Springer

Matthew A. Pons
North Central College
Naperville, IL, USA

ISBN 978-1-4939-4649-5   ISBN 978-1-4614-9638-0 (eBook)
DOI 10.1007/978-1-4614-9638-0
Springer New York Heidelberg Dordrecht London

Mathematics Subject Classification (2010): 26-01

Printed on acid-free paper

Springer is part of Springer Science+Business Media (www.springer.com)

For Ann and Truman

Preface

Like many mathematical textbooks, this project evolved from a set of course notes. These notes were developed and refined over a 5-year period and represent my interpretation of a first course in analysis. Much of the material comes from the standard undergraduate canon including an axiomatic or constructive exploration of the real number system, numerical sequences and series, limits and continuity, differentiation, sequences and series of functions, and some introductory form of integration. In addition I have included material on measure theory and the Lebesgue integral, and a brief invitation to functional analysis. Although these advanced topics are not typically introduced until graduate study, I find that with the proper presentation they are perfectly manageable for undergraduates wishing to delve deeper into the subject matter. My primary reason for the inclusion of this advanced material is to prepare students for graduate study, and to also help them experience the evolution of mathematics. While each chapter covers a topic central to a beginning course in real analysis, the last section in each chapter introduces a topic from functional analysis which is derived from the core chapter content in a natural way; these sections are developed alongside a concrete topic in order to provide grounding. This takes the shape of a particular idea, complete normed linear spaces, eigenvalues, invariant subspaces, or focuses on a particular example or class of examples, ℓ^p spaces, continuous function spaces, and L^p spaces over general measure spaces. Taken as a whole, these sections provide a nice connection to linear algebra and point to a thriving area of mathematical research. The Epilogue then provides a brief road map to further points of interest in the realm of functional analysis.

When I began this project, I first mapped out the sections within each chapter. This led to nine chapters which built on each other, but seemed to lack a sense of harmony. As a reworking, I began to think about the overall story line of the text, of each chapter, and then of each section. The goal was to tie these story lines together in a cohesive manner that conveys the story of real analysis. This activity led to five core beliefs which have influenced every word written here.

- An introductory text should be written for students. There are a great many analysis texts available on the market. Some classical texts are too difficult for today's audience (which has changed drastically in the last few decades) and many of the newer expositions are too watered down. Others seem to strike a balance between rigorous exposition and intuitive reasoning, but fall short when it comes to adequate exercises and examples. In thinking about all of this, my goal in writing this text was to provide a thorough treatment of the necessary subject matter without sacrificing rigor, while shaping the presentation so that every student required to take a course in real analysis can succeed. Examples are frequent, discussion of new ideas motivates definitions, theorems, and proofs before formal statements and arguments are given, and the exercise sets are detailed and thorough. The discussion portions of the text are meant for the newcomer and there will inevitably be instances where an instructor thinks I'm belaboring a point. And while that may be the case, these explanations have arisen out of conversations with students about the material, both in the classroom and in the office hours. Undergraduates think differently about new material than a professional mathematician and my audience is comprised of undergraduates.
- A first course in analysis should, to some degree, focus on providing a theoretical treatment of the material from the first-year calculus sequence. I see this course as a means of coming full circle. Students have experienced calculus, their intuition has been developed, and they have been exposed to the many applications of differentiation and integration. To complete this experience, it is necessary then that they should understand why the things they have taken on faith are in fact true. To do this requires a thorough study of the concept of a limit, and it is necessary that they understand why it is the underlying construct for every topic in analysis. This can take many forms and I have chosen to focus on sequential limits. So, while I do strive to provide a concrete foundation for the differential and integral calculus, I do not believe that a text should be confined to developing this material in the same manner or order as it is presented in the first-year calculus sequence.
- Examples and exercises are the keys to understanding theory. The sections are written so as to provide most of the details for the ideas under consideration, but the reader is encouraged time and time again to pick up a pencil and check a calculation or complete the details of an argument. That being said, core ideas are spelled out in detail and major results are proved in full generality. It is my strong belief that core results should not be left as exercises as these are typically too difficult for many students at the beginner level.
- Students deserve to see how their mathematics courses relate to each other; this in turn leads to an understanding of how mathematics evolves as a discipline. Mathematics is at its most powerful when it incorporates knowledge from its various branches, and I find analysis to be particularly rich in this area as it depends so heavily on the algebra, topology, and geometry of the underlying domain space. The study of operators on a Hilbert space may be the most ideal example of such a lush and vibrant area of inquiry.

• The story of analysis is not over, so a text should not attempt to wrap it up in a tidy package. This again is where the introduction of functional analysis comes into play. The functional topics included are a generalization of the basic analysis topics they accompany and these sections are designed to produce more questions than they necessarily answer. This is the beauty of mathematics and any academic discipline. There will always be questions to ask and my point here is to help the students see that new questions are being asked not only about new topics but about old topics as well. When taken as a whole, these culminating sections provide a basic introduction to very abstract material.

As for prerequisites, the text assumes that students have experienced calculus up through sequences and series, which typically comes after differentiation and integration, and that they have had a formal course in proof writing, including basic logic and set theory, proof techniques (induction, proof by contradiction, and proof by contrapositive), basic function theory, cardinality, and some exposure to the nature of quantifiers. There are several instances where basic topics from linear algebra come into play, but these are only present in the culminating sections of each chapter and do not affect the core material of the chapters; these can be easily supplemented to the student who has not had a formal linear algebra course.

To the Instructor

This text is designed for use over the course of two semesters or quarters. I currently teach the majority of the material over two 10-week terms covering roughly Chaps. 1–5 and 6–9 in the two respective sessions. As mentioned earlier, the text is written so that each chapter culminates with the introduction of a topic from functional analysis so that the students can see how the topics evolve from the content of that particular chapter. However, I will be the first to admit that the natural order will be strenuous for most students. To prepare the students for the challenge I walk a smoother path.

I. *First Term*: Chapters 1–5 excluding the last section in each chapter;
II. *Second Term*:

    a. Chapters 6–7 excluding the last section in each chapter;
    b. Sections 1.5, 2.5, 3.4, 6.5, 7.5;
    c. Chapters 8–9;
    d. Sections 4.5 and 5.6 (if time allows).

For a particularly strong group or as an independent reading project, the natural order may be more manageable. And also please keep in mind that the culminating sections in each chapter are only meant for enticement; in fact, they are actually meant to raise just as many questions as they answer. The material in the last sections is not relevant to the core material in subsequent chapters, so doing things a little out of order should cause no confusion. However, be aware that I find the material in Sect. 7.5 to be the driving force behind introducing the Lebesgue integral at this level. I have found that the material can be introduced along a variety of paths, each of which has its own set of advantages; this provides some room for creating a course that suits a variety of needs depending on the level of the students and the time span of the course.

There are several places where I have varied from tradition. Some of these are simply personal preferences, but all have a pedagogical intent. A few of these are detailed below.

- I do not include a formal chapter on topology of the real number system. All of the topics typically outlined in such a chapter appear throughout the text as necessary. I find that incorporating this material into the text at various places allows for a smoother understanding rather than singling out the study of set structures in an isolated chapter. In Chap. 1 there is an introductory section with a discussion of open, closed, and compact sets. This initial discussion centers around open sets; closed sets are then defined as complements of open sets. The idea of a limit point is introduced in the second chapter where we revisit the notion of set structures via sequential criteria. Compact sets are therefore introduced in the first chapter as closed and bounded subsets of the real line and the characterizations in terms of sequential compactness and open cover criteria are given as equivalent notions. This is very atypical, but the closed and bounded criteria is easier to digest at first, though I do feel that this may detract from the importance of the Heine–Borel theorem.
- My basic treatment of most of the constructs of analysis is presented via a sequential approach. On the whole, this provides a unifying theme to the text and places sequences at the center of all our work. This is particularly relevant to our discussion of functional limits where the sequential definition is presented before the more traditional $\varepsilon - \delta$ definition. When discussing integrals, the sequential approach is presented as a secondary characterization.
- The last two chapters of the text provide an introduction to measure theory and Lebesgue integration. This material is not typical of an undergraduate text and these chapters are more dense than the earlier chapters, but I have taken care to present the material in a manner that is appropriate for the beginner.

As a final comment, the exercise sets are designed to be completed in full meaning that many of the problems are referenced later in the text. This does not mean that every student has to do every problem for submission. Rather, I find it helpful to create homework sets by selecting a few problems from a section without worrying too much at the future. Then, if I need a particular problem later in the course, I do not hesitate to assign it with a later homework set. This gives the student the opportunity to revisit early topics and I find repetition key at this point in their development. However, I do make it clear to students that the homework sets represent the minimal amount of preparation for exams and encourage them to investigate all the problems in each section.

To the Student

Be aware that what you are about to undertake represents several hundred years worth of mathematical inquiry and the topics are difficult. I don't say this to deter you, but my point is that you should expect to devote a significant amount of time to mastering the ideas covered here. There are a few keys to doing this. First, read the text carefully. Second, read the text often. Third, reread the text carefully. Mathematics takes time to sink in and allowing yourself ample time is necessary to fully digest the material. When you read, have a paper and pencil handy. You need to be involved with the ideas and sketching out the details of an argument or checking a calculation is the best way to invest yourself in the subject matter. I have devoted much time to writing clear and concise proofs, to developing and motivating ideas before stating definitions and theorems, and before giving proofs; I have spent many hours crafting problem sets that emphasize the key techniques used in this area of mathematics. All of this is to say that the text is designed *for you*. Your instructor has already learned the material and this text is not designed as a guide for him/her. It is designed as a guide *for you*.

Also, while this text is aimed at the beginner, there will inevitably be a moment when you need some topic clarified. In this moment, ask for help. I have strived to convey my thoughts clearly but this doesn't always translate as effectively as an in-person clarification. This is the nature of writing. I have one chance to express a thought in an effective manner, but that may not be the best explanation for your particular point of confusion. So, while I have been meticulous about details, it may be the case that I have inadvertently glossed over a point that causes you strife. In this moment, ask for help.

If you haven't already picked up on the fact that I am a believer that repetition is a key pedagogical technique, let me repeat. I believe in repetition. The only way to learn is to do and not just once. You should commit the definitions and major statements to memory. As a student I did this by writing out the statements and reading them several times a day. This way it doesn't feel like memorization but you have to find what works for you. For problems, you should work as many exercises as possible. In each section there are multiple problems that get at the same point or technique, and this variety is intended to be repetitive so as to emphasize the importance of a concept or technique.

You should also be aware of the fact that this is a writing course. Your homework sets should look more like short papers than like a calculus assignment. The proofs given in the text then have two purposes. They are included so as to demonstrate the validity of claims made. But they are also present to act as examples of good mathematical writing as you attempt to write your own proofs. Below are a few things to keep in mind.

- The words "show," "verify," "explain," "discuss," etc. all have the same meaning in this context. That is, they are all asking you to prove a specified statement. This may sometimes be just a sentence, but a proof nonetheless.

- Developing a proof is actually a two-step process. Don't expect to read a problem and immediately sit down and write out a proof. First, you need to sketch out the ideas and determine a logical path that takes you from hypothesis to conclusion. I recommend a notebook, almost like a journal, to keep your sketch work in. The second phase of this is then the writing process. You want to translate the ideas from the sketching phase into a clear argument and this presents its own set of challenges. This technique is demonstrated throughout the text as I have included much of the "scratch work" necessary for building a proof before presenting a formal argument. This has added substantial length to the text, but is present to aid you in understanding how mathematicians work. This is analogous to how writers, artists, musicians, film makers, and other creative individuals construct masterpieces.
- Your proofs should consist only of complete sentences with proper punctuation and grammar. And every problem, even if it is simply a computational exercise, should be written in this fashion (there are only a handful of computational exercises in the text). For some problems, e.g. those asking for examples with a specified property, a short explanation is all that is necessary, but this should always be included. As a simple rule of thumb, always supply proof. If the directions of a problem are unclear, ask your instructor.
- Do not abuse symbology, i.e. your proof should have more words than symbols. In fact, you should only use symbols when absolutely necessary. This is not a strict rule and you should consult with your instructor for their particular preference here. For example, I find it perfectly acceptable to write "$x \in A$" or "x in A," but I would be put off by the use of "$\forall \varepsilon > 0, \exists \delta > 0$" as a substitute for the phrase "for every $\varepsilon > 0$, there is a $\delta > 0$" in a formal proof.
- Read a section clearly and carefully before beginning a problem set. Often there are hints in the section for problems. In particular, proof techniques in a section are likely relevant to the problems in that section. Also, mathematics builds on itself, as does this text. Referencing previously completed homework exercises will likely occur and this is expected. To do this effectively, you have to remember things you have previously encountered. In addition, the problems are designed to be completed using *only* ideas from the current section or previous sections. This is for your own development even though there may be a later theorem which makes the argument extremely simple.
- The problems are not easy (though they vary in difficulty level) and you want to allow yourself plenty of time.
- Reread your work before submitting it for assessment. This will help you catch typos/errors and will also help you to solidify your arguments. If upon rereading you do not understand your argument, then there is a high probability that your instructor will not understand your argument. This is an encouragement to work early and often.

Finally, let me take a moment to indicate what makes an analysis course slightly different from your previous mathematics courses. In most areas of study, there are certain obstacles which make that particular discipline challenging. From my perspective, analysis presents two such obstacles. First, you have been trained to understand equality exceptionally well. However, analysis is about inequality. One reason for this is that it is often hard to tell when two quantities which are allowed to vary are equal, and the task is made easier by contenting ourselves with understanding when one quantity is less than another. For example, we will see in the first chapter that a real number a is equal to 0 if and only if $|a| < \varepsilon$ for every $\varepsilon > 0$. This should make intuitive sense as 0 is the unique real number which "separates" the positive and negative real numbers, and the statement above demonstrates that inequality can often tell us something about equality. Also, many of the constructs of analysis will deal exclusively with inequality, sequence convergence for example, as these ideas are really about being close to a fixed value rather than being equal to that fixed value.

The second obstacle will appear in the nature of our proofs. As we will deal primarily with functions, we will most often be working with at least two quantities simultaneously, a domain point and a range point. The goal then will be to *force* some outcome in the range by *controlling* the corresponding domain quantity. This will require us to make very explicit choices with respect to domain values and therein lies the challenge. The nature of such arguments may seem circular at first and the key is practice.

Acknowledgments

Like any significant project, there are many folks to whom I owe a significant amount of gratitude. First let me thank my instructors, Mark McClure, Barbara MacCluer, Larry "LT" Thomas, and Tom Kriete, who introduced me to the beauty and precision of analysis. An extra thank you to Barbara who emphasized clear and concise exposition; as an instructor and a mentor, she has impacted my career in ways too numerous to list.

I owe a tremendous debt of gratitude (and time) to my colleague and friend, Robert Allen. He encouraged this project from its initial phase, has acted as a sounding board over years of planning, writing, revising, and repeating, and read the entire text. I would have faltered many times without his support. He is also responsible for the wonderful illustrations throughout the text. These were based on sketches that I provided, but he infused them with clarity and a professional look that I would never have been able to achieve. Moreover, he taught from the first draft of the manuscript and provided many insights which helped me refine the message of the text.

There are many students who contributed to this project in a number of different ways. First, thanks to Luke Eichelberger, Christina Lorenzo, and Alyssa De Chirico who encouraged me to begin thinking about a textbook. Also, thanks to the students in Real Analysis I and II at North Central College (2010–2013) and at the University of Wisconsin at La Crosse (2012–2013) who used several drafts and helped me refine the style and presentation of the text. And to Maggie Wieczorek, who took

Real Analysis I and II as an independent study; thanks for reading every single word...literally.

I would also like to thank my colleagues in the math department at North Central College (Rich, David, Linda, Mary, Neil, and Katherine) who have encouraged me throughout the duration of the project and who have allowed me the freedom to shape our analysis courses into a more rigorous student experience. I am also grateful to the Office of Academic Affairs and the Faculty Professional Development Committee at North Central College for supporting this project in the form of summer writing grants and a junior faculty enhancement award.

Finally, I would like to thank my editors at Springer, Meredith Rich, Eve Mayer, and Marc Strauss, for their initial interest in the project and their advice over the course of writing and production.

With all of that, I hope that you, the reader, will find this text as engaging to read as it has been for me to write. And while many folks have contributed their efforts to ensure that the text is error free, I take complete responsibility for any remaining errors, oversights, or shortcomings.

Naperville, IL, USA   Matthew A. Pons

Contents

List of Figures

Chapter 1
The Real Numbers

To begin our study we must take a rigorous approach to understanding the real number system. This set has been built on intuition throughout your mathematical career and we seek to develop a concrete foundation to support this intuitive understanding before progressing to a study of functions of a real variable. In order to make sure we have a firm baseline established, the opening section provides a list of necessary preliminaries. We then consider general ordered fields which pave the way for an axiomatic development of the real numbers. Along the way we investigate the behavior of the natural and rational numbers, and we also develop a catalog of set structures available among the subsets of the real numbers.

1.1 Preliminaries

The goal of this section is to provide the reader with a list of prerequisites and to provide a guide for notation to be used throughout the text. We are assuming that all of this material has been encountered in previous courses and so there are very few proofs in this section. The exercises accompanying this material are to assist the reader in brushing up on the basics.

The most fundamental objects in the study of mathematics are sets and functions and these are our main focus here. We begin with the basics of set theory and then provide a brief discussion of the familiar number sets: the natural numbers, the integers, and the rational numbers. From there we move to a brief exposition on functions and their basic properties. We then use this idea to formalize the concept of cardinality.

M.A. Pons, *Real Analysis for the Undergraduate: With an Invitation to Functional Analysis*,   1
DOI 10.1007/978-1-4614-9638-0_1, © Springer Science+Business Media New York 2014

Sets and Operations

Weakly defined, a *set* is a collection of objects. For our purposes, there will always
be some *universal set* X and we will be interested in the objects or *elements* (the
term *points* is also used) of X. Typically a lowercase letter, say x, will represent
an element of X in which case we write $x \in X$. We will also work with smaller
collections (though possibly all of X) of elements of X. For notation, we write
$A \subseteq X$ to mean that A is such a smaller set of elements of X, or a *subset* of X. This
distinction breaks the elements in X into two categories, those in A and those not in
A. The symbols

$$x \in A \quad \text{and} \quad x \notin A$$

will indicate these situations, respectively. The *empty set* is the subset of X which
contains no elements and is denoted by the symbol $\emptyset$.

In order to determine membership in a subset A, we have to have some means
of *defining* subsets; in other words, given an element $x \in X$, we have to be able
to determine if $x \in A$ or $x \notin A$. Thus the elements of a subset are often defined
by having some specific property P. After classifying the subset A in this way, we
represent A with the set-builder notation

$$A = \{x \in X : x \text{ has property } P\};$$

some texts use a vertical bar rather than a colon. The first expression in the braces
gives the general location of the element x and the second expression provides the
specific condition the element must meet to be considered an element of A.

Example 1.1.1. For a simple example, let X be a set consisting of six marbles
numbered 1 through 6. Suppose that the marbles 1, 3, 4 are colored red, marbles
2 and 5 are green, and marble 6 is blue. We then represent the set of red marbles as

$$A = \{x \in X : x \text{ is red}\}.$$

In this example, we could also represent A simply by listing the marbles which are
red $A = \{1, 3, 4\}$ but listing leads to difficulties when considering infinite sets.

If $A, B \subseteq X$, then we say $A \subseteq B$ if all the elements of A are also in B. We say
that two sets A and B are *equal* if they contain the same elements. In practice this
is shown by demonstrating both inclusion statements

$$A \subseteq B \quad \text{and} \quad B \subseteq A.$$

We also define two operations on pairs of sets, the *union* and *intersection*,

$$A \cup B = \{x \in X : x \in A \text{ or } x \in B\}$$

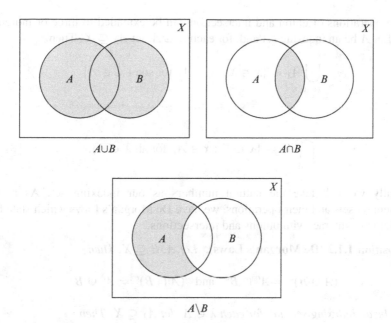

Fig. 1.1 Examples of Venn diagrams

and

$$A \cap B = \{x \in X : x \in A \text{ and } x \in B\}.$$

The set of elements not in a subset A is represented by the *complement*,

$$A^c = \{x \in X : x \notin A\}.$$

The operation of *set difference* is derived from intersections and complements and is defined by

$$A \setminus B = \{x \in X : x \in A \text{ and } x \notin B\} = A \cap B^c.$$

Figure 1.1 provides venn diagrams for the union, intersection, and set difference of two sets.

One final set operation is the *symmetric difference*. Like the set difference, this operation is built using the basic operations,

$$A \triangle B = (A \cup B) \setminus (A \cap B).$$

You are asked to explore this operation in the exercises.

The operations of union and intersection can be extended to three or more sets in X. Let Λ be an indexing set and, for each $\lambda \in \Lambda$, let $A_\lambda \subseteq X$. Then

$$\bigcup_{\lambda \in \Lambda} A_\lambda = \{x \in X : x \in A_\lambda \text{ for some } \lambda \in \Lambda\}$$

and

$$\bigcap_{\lambda \in \Lambda} A_\lambda = \{x \in X : x \in A_\lambda \text{ for all } \lambda \in \Lambda\}.$$

Typically we will take the natural numbers as our indexing set. As a final comment on sets and their operations, we have De Morgan's Laws which state how complements interact with unions and intersections.

Proposition 1.1.2 (De Morgan's Laws). *Let $A, B \subseteq X$. Then*

$$(A \cup B)^c = A^c \cap B^c \quad \text{and} \quad (A \cap B)^c = A^c \cup B^c.$$

Let Λ be an indexing set and, for each $\lambda \in \Lambda$, let $A_\lambda \subseteq X$. Then

$$\left(\bigcup_{\lambda \in \Lambda} A_\lambda\right)^c = \bigcap_{\lambda \in \Lambda} A_\lambda^c \quad \text{and} \quad \left(\bigcap_{\lambda \in \Lambda} A_\lambda\right)^c = \bigcup_{\lambda \in \Lambda} A_\lambda^c.$$

To be more formal about the notion of an operation, we have the following definition.

Definition 1.1.3. Given a set X, a *binary operation* on X is a rule which associates each pair of elements $x, y \in X$ with another element $z \in X$.

In the definition above, the term binary indicates that the operation defines a relationship between *two* elements in X. The most common binary operations on number sets are addition and multiplication which arise quite naturally, but we will study other operations as we go along, such as the union and intersection just defined for general sets. Be aware that there are more general types of operations.

Number Systems

The natural numbers are the most basic number system and play a very prominent role in the study of analysis. We will use the symbol $\mathbb{N}$ to denote this set whose members are listed

$$\mathbb{N} = \{1, 2, 3, \ldots\}.$$

There are two binary operations, addition and a multiplication, defined on the natural numbers both of which are commutative and associative. There is a multiplicative identity, but this number set has no additive identity and there are no additive or multiplicative inverses. There is also an order on $\mathbb{N}$, that is, a means to determine when one natural number is greater than another. The symbol $n < m$ indicates that n is less than m. The following statements represent the most basic properties of $\mathbb{N}$. The first is one of the defining characteristics of this set.

Proposition 1.1.4 (Principle of Induction). *Let* $S \subseteq \mathbb{N}$. *Suppose that*

(a) $1 \in S$;
(b) *if* $n \in S$, *then* $n + 1 \in S$.

Then $S = \mathbb{N}$.

Example 1.1.5. It is assumed that induction is familiar territory but we outline the general strategy and provide an example of application. Let P be a property and define $S \subseteq \mathbb{N}$ by

$$S = \{n \in \mathbb{N} : n \text{ has property } P\}.$$

We then use induction to show that every natural number has the property P. To do this, we *show* that 1 has property P; this must be exhibited explicitly and is often referred to as the *base case*. Then we *assume* that an arbitrary natural number n has property P. This is called the *inductive hypothesis*. Next, we demonstrate, again explicitly, that $n + 1$ has property P. If this implication is true, then the above statement indicates that every natural number has property P and hence $S = \mathbb{N}$.

For an example, we will show that for every $n \in \mathbb{N}$, the summation identity

$$1 + 2 + \cdots + n = \frac{n(n + 1)}{2}$$

holds true. For a fixed natural number, let P be the property that the summation identity holds.

Proof. For the base case, we show that $n = 1$ satisfies the rule above. The statement is trivial in this case (as is usual) since

$$1 = \frac{1(2)}{2}.$$

For our inductive hypothesis, assume that for some $n \in \mathbb{N}$ we have

$$1 + 2 + \cdots + n = \frac{n(n + 1)}{2}.$$

Now consider $n + 1$. Some arithmetic shows that

$$1 + 2 + \cdots + n + (n + 1) = \underbrace{1 + 2 + \cdots + n}_{} + (n + 1) = \frac{n(n + 1)}{2} + (n + 1)$$
$$= \frac{n(n + 1)}{2} + \frac{2(n + 1)}{2} = \frac{n(n + 1) + 2(n + 1)}{2}$$
$$= \frac{(n + 1)(n + 2)}{2}.$$

Notice that we used the induction hypothesis in the second equality. This is the power of induction—the $n + 1$ case depends on the previous case, just like climbing a ladder. We see then that

$$1 + 2 + \cdots + n + (n + 1) = \frac{(n + 1)(n + 2)}{2}$$

which is exactly the summation identity for $n + 1$. Thus we conclude that the identity holds for all $n \in \mathbb{N}$. $\square$

The next statement is often taken for granted but provides a nice contrast between the natural numbers and the other number systems we encounter.

Theorem 1.1.6 (Well-Ordering Principle). *Let $A \subseteq \mathbb{N}$ be nonempty. Then there is an $n \in A$ such that $n \leq m$ for every $m \in A$.*

One last result for the natural numbers shows that the prime numbers are indivisible quantities. This will be of particular relevance when we discuss roots of natural numbers.

Lemma 1.1.7 (Euclid's Lemma). *Let $a, b, p \in \mathbb{N}$ with p a prime number. If p divides the product ab, then p divides a or p divides b.*

The integers are the next number set,

$$\mathbb{Z} = \{\ldots, -3, -2, -1, 0, 1, 2, 3, \ldots\}.$$

These numbers arise as a solution to the need for subtraction in the natural number system. As with the natural numbers, there are two operations, addition and multiplication, which are derived from their natural number ancestors. Here we do have an additive identity and additive inverses, but we still lack multiplicative inverses.

The rational numbers solve the problem of multiplicative inverses and provide us with the concept of division,

$$\mathbb{Q} = \left\{ \frac{p}{q} : p \in \mathbb{Z} \text{ and } q \in \mathbb{N} \right\}.$$

One drawback to the notation above is that there are repeated elements, e.g. 1/2 and 2/4. To alleviate this issue, we will often also require that p and q have no common factors in order to get a true representation for the rational numbers. This set has the most structure of the number systems we have considered thus far. There are two operations, again derived from the addition and multiplication on the natural numbers, and there are additive and multiplicative identities and inverses. Such a set structure is called a *field*; we provide more details on this in the next section.

The rational number system also has an order $<$ which satisfies:

(i) if $x, y \in \mathbb{Q}$, then exactly one of the following holds: $x < y$, $y < x$, or $x = y$;
(ii) if $x, y, z \in \mathbb{Q}$ with $x < y$ and $y < z$, then $x < z$.

As is customary, we use the symbol $y > x$ to mean $x < y$ and $x \leq y$ to mean $x < y$ or $x = y$. To get a sense for how this works, if $x \in \mathbb{Q}$, we say that $x > 0$ if x can be written as p/q where $p, q \in \mathbb{N}$. For example, $1/2$, $2 = 2/1$, and $37/11$ are all greater than 0. This is then extended to an order on $\mathbb{Q}$ by defining $x < y$ provided $y - x > 0$.

Be aware that while we are assuming familiarity with these systems, they can all be developed in a rigorous fashion from the basics of set theory. Our goal here, however, is to understand the larger class of real numbers and these three sets provide a starting point. A truly detailed exposition beginning with the construction of the natural numbers would require a much lengthier text. Before moving on to functions we highlight one fault of the rational numbers. This result was discovered by the Greeks circa 500 BCE and provides part of our motivation for moving to the real number system.

Theorem 1.1.8. *There is no rational number whose square is 2.*

Proof. To begin let us restate the theorem as follows:

If r is rational, then $r^2 \neq 2$.

In this form, we see a clear hypothesis and conclusion. We assume that $r \in \mathbb{Q}$. If this was our only assumption, we would be left with a difficult choice as to how to move forward, but in such cases proof by contradiction is a valuable tool. With this in mind, we also assume to the contrary that $r^2 = 2$. Using our hypothesis that r is rational, we can represent it by a ratio

$$r = \frac{p}{q},$$

where $p \in \mathbb{Z}$ and $q \in \mathbb{N}$. Furthermore, we can assume that p and q have no common factors. With this, we apply our second hypothesis to this ratio and we obtain

$$\left(\frac{p}{q}\right)^2 = 2$$

which is equivalent to

$$p^2 = 2q^2. \tag{1.1}$$

From this, we conclude that p^2 is an even number. By Euclid's Lemma, we are also assured that p is an even number, i.e. there is a positive integer s with

$$p = 2s.$$

If we substitute this into Eq. (1.1), we see that

$$(2s)^2 = 2q^2$$

or

$$2s^2 = q^2.$$

However, this shows that q^2 is even and Euclid's Lemma dictates that q must also be even. Now the problem is apparent. We know that p and q have no common factors, but we have shown that both numbers are even. This is a contradiction and thus it must be the case that there is no rational number whose square is 2. $\square$

Functions

Every topic that we consider in this book builds on the idea of a function which is a means of relating elements in two sets. Our sets will naturally consist of numbers, though the definition of a function is stated in a more global context.

Definition 1.1.9. Let X and Y be sets. A *function* $f : X \to Y$ is a rule that assigns to each element of X a unique element of Y. We call X the *domain* of f and Y the *codomain*. If $x \in X$, we use the symbol $f(x)$ to represent the point in Y associated with x. On occasion we will use the symbol $x \mapsto f(x)$ which means that x maps to $f(x)$. If $y \in Y$ is associated with the point $x \in X$, then we write $y = f(x)$.

Mathematicians have studied functions for centuries and thus it is interesting to note that this formal definition was not provided until the 1830s by Dirichlet. Early mathematicians studied functions that were given by an algebraic relation such as $x \mapsto x^2$ or $x \mapsto \sin x$. As mathematics itself evolved, the definitions for such maps became more complicated and such formulaic identifications were realized to only represent a small class of possible functions. This new definition for what qualifies as a function broadened the scope for mathematical inquiry.

Thinking about the definition, be careful with the uniqueness statement. While each element in X is assigned a unique element in Y, it is not necessarily the case that each $y \in Y$ is associated with a unique point in X. For the given statement, the requirement is that each point of x maps to exactly one point of Y; we will use the phrase *well-defined* to signify this. For a point $y \in Y$, however, it may be the case that no point of x maps to y or it may be that there are multiple points in X

which map to y. To clarify the first of these two scenarios, we define the *range* of a function f to be the set

$$f(X) = \{y \in Y : y = f(x) \text{ for some } x \in X\}$$

and thus $f(X) \subseteq Y$. For example, define $f : \mathbb{N} \to \mathbb{N}$ by $f(n) = n^2$. Then the domain of f is $\mathbb{N}$ since n^2 is a well-defined natural number. The codomain of f is also $\mathbb{N}$, but the range of f is the set of perfect squares $\{1, 4, 9, \ldots\}$. The codomain provides us with a target set for the function f, but the range gives more specific information about what the function actually maps to.

We also single out the particular situations when each point in Y is mapped to by a point of X and when all points in the range are mapped to by a unique point of X.

Definition 1.1.10. Let X, Y be sets and $f : X \to Y$. We say that f maps X *onto* Y if for each $y \in Y$, there is a point $x \in X$ such that $f(x) = y$. We say that f is *one-to-one* if each y in $f(X)$ is mapped to by a unique element of X. This second property is better understood as the conditional statement

$$\text{if } f(x) = f(y), \text{ then } x = y.$$

With these definitions it should be clear that $f : X \to Y$ always maps onto $f(X)$ and the definition above is an effort to identify situations when $Y = f(X)$. Often the term *surjective* is used as a synonym for onto and *injective* for one-to-one. A function that is both one-to-one and onto is called *bijective* or a bijection.

Given sets X, Y, Z and functions $f : X \to Y$ and $g : Y \to Z$, we define a new function $h : X \to Z$ called the *composition* of f and g by

$$h(x) = g(f(x));$$

we usually forgo the use of h and simply write $g \circ f$ so that the previous line reads

$$(g \circ f)(x) = g(f(x)).$$

When a function $f : X \to Y$ is one-to-one and onto Y, it is possible to define a new function called the *inverse* of f, denoted $f^{-1} : Y \to X$, by the rule

$$f^{-1}(y) = x \text{ if and only if } f(x) = y.$$

Since f is onto we know that every $y \in Y$ is mapped to by some x in X and the fact that f is one-to-one provides the means to conclude that f^{-1} is well defined. Furthermore, f^{-1} satisfies the equations

$$f^{-1}(f(x)) = x \quad \text{for all } x \in X$$

and

$$f(f^{-1}(y)) = y \quad \text{for all } y \in Y.$$

It is also interesting to note that when such a function exists, then it is unique.

Example 1.1.11. Define a function f by

$$f(x) = \frac{x}{x-2}.$$

We want to consider this as a function on $\mathbb{Q}$. For such an algebraic definition, we take the domain to be the largest possible set of numbers for which the function is defined (typically this is done over the real numbers but we will focus on the rationals for the moment). Thus we see that the natural domain of this function is the set

$$A = \{x \in \mathbb{Q} : x \neq 2\}.$$

Certainly $f(x)$ is rational if $x \in \mathbb{Q}$, and we have $f : A \to \mathbb{Q}$. Next we want to identify the range of f. In general, identifying the range of a function is trickier than identifying its domain. First let us show that f is one-to-one. Letting $x_1, x_2 \in A$, we assume that

$$\frac{x_1}{x_1 - 2} = \frac{x_2}{x_2 - 2}.$$

Algebraic manipulation shows that

$$x_1 x_2 - 2x_1 = x_1 x_2 - 2x_2$$

and it follows that $x_1 = x_2$. Now, to find the range, let $y \in \mathbb{Q}$ and assume there is an $x \in A$ with $f(x) = y$. Then we can solve the equation

$$\frac{x}{x-2} = y$$

to find that

$$x = \frac{2y}{y-1}.$$

This shows that y is in the range of f provided $y \neq 1$,

$$f(A) = \{y \in \mathbb{Q} : y \neq 1\}.$$

It also shows that the function

$$f^{-1}(y) = \frac{2y}{y-1}$$

is the inverse of f.

For a moment, let $f : X \to Y$ and let's consider f as a function on the subsets of X, rather than as a function on the points of X. Let $A \subseteq X$ and $B \subseteq Y$. We can then restrict the notion of range to subsets of the domain by defining $f(A) = \{y \in Y : f(x) = y$ for some $x \in A\}$. When a function is viewed in this more general manner rather than as a function acting on particular domain points, we can always consider f^{-1}, whether or not f is one-to-one, if we define $f^{-1}(B) = \{x \in X : f(x) = y$ for some $y \in B\}$. The set $f^{-1}(B)$ is called the *inverse image of B* (other terms include the *pre-image* or *pull back of B*).

Example 1.1.12. Define $f : \mathbb{N} \to \mathbb{N}$ by $f(n) = n^2$. If $A = \{1, 2, 3, 4, 5, 6\}$, then

$$f(A) = \{1, 4, 9, 16, 25, 36\}.$$

Also, if $B = \{3, 4, 5, 6\}$, then

$$f^{-1}(B) = \{2\}$$

whereas $f^{-1}(B) = \emptyset$ if $B = \{3, 5\}$.

Later we will have need to consider the inverse image of unions and intersections of sets, and the following proposition indicates that these operations commute with the inversion. We leave the proof as an exercise.

Proposition 1.1.13. *Let $f : X \to Y$ where X and Y are sets. If $\{A_n : n \in \mathbb{N}\}$ is a collection of subsets of Y, then*

$$f^{-1}\left(\bigcup_{n=1}^{\infty} A_n\right) = \bigcup_{n=1}^{\infty} f^{-1}(A_n) \quad \text{and} \quad f^{-1}\left(\bigcap_{n=1}^{\infty} A_n\right) = \bigcap_{n=1}^{\infty} f^{-1}(A_n).$$

Cardinality

We have a natural intuition when it comes to the size of finite sets, but this is challenged when we move to infinite sets. Naively one would say that there are fewer even natural numbers than there are natural numbers since one set is a strict subset of the other. This manner of thinking muddles the issue though. For instance, one way to interpret the idea of size for two finite sets is to line the objects up next to each other and determine which line is longer. If they have the same length, then the sets have the same size. Such a demonstration is actually the construction of a bijective function between the two sets.

$$x_1 \ x_2 \ x_3 \ \ldots \ x_n$$
$$\updownarrow \ \updownarrow \ \updownarrow \qquad \updownarrow$$
$$y_1 \ y_2 \ y_3 \ \ldots \ y_n$$

Returning to the even natural numbers then, it is possible to line them up in a similar manner with the natural numbers, except that the lines never terminate.

$$1\ 2\ 3\ \ldots\ n\ \ldots$$
$$\updownarrow \updownarrow \updownarrow \qquad \updownarrow$$
$$2\ 4\ 6\ \ldots 2n\ \ldots$$

We would therefore conclude that there are just as many even natural numbers as there are natural numbers *even though* one set is a strict subset of the other. The concept of cardinality formalizes this notion.

Definition 1.1.14. Let A and B be sets. We say that A and B have the same *cardinality* if there is a bijection $f : A \to B$. If such a bijection exists, we symbolize this by $\text{card}(A) = \text{card}(B)$.

For the moment you should take caution with the notation above. The equal sign in the expression $\text{card}(A) = \text{card}(B)$ is not the same notion of equality that you are accustomed to and is merely a shorthand way of indicating that such a bijective function exists. It is therefore not obvious that it obeys the familiar rules we associate with equality. For instance, if A, B, C are sets such that $\text{card}(A) = \text{card}(B)$ and $\text{card}(B) = \text{card}(C)$, must it be true that $\text{card}(A) = \text{card}(C)$? Simply writing $\text{card}(A) = \text{card}(B) = \text{card}(C)$ is a tempting method of proof, but this is relying on what you know about equality of numbers which is not what we have here as cardinalities are not necessarily finite quantities. To show that A and C have the same cardinality, we must exhibit a bijective function from A to C. This of course will depend on the fact that we can construct bijections between A and B, and between B and C; this is requested in the exercises.

To classify the cardinalities of sets we have the following definition.

Definition 1.1.15. A set A is a finite set with cardinality n if $\text{card}(A) = \text{card}(B)$ where $B = \{1, 2, \ldots, n\}$. The empty set has cardinality 0. A set A is *countable* if $\text{card}(A) = \text{card}(\mathbb{N})$ or if A is finite, and *uncountable* otherwise.

The definition indicates that the natural numbers form the basis for the smallest type of infinite sets. Furthermore, it is easy to see that a set A is countable if there is a one-to-one function $f : A \to B$ where $B \subseteq \mathbb{N}$. We use the term countable for these sets to indicate that we can *enumerate* them or write them out similar to the way we list the elements of $\mathbb{N}$. If A is countable, then we can write

$$A = \{a_1, a_2, a_3, \ldots\}$$

where we do not necessarily know the order of the elements but we are assured that every element of A is represented exactly once in this list. If A is finite, then the list terminates. Otherwise, it continues indefinitely.

Proposition 1.1.16. *(a) Let A_1 and A_2 be countable sets. Then $A_1 \cup A_2$ is countable.*
(b) For each $n \in \mathbb{N}$, let A_n be a countable set. Then $\bigcup_{n=1}^{\infty} A_n$ is countable.

Proof. Using the idea of an enumeration, let us write

$$A_1 = \{a_{11}, a_{12}, a_{13}, \ldots\}$$

and

$$A_2 = \{a_{21}, a_{22}, a_{23}, \ldots\}.$$

Then we can define a function $f : A_1 \cup A_2 \to \mathbb{N}$ by $f(a_{ij}) = 2^i 5^j$ where i is 1 or 2 and $j \in \mathbb{N}$; there may also be restrictions on j if either A_1 or A_2 is finite. Notice that this identification covers all points of $A_1 \cup A_2$. Furthermore, the function f is one-to-one by the uniqueness of prime factorizations. The fact that

$$B = \{2^i 5^j : i = 1, 2 \text{ and } j \in \mathbb{N}\} \subseteq \mathbb{N}$$

guarantees us that $A_1 \cup A_2$ is countable.

The exercises ask you to extend this to part (b). $\quad\square$

Proposition 1.1.17. *The rational numbers are countable.*

Proof. To prove this statement we will break $\mathbb{Q}$ into three sets and show that each of these is countable. We will then apply the previous proposition. First, let $\mathbb{Q}^+$ denote the set of all positive rational numbers, that is,

$$\mathbb{Q}^+ = \left\{ \frac{p}{q} : p, q \in \mathbb{N} \text{ with no common factors} \right\}.$$

To see that this set is countable we use the idea from the previous proof and we define $f : \mathbb{Q}^+ \to \mathbb{N}$ by $f(p/q) = 2^p 5^q$. We know that this function is one-to-one and thus $\mathbb{Q}^+$ is countable.

Similarly, we define the set $\mathbb{Q}^-$ by

$$\mathbb{Q}^- = \{r \in \mathbb{Q} : r < 0\} = \left\{ \frac{-p}{q} : p, q \in \mathbb{N} \text{ with no common factors} \right\}.$$

Here we ignore the negative sign and define a function $g : \mathbb{Q}^- \to \mathbb{N}$ by $g(-p/q) = 2^p 5^q$ which shows that $\mathbb{Q}^-$ is also countable. The observation that

$$\mathbb{Q} = \{0\} \cup \mathbb{Q}^+ \cup \mathbb{Q}^-$$

with Proposition 1.1.16 verifies that $\mathbb{Q}$ is countable. $\quad\square$

Exercises

Exercise 1.1.1. Let $A, B, C \subseteq X$. Show each of the following set equalities.

(a) $A \cap (B \cup C) = (A \cap B) \cup (A \cap C)$
(b) $A \cup (B \cap C) = (A \cup B) \cap (A \cup C)$
(c) $A \setminus (B \cup C) = (A \setminus B) \cap (A \setminus C)$
(d) $A \setminus (B \cap C) = (A \setminus B) \cup (A \setminus C)$

Exercise 1.1.2. If $A, B \subseteq X$, show that $A \cup B = A \cup (B \setminus A)$ and $B = (A \cap B) \cup (B \setminus A)$.

Exercise 1.1.3. Draw a Venn diagram representing the symmetric difference of two sets.

Exercise 1.1.4. Let $A, B \subseteq X$. Show each of the following statements.

(a) $A \Delta \emptyset = A$
(b) $A \Delta X = A^c$
(c) $A \Delta B = (A \setminus B) \cup (B \setminus A)$
(d) If $A \subseteq B$, then $A \Delta B = B \setminus A$

Exercise 1.1.5. Let $A, B, C, D \subseteq X$. Show each of the following statements.

(a) $A \Delta C \subseteq (A \Delta B) \cup (B \Delta C)$
(b) $(A \cup B) \Delta (C \cup D) \subseteq (A \Delta C) \cup (B \Delta D)$
(c) $A \Delta B = A^c \Delta B^c$
(d) $(A \setminus B) \Delta (C \setminus D) \subseteq (A \Delta C) \cup (B \Delta D)$

Exercise 1.1.6. Let $A \subseteq X$ and for each $n \in \mathbb{N}$, let $B_n \subseteq X$. Show each of the following set equalities.

(a) $A \cap \left(\bigcup_{n=1}^{\infty} B_n \right) = \bigcup_{n=1}^{\infty} (A \cap B_n)$

(b) $A \cup \left(\bigcap_{n=1}^{\infty} B_n \right) = \bigcap_{n=1}^{\infty} (A \cup B_n)$

Exercise 1.1.7. If $A_n \subseteq X$ for $n = 1, 2, 3, \ldots$, show that

$$\bigcap_{n=1}^{\infty} A_n = A_1 \setminus \left(\bigcup_{n=1}^{\infty} (A_1 \setminus A_n) \right).$$

Exercise 1.1.8. Let $y_1 = 1$ and, for each $n \in \mathbb{N}$, define $y_{n+1} = (3y_n + 4)/4$.

(a) Use induction to show that $y_n < 4$ for all $n \in \mathbb{N}$.
(b) Use induction to show that $y_n \leq y_{n+1}$ for all $n \in \mathbb{N}$.
(c) What happens to the statements in (a) and (b) if we let $y_1 = 8$? Provide formal statements and proof.

Exercise 1.1.9. Show that the summation identity

$$1 + 4 + 9 + \cdots + n^2 = \frac{n(n+1)(2n+1)}{6}$$

holds for all $n \in \mathbb{N}$.

Exercise 1.1.10. (a) Modify the proof of Theorem 1.1.8 to show that there is no
rational number which satisfies $r^2 = 3$.
(b) Let p be a prime number. Extend the above result to show that there is no
rational number which satisfies $r^2 = p$.

Exercise 1.1.11. For sets A and B we define a relation $A \sim B$ if $\text{card}(A) = \text{card}(B)$. Show that this is an equivalence relation.

Exercise 1.1.12. Let A be an uncountable set and B a countable set with $B \subseteq A$.
Show that $A \setminus B$ is uncountable.

Exercise 1.1.13. Verify that the following pairs of sets have the same cardinality
by constructing a bijection between the given sets.

(a) $\mathbb{N}$ and $\mathbb{N} \cup \{0\}$
(b) $\mathbb{Q}$ and $\mathbb{Q} \cup \{\pi, e, \sqrt{2}\}$
(c) $\mathbb{N}$ and $\mathbb{Z}$

Exercise 1.1.14. Write $\mathbb{Q}$ as a countable union of finite sets.

Exercise 1.1.15. Prove part (b) of Proposition 1.1.16.

1.2 Complete Ordered Fields

Our definition of the real number system will consist of three ingredients and is built
around the rational numbers. Our goal is to take the existing structures in $\mathbb{Q}$ and
extend them in such a way that we overcome the issue presented in Theorem 1.1.8.
We will first investigate the algebraic structure that we desire of the real number
system. Next we will focus on an order for the system, and the final step will concern
an analytic property that distinguishes it from the rational number system.

Algebra is the area of mathematics which examines set structures. Given a set X,
we define operations on the elements in the set, e.g. addition on the natural numbers,
multiplication on the integers, etc., and the purpose of algebra is to understand the
resulting structure and relationships between the elements that arise from these
operations. These constructions range from a set and one operation, e.g. monoids
and groups, to more complicated objects consisting of a set with two or more
operations, rings, fields, vector spaces, algebras, and so on.

The algebraic structure we focus on here is that of a *field*, which concerns a
set and two binary operations. We use the typical symbols $+$ and $\cdot$ to represent

an addition and multiplication, and, for $x, y \in X$, $x + y$ for the sum and xy for the product. It is sometimes necessary to use the more formal symbol $x \cdot y$ for the product but we will typically use the shorter form. The definition of a field then states the axioms satisfied by the operations and the elements of X. Notice that the definition below does not include closure axioms, which are more common in linear or abstract algebra texts, because we have incorporated these into our definition of an operation.

Definition 1.2.1. A *field* is a set X and two binary operations (see Definition 1.1.3), addition $+$ and multiplication $\cdot$, which satisfy:

Addition Axioms

(a) *Commutativity*: $x + y = y + x$ for every $x, y \in X$;
(b) *Associativity*: $(x + y) + z = x + (y + z)$ for every $x, y, z \in X$;
(c) *Additive identity* or *zero*: there is an element, denoted 0, in X such that $x + 0 = x$ for every $x \in X$;
(d) *Additive inverse*: for each $x \in X$, there is an element, denoted $-x$, in X such that $x + (-x) = 0$;

Multiplication Axioms

(e) *Commutativity*: $xy = yx$ for every $x, y \in X$;
(f) *Associativity*: $(xy)z = x(yz)$ for every $x, y, z \in X$;
(g) *Multiplicative identity*: there is an element, denoted 1, in X such that $1x = x$ for every $x \in X$;
(h) *Multiplicative inverse*: for each nonzero $x \in X$, there is an element, denoted x^{-1}, in X such that $xx^{-1} = 1$;
(i) *Distributivity*: $x(y + z) = xy + xz$ for every $x, y, z \in X$.

The operational axioms above are meant to endow the field X with a reasonable degree of functionality and these properties are all extracted from the rational number system. We will therefore find it imperative that the real number system which we are in the process of constructing also exhibit these qualities. As a comment on notation, we will use the symbol $x - y$ to represent the addition of an element and an additive inverse, $x + (-y)$.

The definition of a field represents the minimal requirements that a collection $(X, +, \cdot)$ must meet to be considered a field. However, there are many more properties which will be of use to us. The most relevant are captured in the next two theorems. The first of these concerns the operation of addition and additive inverses in a field. In the statement below, -1 refers to the additive inverse of the multiplicative identity 1. It is likely that you will think of 1 and -1 as their rational number counterparts, but it is good to keep in mind that momentarily we are not assuming any relationship between the field X and $\mathbb{Q}$.

Theorem 1.2.2. *Let $(X, +, \cdot)$ be a field and let $x, y, z \in X$.*

(a) the additive identity is unique;
(b) the additive inverse of an element x is unique;

(c) *if* $x + z = y + z$, *then* $x = y$;
(d) $0 \cdot x = 0$;
(e) $-(x + y) = (-x) + (-y)$;
(f) $-(-x) = x$;
(g) $(-1)x = -x$;
(h) $(-1)(-1) = 1$.

Proof. We will prove several parts of the theorem to demonstrate some of the algebraic techniques involved and will leave the others as an exercise.

For (a) we will employ the typical mathematical approach for demonstrating that an object is unique and assume that there are two zeros in our fields. We will denote these by 0 and $0'$. Our goal is now to use the definition of a field to show that these two zeros are in fact equal. Appealing to the fact that both 0 and $0'$ satisfy the definition of a zero, the commutativity of addition shows that

$$0 = 0 + 0' = 0' + 0 = 0'.$$

For (c), we assume that $x + z = y + z$ for $x, y, z \in X$. To see that $x = y$, observe that

$$x = x + 0 = x + (z - z) = (x + z) + (-z) = (y + z) + (-z) = y + (z - z) = y + 0 = y$$

where we have used the zero and additive inverse properties, and the associativity of addition in X.

To prove (d), first notice that $0 = 0 + 0$ and thus by the distributive property,

$$0 \cdot x = (0 + 0)x = 0 \cdot x + 0 \cdot x.$$

With this we can use the additive identity once more to generate the equation

$$0 + 0 \cdot x = 0 \cdot x + 0 \cdot x.$$

Part (c) now indicates that $0 \cdot x = 0$.

Part (f) of the statement is actually reiterating the fact that additive inverses are unique by insisting that the additive inverse of $-x$ is x. To prove the statement, note that $x + (-x) = 0$ not only implies that $-x$ is the additive inverse of x, but it also indicates that the additive inverse of $-x$ is x. Hence $-(-x) = x$.

Finally, (h) follows from (f) and (g) since

$$(-1)(-1) = -(-1) = 1. \qquad \square$$

Part of the point of demonstrating so many parts of the proof above is to demonstrate a common theme. Each component of the proof produces a string of equalities that, when taken as whole, gives the desired conclusion. There are of course other proof techniques, but getting used to this form now will aid in

understanding as we encounter more complicated arguments and will also enable
you to produce concise and effective proofs.

The following theorem concerns the multiplication and multiplicative inverses
in a field. We use the standard exponential notation $x \cdot x = x^2$, and for $n \in \mathbb{N}$,
$\underbrace{x \cdot x \cdots x}_{n} = x^n$.

Theorem 1.2.3. *Let $(X, +, \cdot)$ be a field and let $x, y, z \in X$.*

(a) the multiplicative identity is unique;
(b) the multiplicative inverse of a nonzero element x is unique;
(c) if $xz = yz$ and $z \neq 0$, then $x = y$;
(d) if $xy = 0$, then $x = 0$ or $y = 0$;
(e) if $x \neq 0$, then $(x^{-1})^{-1} = x$;
(f) $(xy)^n = x^n y^n$;
(g) $-(xy) = (-x)y = x(-y)$;
(h) $(-x)(-y) = xy$;
(i) if $x, y \neq 0$, then $(xy)^{-1} = x^{-1}y^{-1}$.

Proof. To prove (b), we come again to a uniqueness proof and we will apply the
same strategy as above. For $x \in X$ with $x \neq 0$, assume that x^{-1} and y are both
multiplicative inverses for x. To see that these two elements are equal, observe that

$$y = y1 = y(xx^{-1}) = (yx)x^{-1} = 1x^{-1} = x^{-1}.$$

where here we have used the multiplicative identity and inverse, and the associativity
of multiplication.

Part (d) will be one of the most important properties that we have at our disposal.
First suppose that $x \neq 0$. In this case, we will show that y must be zero. Notice that

$$y = 1y = (xx^{-1})y = (x^{-1}x)y = x^{-1}(xy) = x^{-1}0 = 0;$$

here we have used the fact that x has a multiplicative inverse since we have
assumed that $x \neq 0$ together with the commutative and associative properties of
multiplication and part (d) of the previous theorem. The statement clearly holds if
$x = 0$.

We will conclude with a proof of (g) which is an easy consequence of
Theorem 1.2.2 (g). First, we can write $-(xy)$ as $(-1)(xy)$. The associativity and
commutativity of multiplication show that

$$-(xy) = (-1)(xy) = (-1x)y = x(-1y).$$

A second application of (g) then confirms that

$$-(xy) = (-x)y = x(-y). \qquad \square$$

The definition and theorems above indicate that we can manipulate expressions in a field according to the rules of algebra that should be second nature at this point in your mathematical development. With this algebraic structure in hand, we turn our attention to order. Many of the ideas we will encounter will require us to determine when one quantity is smaller than another given quantity and hence a thorough understanding of what it means to have an order on a field is an absolute necessity.

Definition 1.2.4. Let $(X, +, \cdot)$ be a field. We say X is an *ordered field* if there is a relation $<$ on X which satisfies the following:

(a) *Trichotomy*: if $x, y \in X$, then exactly one of the following holds: $x < y, y < x$, or $x = y$.
(b) *Transitivity*: if $x, y, z \in X$ with $x < y$ and $y < z$, then $x < z$;
(c) *Addition*: if $x, y, z \in X$ with $x < y$, then $x + z < y + z$;
(d) *Multiplication*: if $x, y \in X$ with $x > 0$ and $y > 0$, then $xy > 0$;
(e) *Non-degeneracy*: $0 \neq 1$.

To be clear, we use $y > x$ to mean $x < y$, and $x \leq y$ to mean that either $x < y$ or $x = y$. As with a field, the axioms above indicate the minimal requirements for an ordered field and the following theorem produces several more important properties.

Theorem 1.2.5. *Let X be an ordered field and let $x, y, z \in X$.*

(a) $x < y$ *if and only if* $0 < y - x$;
(b) $x > 0$ *if and only if* $-x < 0$;
(c) *if* $x \neq 0$, *then* $x^2 > 0$;
(d) $-1 < 0 < 1$;
(e) *if* $x < y$ *and* $z > 0$, *then* $xz < yz$;
(f) *if* $x < y$ *and* $z < 0$, *then* $yz < xz$;
(g) *if* $x > 0$, *then* $x^{-1} > 0$;
(h) *if* $0 < x < y$, *then* $0 < y^{-1} < x^{-1}$.

Proof. We begin with (a). Here we must prove two statements. We first assume that $x < y$. Then, according to the axioms of an ordered field, we can add an element to both sides and not change the order. Adding $-x$ we obtain

$$x - x < y - x$$

which implies that $0 < y - x$. Similarly, if we assume that $0 < y - x$, we can add x to both sides and we obtain $x < (y - x) + x$. The associativity of addition in a field guarantees us that $x < y$.

For (b) we employ a similar technique. Assuming that $x > 0$, we add $-x$ to both sides and we have

$$x - x > 0 - x;$$

this implies that $-x < 0$ by the additive identity and inverse properties of a field. The reverse argument follows identically.

Moving on to (c), we consider two cases. If $x > 0$, then ordered field axiom (d) immediately implies that $x^2 = xx > 0$. On the other hand, if $x < 0$, then by part (b) we know that $-x > 0$. The previously mentioned ordered field axiom now implies that $(-x)(-x) > 0$. However, Theorem 1.2.3 (g) indicates that $(-x)(-x) = xx = x^2$ and thus upon substituting, $x^2 > 0$.

To prove (d), we first show the upper inequality. By (c) and the properties of the multiplicative identity, we know that $1 = 1 \cdot 1 = 1^2 > 0$. Part (b) of this theorem now shows that $-1 < 0$.

We conclude with a proof of (g) and suppose $x > 0$. Towards a contradiction, assume also that $x^{-1} < 0$. Then (f) implies that $xx^{-1} < 0x^{-1}$, but this is a contradiction since $1 = xx^{-1} < 0x^{-1} = 0$. Thus it must be the case that $x^{-1} > 0$.   □

The final idea necessary to define the real numbers is an analytic property. The rational numbers are an ordered field and thus requiring that the real numbers be an ordered field doesn't do much as far as distinguishing them from the rational numbers. Again, our hope is that this third property will alleviate the problem encountered in Theorem 1.1.8. Thinking visually, if we were to plot out the rational numbers on a number line, there would be gaps. This is the issue we hope to correct by introducing the real number system. The following circle of ideas supplies this third component.

Definition 1.2.6. Let X be an ordered field and let $A \subseteq X$ be nonempty. We say A is *bounded above* if there is a point $M \in X$ such that $a \leq M$ for every $a \in A$. In this case we call M an *upper bound* for A. Similarly, we say A is *bounded below* if there is a point $m \in X$ such that $m \leq a$ for every $a \in A$. Here we call m a *lower bound* for A.

Example 1.2.7. As a first example, consider the set $A = \{1, 3, 7\} \subseteq \mathbb{Q}$. It is clear that this set is bounded above by 7. Of course this is not the only upper bound. Indeed 8, 33/2, 92, and many other rational numbers are upper bounds. Concretely, any rational number M which satisfies $M \geq 7$ will serve as an upper bound. Similarly, any rational number m which satisfies $m \leq 1$ will act as a lower bound.

Next consider $\mathbb{N} \subseteq \mathbb{Q}$. Certainly 1 is a lower bound, as is any rational number less than 1. Is this set bounded above? Your intuition should lead you to believe that it is not and we will prove this fact in the next section.

For a third example, let $A = \{x : 0 < x < 3\} \subseteq \mathbb{Q}$. Keep in mind that this is not the interval $(0, 3)$ as we have not encountered the real numbers yet. Rather, this is the collection of all rational number strictly between 0 and 3. The set is certainly bounded above by 3 and below by 0. For a more interesting concept, consider the fact that in the previous two examples, when an upper/lower bound existed, there was an element *in* the set that served as an upper/lower bound. Is there an element in the set A which is an upper bound? The value 3 seems like a reasonable candidate, but it is not in A. For any $a \in A$, notice that the point $(a + 6)/3 \in A$; this follows from the order properties in $\mathbb{Q}$. Furthermore $a < (a + 6)/3$. To see this, notice that $a < 3$ implies that $2a < 6$, which in turn implies that $3a < a + 6$. Multiplying both sides by $\frac{1}{3}$, we arrive at $a < (a + 6)/3$. Thus no matter what a we choose, there is an element in A which is greater than a. Therefore no element of A is greater than all the others.

As a final comment, notice that 3 is smaller than all the other upper bounds. This is really just a restatement of the previous comment that no rational number in A is an upper bound, and hence no rational number less than 3 is an upper bound. But again, 3 is not in the set. The point here is that the idea of a smallest or least upper bound should not be confused with the more common notion of a *maximum* which is the largest element of a set or an upper bound which is in the set.

We next provide a rigorous definition for a smallest or least upper bound. It turns out that this is the property we need to complete our definition of the real number system.

Definition 1.2.8. Let X be an ordered field and let $A \subseteq X$ be nonempty and bounded above. We say A has a *least upper bound* or a *supremum* if there is a point in X, denoted $\sup(A)$, such that

(a) $\sup(A)$ is an upper bound for A;
(b) if b is another upper bound for A, then $\sup(A) \leq b$.

One of the first questions that should come to mind is whether or not a set can have more than one supremum. Thankfully this is not the case and, when a supremum exists, it is unique; see Exercise 1.2.6. There you will also consider the equivalent idea of greatest lower bound or infimum of a set.

Example 1.2.9. Returning to the set $A = \{x : 0 < x < 3\} \subseteq \mathbb{Q}$, we see that 3 is the supremum; see the comments above. By similar reasoning, 0 is the infimum and neither of these are in the set A. The concept of infimum and supremum have simply provided a generalization of the maximum and minimum of a set.

For the set $A = \{1, 3, 7\} \subseteq \mathbb{Q}$, notice that 1 is the infimum and 7 is the supremum.

Lemma 1.2.10. *Let X be an ordered field and let $A \subseteq X$ be nonempty and bounded above. If $M \in X$ is an upper bound for A, then $M = \sup(A)$ if and only if for every $\varepsilon \in X$ with $\varepsilon > 0$, there is a point $a \in A$ such that $M - \varepsilon < a$.*

The idea here is a restatement of part (b) of the definition of supremum and it says that M is the supremum if no point less than M is an upper bound for A; the term $M - \varepsilon$ is interpreted as a point of X satisfying $M - \varepsilon < M$ as in Exercise 1.2.5. Thus there must be a point of A which satisfies $M - \varepsilon < a \leq M$. Our investigation of the third set in Example 1.2.7 exhibits this style of argument. There we argued that 3 was the least upper bound by showing that no rational number less than 3 was an upper bound. Figure 1.2 illustrates these two ideas side by side. Thinking of the ordered field X illustrated by a number line, the left-hand image illustrates that *every* point to the right of M is an upper bound for A, while the right-hand image shows that *no* point to the left of M is an upper bound.

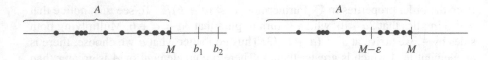

Fig. 1.2 Supremum of a set

Proof. We first assume that $M = \sup(A)$. With this hypothesis we have that M satisfies the definition of supremum above. Now, let $\varepsilon > 0$ in X and consider the element $M - \varepsilon$. By Exercise 1.2.5, we know that $M - \varepsilon < M$. This shows that $M - \varepsilon$ is not an upper bound for A since M is the least upper bound. Thus, by negating the definition of upper bound, we find that there must be a point $a \in A$ with $M - \varepsilon < a$.

For the converse, we assume that for every $\varepsilon > 0$ in X, there is a point $a \in A$ with $M - \varepsilon < a$. Our goal is now to show that M satisfies the second part of the definition of supremum; keep in mind that we are assuming M is an upper bound in the statement of the lemma. To this end, let b be an upper bound for A. We will be done if we can show that $M \leq b$. By order axiom (a), it suffices to rule out the possibility that $b < M$. Hence we assume $b < M$ and seek a contradiction. With this assumption, we know that $M - b > 0$. Here is where our hypothesis comes into play. Setting $\varepsilon = M - b$, we know that there is a point $a \in A$ with $M - \varepsilon < a$. Notice also that our choice of ε implies that $M - \varepsilon = M - (M - b) = b$. These two statements together show that $b < a$ which contradicts the fact that b is an upper bound for A; hence it must be the case that $M \leq b$. Therefore we conclude that $M = \sup(A)$. $\qquad\square$

Definition 1.2.11. We say an ordered field X is *complete* if every set $A \subseteq X$ that is nonempty and bounded above has a least upper bound in X.

Always keep in mind that the definition states that the supremum of a set exists as an element of the larger set X, but that this point is not required to be *in* the set A. Furthermore, the definition above is equivalent to requiring that every nonempty set that is bounded below has a greatest lower bound in X.

Example 1.2.12. We know that the rational numbers form an ordered field; however, it is not complete. To see this, we must produce a set $A \subseteq \mathbb{Q}$ which is not empty and bounded above, but which has no supremum *in* $\mathbb{Q}$. For an example of such a set, take

$$A = \{a \in \mathbb{Q} : a^2 < 2\}.$$

We claim that 2 is an upper bound. To see this, notice that for any $a \in A$, it must be the case that $a^2 < 2 < 4$, according to the order on $\mathbb{Q}$. This immediately implies that $a < 2$ for all positive a in A by Exercise 1.2.5(b); if a is negative or zero, it is trivially true that $a < 2$. Thus 2 is an upper bound.

Next we consider computing the supremum of A. Clearly such a value would need to be positive since $0 \in A$. Notice that the rational number $3/2$ is also an upper bound for A. In fact, any *positive* rational number r with $2 < r^2$ will be an upper bound here, as this inequality implies that $a < r$ for all $a \in A$. From this observation it is evident that the rational number line is split into two disjoint collections,

$$\mathbb{Q} = \{a \in \mathbb{Q} : a^2 < 2\} \cup \{r \in \mathbb{Q} : 2 < r^2\}.$$

With this it is easy to see that the least upper bound M would necessarily need to satisfy $M^2 = 2$. However, there is no rational number with this property! Thus $\mathbb{Q}$ is not complete. In the next section, we will show that there is a *real number* with this property; the construction of such a number will revolve around the sets above.

Exercises

Exercise 1.2.1. Prove parts (b), (e), and (g) of Theorem 1.2.2.

Exercise 1.2.2. Prove parts (a), (c), (e), (f), (h), and (i) of Theorem 1.2.3.

Exercise 1.2.3. Let X be an ordered field and let $x \in X$.

(a) Show that $-0 = 0$.
(b) If $x \neq 0$, show that $(-x)^{-1} = -x^{-1}$.

Exercise 1.2.4. Prove parts (e), (f), and (h) of Theorem 1.2.5.

Exercise 1.2.5. Let X be an ordered field and let $x, y, z, w \in X$.

(a) Let $x, y > 0$. If $x < y$, show that $x^2 < y^2$.
(b) Let $x, y > 0$. If $x^2 < y^2$, show that $x < y$.
(c) Let $x, y > 0$. If $x < y$ and $n \in \mathbb{N}$, show that $0 < x^n < y^n$.
(d) If $x \leq y$ and $x \geq y$, show that $x = y$.
(e) If $x < y$ and $z < w$, show that $x + z < y + w$.
(f) If $0 < x < y$, show that $xy^{-1} < 1$.
(g) If $\varepsilon > 0$ in X, show that $x < x + \varepsilon$ for every $x \in X$.
(h) If $\varepsilon > 0$ in X, show that $x - \varepsilon < x$ for every $x \in X$.

Exercise 1.2.6. Let X be an ordered field and let $A \subseteq X$ be nonempty and bounded above. If $\sup(A)$ exists, show that it is unique.

Exercise 1.2.7. Let X be an ordered field and let $A \subseteq X$ be nonempty and bounded below.

(a) Provide a definition for the greatest lower bound or infimum of A.
(b) Show that $\inf(A)$ is unique when it exists.
(c) Provide a version (including a proof) of Lemma 1.2.10 for infima.

Exercise 1.2.8. Let X be a complete ordered field. Show that every set $A \subseteq X$ that is nonempty and bounded below has a greatest lower bound in X.

Exercise 1.2.9 (Binomial Expansion Theorem). For nonnegative integers n and k we define the binomial coefficient $\binom{n}{k}$ (read "n choose k") by

$$\binom{n}{k} = \frac{n!}{k!(n-k)!}, \quad k = 0, 1, 2, \ldots, n,$$

where $0! = 1$, and $\binom{n}{k} = 0$ for $k > n$.

(a) Show that

$$\binom{n}{k-1} + \binom{n}{k} = \binom{n+1}{k}$$

for $k = 1, 2, \ldots, n$.

(b) Let X be a field. For $x \in X$, show that

$$(1 + x)^n = \sum_{k=0}^{n} \binom{n}{k} x^k$$

for all $n \in \mathbb{N}$.

(c) Use (b) to show that

$$(a + b)^n = \sum_{k=0}^{n} \binom{n}{k} a^k b^{n-k}$$

for all $a, b \in X$.

1.3   The Real Number System

Our goal in this text is to study functions defined on the real numbers. At this point we are ready to define the real numbers but first let us take a moment to consider in more detail why we are in need of this set. The development of the theory of calculus grew out of a desire to model natural phenomena with functions in an effort to predict behavior. This led to the concepts of differentiation and integration, topics that you are somewhat familiar with from your experience in calculus. These two topics both depend heavily on the fact that the real number system is an interval, a term we will define in the next section. If we were to only consider the rational number system, then we lose this property for there are no intervals in $\mathbb{Q}$; we can define an interval in $\mathbb{Q}$ but such a set would be drastically different from the continuum visual that we intuitively associate with intervals. For example, the set

$$\{x \in \mathbb{Q} : 0 \leq x \leq 4\}$$

has breaks where $\sqrt{2}$ and π would appear, and many other gaps as well. This has strong ramifications for differentiation and integration. If we take $\mathbb{Q}$ as our base number system, then we cannot even define the Riemann integral. There are, however, other integration theories and one result will demonstrate that if we consider a function $f : \mathbb{Q} \rightarrow \mathbb{Q}$, then it necessarily follows that the integral of f is zero. Hence the resulting theory of integration would be rather boring in this context.

Another motivating factor is the desire to have every naturally occurring length correspond to a number of some sort and this is certainly not the case for the rational numbers; this issue was discovered by the Greeks who had proved Theorem 1.1.8. This result along with the Pythagorean Theorem led to distress, for it is also true that the length of the hypotenuse, call it c, of a right triangle with side lengths 1 must satisfy $c^2 = 2$. This is an issue that the Greeks were unable to resolve as they associated length with rational number.

Definition 1.3.1. We define the set $\mathbb{R}$ to be a complete ordered field containing $\mathbb{Q}$ as an ordered subfield. The irrational numbers are then defined to be $\mathbb{I} = \mathbb{R} \setminus \mathbb{Q}$.

Two concerns should immediately arise. First, do we have any reason to believe that such a set exists? The approach we are taking is an axiomatic approach. We have laid out the rational number system (in minimal detail) and the axioms for a complete ordered field, and we now declare that $\mathbb{R}$ exists and satisfies the above definition, but we have given no indication as to why such a declaration is valid. If it were not, then your entire experience with number systems up to this point in your life has been a lie! Do not let this alarm you. There is another approach to generating the real number system, a constructive approach in which a set and operations are constructed and an order is defined. However, this is a very involved process and is best appreciated (in this author's opinion) only after a certain level of sophistication has been acquired. The founders of calculus and analysis worked as we have to this point with only an intuitive/axiomatic definition of the real numbers. They laid the foundations of the subject with care, but took certain liberties with regard to the number system in which they worked. Eventually, mathematicians realized a need to imbue the real numbers with a rigorous definition and four distinct constructions were presented in the 1870s. This work came well after much of what we will study here and this is an indication that these earlier mathematicians had taken great care to understand the real numbers even without an explicit formulation for the set; this bed of knowledge was developed as an extension of the well-understood set of rational numbers. To see two of these constructions, see Sect. 8.4 of [1] and Chap. 2 of [30].

The second possible concern arises after we have settled the issue of the existence of such a set and we now ask if the set is unique. Is it possible that there are distinct sets $\mathbb{R}$ and $\mathbb{R}'$ which satisfy the definition above? And, if so, which do we take as our number system? Does the choice matter? This worry is particularly relevant in light of the comments above which indicate that there are four distinct constructions of the real numbers. In mathematics, it is often the case that different objects, two complete ordered fields in our case, have the same *structure*. Verifying this involves showing that the elements in the sets are in 1-to-1 correspondence with each other and that the operations and the order behave similarly; this similarity is summed up by the phrase "there is an order preserving field isomorphism." With an understanding of isomorphisms it is a straightforward, if somewhat involved, process to show that if $\mathbb{R}$ and $\mathbb{R}'$ both satisfy the definition above, then they are isomorphic, meaning that structurally they are the same, though the appearance of the elements may be different.

It is not standard practice to assign the letter $\mathbb{I}$ for the irrational numbers, but we will adopt this convention for the sake of referencing. The sets $\mathbb{N}$, $\mathbb{Z}$, $\mathbb{Q}$, and $\mathbb{R}$, all have common symbolic representations, but these sets have sound mathematical structures. We will see at the end of this section that the irrational numbers do not share this trait; this provides some evidence as to why this set is typically denoted simply by $\mathbb{R} \setminus \mathbb{Q}$ and is not given the honor of a formal symbol.

Now that we have the real number system established as a complete ordered field, it immediately follows that all of the results from the previous section hold. We will typically not reference every field and order property that we use in order to keep our proofs tidy, so it is imperative that you are familiar with all the properties of ordered fields. Furthermore it is also of the utmost importance that you use these properties as stated; obtaining proofs through incorrect manipulation is as good as no proof at all.

One result of the requirement that $\mathbb{Q}$ be a subfield of $\mathbb{R}$ indicates that the additive and multiplicative identities in $\mathbb{R}$ are the same as those in $\mathbb{Q}$. Thus the 0 and 1 elements of $\mathbb{R}$ are the familiar elements of $\mathbb{Q}$. This implies that -1 is also the same in $\mathbb{R}$ as its rational counterpart. These observations will be important when we consider the set of irrational numbers at the end of this section. As a comment on notation, as in $\mathbb{Q}$, we will use the standard fractional notation and write a/b for ab^{-1}, provided $b \neq 0$.

Example 1.3.2. Most of this text concerns functions which map $\mathbb{R}$ into $\mathbb{R}$ and properties of such functions. Here we introduce the most important function of the text, the *absolute value function*,

$$|x| = \begin{cases} x, & x > 0; \\ 0, & x = 0; \\ -x, & x < 0. \end{cases}$$

You should be familiar with this function and it will form the basis for many of our arguments as we proceed. The reason for this is the fact that for $x, y \in \mathbb{R}$, we interpret the quantity $|x - y|$ to represent the distance between x and y, just as

$|x|$ represents the distance from x to 0. Moreover, this function gives us a way to determine when two real numbers are equal. Notice that $|x| = 0$ if and only if $x = 0$ and thus if $|x - y| = 0$, we have $x - y = 0$, or $x = y$. The following proposition details the most important features of this function.

Proposition 1.3.3. *Let* $x, y \in \mathbb{R}$ *and let* $\varepsilon > 0$. *Then*

(a) $|x| = |-x|$;
(b) *if* $x < 0$, *then* $x = -|x|$;
(c) $|xy| = |x||y|$;
(d) $|1/y| = 1/|y|$ *provided* $y \neq 0$;
(e) $|x/y| = |x|/|y|$ *provided* $y \neq 0$;
(f) $-|x| \leq x \leq |x|$;
(g) $|x| < \varepsilon$ *if and only if* $-\varepsilon < x < \varepsilon$;
(h) $|x + y| \leq |x| + |y|$;
(i) $x = 0$ *if and only if* $|x| < \varepsilon$ *for all* $\varepsilon > 0$.

Proof. For (a) we consider three cases. First, if $x = 0$, then $0 = -0$, and thus both sides of the desired equality are 0 by the definition of absolute value. Next, if $x > 0$, then $|x| = x$. Also, $-x < 0$, and hence $|-x| = -(-x) = x$. Combining these two equalities shows that

$$|x| = x = -(-x) = |-x|$$

as desired. Finally, if $x < 0$, then $|x| = -x$. It is also then true that $-x > 0$, and therefore $|-x| = -x$. This shows that $|x| = -x = |-x|$ completing the proof.

To show that (b) holds, notice that $|x| = -x$ whenever $x < 0$ by definition of absolute value. Multiplying both sides of this equality by -1 shows that $x = -|x|$.

To confirm (c) we must consider several cases based on the definition of absolute value. First, if either x or y is 0, then both sides of the desired equation are 0. For a second case, consider what happens when $x, y > 0$. In this case $xy > 0$ and we immediately see that

$$|xy| = xy = |x||y|$$

by the definition. A similar thing happens when $x, y < 0$. Here we have that $xy > 0$, and thus

$$|xy| = xy = (-x)(-y) = |x||y|.$$

The remaining case concerns the situation when one of the two numbers is positive and the other is negative. Without loss of generality we may assume that $x < 0$ and $y > 0$. Then $xy < 0$ and

$$|xy| = -(xy) = (-x)y = |x||y|.$$

To understand (d), notice that part (c) implies $|y||1/y| = |y(1/y)| = |1| = 1$. This shows that $|y|$ and $|1/y|$ are multiplicative inverses and thus

$$|1/y| = (|y|)^{-1} = 1/|y|.$$

For (i), we first suppose that $x = 0$. Then we have that $|x| = 0$ by definition of absolute value and it is immediately clear that $|x| = 0 < \varepsilon$ for every $\varepsilon > 0$.

For the second implication, we assume that $|x| < \varepsilon$ for every $\varepsilon > 0$. If $x \neq 0$, then $|x| > 0$ by definition of absolute value. Choose $\varepsilon = |x|/2$. Then, by hypothesis we have

$$|x| < \varepsilon < |x| \quad \text{or} \quad |x| < \frac{|x|}{2} < |x|$$

which cannot be true by the trichotomy property of our order. Thus we conclude that x must be equal to 0.

The proofs of the remaining parts are left as Exercise 1.3.1   □

Property (h) above will be our most commonly used tool and is called the *triangle inequality*. To see where the name comes from and why this is important, consider three points $r, s, t \in \mathbb{R}$. Then we can write

$$|r - s| = |r + 0 - s| = |r + (-t + t) - s| = |(r - t) + (t - s)|.$$

If we now consider the real numbers $x = r - t$ and $y = t - s$, then $x + y = r - s$ and the triangle inequality shows that

$$|(r - t) + (t - s)| = |r - s| = |x + y| \le |x| + |y| = |r - t| + |t - s|;$$

the use of the labels x and y is meant to clarify the use of the inequality but we will typically forgo such substitutions and write

$$|r - s| = |(r - t) + (t - s)| \le |r - t| + |t - s|.$$

The reason this is important for our work is that we will often want to control the size of a quantity of the form $|r - s|$. However, the information at our disposal will typically involve relationships between r and s, and some other parameter t. Thus by controlling the size of $|r - t|$ and $|t - s|$, we will also be able to control the size of the desired quantity $|r - s|$. To understand the name given to this inequality, imagine that the points r, s, t lie in a two-dimensional plane. Then these three points determine a triangle and the quantities $|r-s|$, $|r-t|$, and $|t-s|$ represent the lengths of the three sides. Thus the inequality states that no single side length exceeds the sum of the other two lengths.

Now we begin a proper investigation of the real numbers. First, to tie up a loose end we show that there is a real number that squares to two, i.e. $\sqrt{2}$ is a real number.

Theorem 1.3.4. *There is a real number whose square is 2.*

Before the proof, a word of caution. The argument we use will seem somewhat technical and, possibly, a little complicated at the moment. This is due in part to the fact that the real numbers are a complicated system. Another reason for this is that we are only at the beginning of our journey. Understanding the proof is the most important thing at the moment, while completely digesting it will come with time. As we will see in later sections, there will be easier ways to assert the existence of roots once we have more machinery in hand.

Proof. The proof here depends solely on the fact that $\mathbb{R}$ is a complete ordered field and some simple estimates. To begin, we let

$$A = \{a \in \mathbb{R} : a^2 < 2\}.$$

We know that A is not empty since $0, 1 \in A$. Also, A is bounded above by 2 and $3/2$ from Example 1.2.12. Thus we are assured that $\sup(A)$ exists; for the sake of convenience, set $x = \sup(A)$. We now seek to show that $x^2 = 2$. To do this, we will rule out the possibilities $x^2 < 2$ and $x^2 > 2$.

First suppose that $x^2 < 2$. We will produce a real number t such that $x < t$ and $t^2 < 2$. This will show that $t \in A$ but will contradict the fact that x is the supremum of A. To choose the point t we will add a tiny increment to x, the motivation for which is obtained from the inequality $x^2 < 2$. From our assumption, we know that $2 - x^2 > 0$, and we set

$$t = x + \frac{2 - x^2}{4}.$$

It is clear that $x < t$ since the second term in the definition of t is positive; the factor of $1/4$ is chosen so as to keep t close to x in order to guarantee that $t^2 < 2$. To compute t^2, we expand and factor to obtain

$$t^2 = \left(x + \frac{2-x^2}{4}\right)^2 = x^2 + \frac{x(2-x^2)}{2} + \frac{(2-x^2)^2}{16} = x^2 + \frac{(2-x^2)}{2}\left(x + \frac{2-x^2}{8}\right).$$

Now, we estimate this quantity. First, we know that $x < 3/2$. Also, since $1 \in A$, we know that $1 < x$ (the inequality is strict since we know x is not rational), which implies that $1 < x^2$. A slight modification produces the inequality $2 - x^2 < 1$. Now, focus on the term in parentheses on the far right of the previous equation. Using the estimates above, we know that

$$x + \frac{2-x^2}{8} < x + \frac{1}{8} < \frac{3}{2} + \frac{1}{8} < 2.$$

Thus

$$t^2 = x^2 + \frac{(2-x^2)}{2}\left(x + \frac{2-x^2}{8}\right) < x^2 + 2\frac{(2-x^2)}{2} = x^2 + 2 - x^2 = 2.$$

This rules out the possibility that $x^2 < 2$.

For the second case, we assume $2 < x^2$. Here we will take a similar tactic and produce a *positive* real number t such that $t < x$ and $2 < t^2$. This will show that t is an upper bound for A but will contradict the fact that x is the smallest of all the upper bounds. Again, we will choose a point that is incrementally close to x. For this case, we set

$$t = x - \frac{x^2 - 2}{4}.$$

It should be clear that $t < x$ since $x^2 - 2 > 0$ in this case. To see that $t > 0$, we again use the fact that $x < 3/2$ which implies that $x^2 < 9/4$. Combining this with the assumption that $x^2 > 2$, we conclude that $x^2 - 2 < 1/4$ or

$$0 < \frac{x^2 - 2}{4} < \frac{1}{16}.$$

From this it follows that

$$t = x - \frac{x^2 - 2}{4} > x - \frac{1}{16} > 1 - \frac{1}{16} = \frac{15}{16} > 0$$

since $x > 1$.

Now, to see that $2 < t^2$, we have

$$t^2 = \left(x - \frac{x^2 - 2}{4}\right)^2 = x^2 - \frac{x(x^2 - 2)}{2} + \frac{(x^2 - 2)^2}{16} > x^2 - \frac{x(x^2 - 2)}{2}$$

where the last inequality comes from the fact that

$$\frac{(x^2 - 2)^2}{16} > 0.$$

To estimate what we have left, notice that $x < 2$ from our previous work in Example 1.2.12. Using the order properties, this implies that $x/2 < 1$, from which we see that $-x/2 > -1$. Using this we see that

$$t^2 > x^2 - \frac{x(x^2 - 2)}{2} > x^2 - (x^2 - 2) = 2.$$

This rules out the possibility that $x^2 > 2$, and thus we conclude that $x^2 = 2$.   □

N, Q, *and* I *as Subsets of* R

The first two results of this subsection allow us to understand the behavior of the natural and rational numbers inside the real number system. Both of these subsystems will be of great importance throughout the text.

Theorem 1.3.5. *Let x and y be positive real numbers. Then there is an $n \in \mathbb{N}$ with $x < ny$.*

The statement here looks as though we should fix x and y in $\mathbb{R}$ and then *produce* a natural number with the desired property. Constructive proofs are appreciated when possible, however, if we consider the negation of the conclusion it reads, "$ny \leq x$ for every $n \in \mathbb{N}$." This is a statement about an upper bound and with this extra bit of information, we will work to obtain a contradiction.

Proof. Let x and y be positive real numbers and suppose that $ny \leq x$ for every $n \in \mathbb{N}$. Then the set $A = \{ny : n \in \mathbb{N}\}$ is bounded above; it is certainly nonempty since $y \in A$. With this we know that $\sup(A)$ exists and we will call it M. We also know that $y > 0$, so $M - y < M$. From here, Lemma 1.2.10 shows that there is an element $ny \in A$ such that $M - y < ny$. But then

$$M < ny + y = (n + 1)y$$

and $(n + 1)y \in A$. This is a contradiction since M is assumed to be the supremum of A. Thus our hypothesis that $ny \leq x$ for every $n \in \mathbb{N}$ must be false and we conclude that there is an $n \in \mathbb{N}$ such that $x < ny$. □

We will typically use this theorem in a simpler form which we state as a corollary.

Corollary 1.3.6 (Archimedean Property).

(a) If x is a real number, then there is an $n \in \mathbb{N}$ with $x < n$.
(b) If y is a real number with $0 < y < 1$, then there is an $n \in \mathbb{N}$ with $1/n < y$.

Proof. For (a), the statement is trivially true if $x \leq 0$. For $x > 0$, simply take $y = 1$ in the preceding theorem. For the second statement, take $x = 1$ in the theorem above. □

Notice that the two parts of the corollary can be interpreted as striving for two extremes. When applying the first statement, we will typically think of x as a large number. The corollary then says that we can find a natural number that exceeds this value. On the other hand, the second statement says that if we have a number y with $0 < y < 1$, a small number, then we can get below it with a number of the form $1/n$. Put another way, *no matter* how small the number y may be, there is a smaller number of the form $1/n$. Also, note that the hypothesis that y satisfy $0 < y < 1$ is not necessary, but it is the only interesting situation as the conclusion of part (b) is trivially satisfied for $y > 1$.

We now turn our attention to the rational numbers. This set has the same cardinality as the natural numbers but they are more interspersed throughout the real number system than the natural numbers and the integers. The following definition gives a concrete meaning to this intuitive statement.

Definition 1.3.7. Let S be a subset of $\mathbb{R}$. We say that S is *dense* (or *dense in* $\mathbb{R}$) if there is a point of S between any two points of $\mathbb{R}$, i.e. given $x, y \in \mathbb{R}$ with $x < y$, there exists $r \in S$ such that $x < r < y$.

Theorem 1.3.8. *The rational numbers are dense in* $\mathbb{R}$.

Our proof will focus on the case for two positive real numbers $0 \leq x < y$. We will then need to make two choices, one for the numerator m and one for the denominator n of our desired rational number and both of these choices will play a specific role in guaranteeing that $x < \frac{m}{n} < y$. The denominator n is chosen so that the incremental step $1/n$ does not exceed the distance between x and y. The numerator m then counts the minimal number of steps of size $1/n$ that we take from 0 to step over x. Then, the choice of n guarantees us that if we stop the moment we pass over x, we will be in between x and y.

Proof. Let x and y be real numbers with $x < y$. The definition above states that we must produce a rational number r with $x < r < y$. We can reduce our work if we consider three distinct cases. First, if $x < 0 < y$, then $r = 0$ is our desired rational number.

For the remaining cases, we consider $0 \leq x < y$ and $x < y \leq 0$. First suppose that $0 \leq x < y$. With this assumption it is clear that we seek a positive rational number r; specifically, we need to find $m, n \in \mathbb{N}$ such that $x < \frac{m}{n} < y$. To choose n, the Archimedean Property guarantees the existence of a natural number n such that $1/n < y - x$ which is equivalent to $x + 1/n < y$. For the choice of m, we appeal to the fact that $\mathbb{N}$ is a well-ordered set and choose m to be the *smallest* natural number such that $nx < m$. This immediately implies that $m - 1 \leq nx < m$ which is equivalent to

$$\frac{m-1}{n} \leq x < \frac{m}{n};$$

this demonstrates half of the desired inequality and to complete this argument we need to show that $\frac{m}{n} < y$. To see this, first note that our choice of m also demonstrates that $m \leq nx + 1$. Dividing both sides of this expression by n and recalling the restrictions on n yields

$$\frac{m}{n} \leq x + \frac{1}{n} < y$$

as desired.

For the final case, suppose that $x < y \leq 0$. Multiplying here by -1 produces the inequality $0 \leq -y < -x$. By the previous case there is a rational number r with

$-y < r < -x$. Another multiplication together with the fact that $\mathbb{Q}$ is a field shows that $-r \in \mathbb{Q}$ with $x < -r < y$, completing the proof.   □

An immediate corollary shows that the irrational numbers also form a dense subset of $\mathbb{R}$. We leave the proof as an exercise.

Corollary 1.3.9. *The irrational numbers are dense in* $\mathbb{R}$.

This corollary provides a segue to a brief discussion of the irrational numbers. Many students feel more comfortable with the rational numbers than with their irrational counterpart, but the fact that these numbers make up a large portion of the real number system should indicate a need for familiarity with this set. First, the irrational numbers do not form a subfield of $\mathbb{R}$. The easiest way to see this is to recognize the fact that there is no 0 element in $\mathbb{I}$. For if there were, then it would necessarily be the same as the 0 element in $\mathbb{R}$, but this particular zero has been claimed by the rational numbers.

This is enough to show that $\mathbb{I}$ is not a field, but it is also instructive to understand which of the other axioms are not satisfied. The purely operational axioms such as commutativity and associativity will be satisfied as they do not involve inclusion statements. As with the zero, there is no multiplicative identity. Of the other axioms, it is easiest at the moment to show that the set is not closed under multiplication. To understand why, observe that $(\sqrt{2})^2 = 2$, two irrational numbers with product in $\mathbb{Q}$. The set is also not closed with respect to addition, it is, however, closed with respect to additive and multiplicative inverses; proofs of these statements are requested in the exercises.

The set is certainly ordered, a trait it inherits from $\mathbb{R}$. This is enough to consider whether the set is complete, but this is rather uninteresting without the field structure, as we lose properties such as Lemma 1.2.10. We leave it as an exercise to come up with a set $A \subseteq \mathbb{I}$ which is nonempty and bounded above, but which has no supremum in $\mathbb{I}$.

A topic which we will consider in the next section investigates the cardinality of $\mathbb{R}$ and $\mathbb{I}$. Intuition may lead you to believe that $\mathbb{I}$ is countable as it is constructed by "filling in the gaps" in the countable set $\mathbb{Q}$. This would then imply that $\mathbb{R}$ is also countable as it is the union of these two sets. We'll call this a cliffhanger!

As a final comment, recall that the irrational numbers are defined by a negated statement, i.e. a real number that is *not* rational. Understanding this tiny detail indicates that, most often, to show a number is irrational we will use a proof by contradiction and assume that the number is rational; this strategy allows us to exploit our understanding of the structure in $\mathbb{Q}$. Keep in mind that we have seen examples of this strategy for $\sqrt{2}$ and $\sqrt{3}$ and, for the moment, these are our only examples of irrational numbers. The exercises will demonstrate several other examples based on roots of natural numbers.

For more information on irrational numbers, the text [22] provides a gentle introduction while the more mature text [23] involves a rigorous investigation of the set.

Exercises

Exercise 1.3.1. Prove parts (e), (f), (g), and (h) of Proposition 1.3.3

Exercise 1.3.2. Let $a, b \in \mathbb{R}$. Show that the following relationships hold:

(a) $|a - b| \leq |a| + |b|$;
(b) $|a| - |b| \leq |a + b|$;
(c) $|a| - |b| \leq |a - b|$;
(d) $\big| |a| - |b| \big| \leq |a - b|$;
(e) $|a + b|^2 + |a - b|^2 = 2|a|^2 + 2|b|^2$.

Exercise 1.3.3. Let $a, b \in \mathbb{R}$.

(a) Show that $a \leq 0$ if and only if $a < \varepsilon$ for every $\varepsilon > 0$.
(b) Show that $a \geq 0$ if and only if $a > -\varepsilon$ for every $\varepsilon > 0$.
(c) Show that $a = b$ if and only if $|b - a| < \varepsilon$ for all $\varepsilon > 0$.
(d) Show that $a \leq b$ if and only if $a < b + \varepsilon$ for every $\varepsilon > 0$.
(e) Show that $a \geq b$ if and only if $a > b - \varepsilon$ for every $\varepsilon > 0$.

Exercise 1.3.4. Let $A \subseteq \mathbb{R}$ be nonempty. We say A is *bounded* if there exists an $M > 0$ such that $|a| \leq M$ for all $a \in A$. Show that A is bounded if and only if A is bounded above and bounded below.

Exercise 1.3.5. Show that the infimum of the set $\left\{ \frac{1}{n} : n \in \mathbb{N} \right\}$ is 0.

Exercise 1.3.6. Compute, without proofs, the suprema and infima of the following sets. Keep in mind that $0 \notin \mathbb{N}$.

$$\text{(a) } \{n \in \mathbb{N} : n^2 < 10\} \qquad \text{(e) } \left\{ \frac{n}{m} : m, n \in \mathbb{N} \text{ with } m + n \leq 10 \right\}$$

$$\text{(b) } \left\{ \frac{n}{m+n} : m, n \in \mathbb{N} \right\} \qquad \text{(f) } \left\{ \frac{4+x}{x} : x \geq 1 \right\}$$

$$\text{(c) } \left\{ \frac{n}{2n+1} : n \in \mathbb{N} \right\} \qquad \text{(g) } \left\{ \frac{\sqrt{x+1}}{x} : x \geq 2 \right\}$$

$$\text{(d) } \left\{ \frac{n+m}{m} : n, m \in \mathbb{N} \right\} \qquad \text{(h) } \{x \in \mathbb{R} : x^2 - x < 6\}$$

Exercise 1.3.7. Assume that $A, B \subseteq \mathbb{R}$ are nonempty, bounded, and satisfy $B \subseteq A$.

(a) Show $\inf(A) \leq \sup(A)$.
(b) Show $\sup(B) \leq \sup(A)$.
(c) Show $\inf(A) \leq \inf(B)$.

Exercise 1.3.8. Assume that $A, B \subseteq \mathbb{R}$ are nonempty and bounded above. Further assume that for each $a \in A$, there is a $b \in B$ such that $a \leq b$.

(a) Show that $\sup(A) \leq \sup(B)$.
(b) What happens to the conclusion above if we assume that for each $a \in A$ there is a $b \in B$ such that $a < b$.

Exercise 1.3.9. Assume that $A \subseteq \mathbb{R}$ is nonempty and bounded above and let $c \in \mathbb{R}$. Define the sets $c + A = \{c + a : a \in A\}$ and $cA = \{ca : a \in A\}$.

(a) Show that $\sup(c + A) = c + \sup(A)$.
(b) If $c \geq 0$, show that $\sup(cA) = c \sup(A)$.

Exercise 1.3.10. Assume that $A, B \subseteq \mathbb{R}$ are nonempty and bounded. Define the sets $-A = \{-a : a \in A\}$ and $A + B = \{a + b : a \in A \text{ and } b \in B\}$.

(a) Show that $\sup(-A) = -\inf(A)$.
(b) Show that $\sup(A + B) = \sup(A) + \sup(B)$ and $\inf(A + B) = \inf(A) + \inf(B)$.

Exercise 1.3.11. Assume that $A \subseteq \mathbb{R}$ is nonempty, bounded, and contains both positive and negative values. Define the sets $A^+ = A \cap [0, \infty)$ and $|A| = \{|a| : a \in A\}$.

(a) Show that

$$\inf(|A|) \leq \inf(A^+) \leq \sup(A^+) = \sup(A) \leq \sup(|A|).$$

(b) If $\sup(A) < \sup(|A|)$, show that $\sup(|A|) = -\inf(A)$.

Exercise 1.3.12. Modify the proof of Theorem 1.3.4 to show that there is a real number whose square is 3.

Exercise 1.3.13. (a) If $t \in \mathbb{I}$, show that $-t$ and $1/t$ are both in $\mathbb{I}$.
(b) Show that $\mathbb{I}$ is not closed with respect to addition.
(c) If $a \in \mathbb{Q}$ and $t \in \mathbb{I}$, show that $a + t$ and at $(a \neq 0)$ are both in $\mathbb{I}$.
(d) If $t \in \mathbb{R}$ with $t^2 \in \mathbb{I}$, show that $t \in \mathbb{I}$.

Exercise 1.3.14. Supply a proof for Corollary 1.3.9. To do this, first consider $x - \sqrt{2}$ and $y - \sqrt{2}$, where $x, y \in \mathbb{R}$ with $x < y$, and Theorem 1.3.8.

Exercise 1.3.15. Let $n \in \mathbb{N}$ and let p, q be prime numbers.

(a) If n is not a perfect square, show that $\sqrt{n}$ is irrational.
(b) Show that $\sqrt{p} + \sqrt{q}$ is irrational.

Exercise 1.3.16. Give an example of a set $A \subseteq \mathbb{I}$ such that A is nonempty and bounded above, but which does not have a supremum in $\mathbb{I}$.

1.4 Set Structures in $\mathbb{R}$

We will be interested in understanding how the domain of a function affects the behavior of the given function. In order to do this, we need to understand the possible set structures available in $\mathbb{R}$ and intervals are the simplest of these.

Definition 1.4.1. Let $A \subseteq \mathbb{R}$. We say that A is an *interval* if whenever $a, b \in A$ and $c \in \mathbb{R}$ with $a < c < b$, it is also true that $c \in A$.

The definition implies that $\mathbb{R} = (-\infty, \infty)$ is an interval, as is the empty set, denoted $\emptyset$, and the singleton $\{a\}$, where $a \in \mathbb{R}$. There are eight other possibilities:

(i) $(a, b) = \{x \in \mathbb{R} : a < x < b\}$ (v) $(a, \infty) = \{x \in \mathbb{R} : x > a\}$

(ii) $[a, b] = \{x \in \mathbb{R} : a \le x \le b\}$ (vi) $[a, \infty) = \{x \in \mathbb{R} : x \ge a\}$

(iii) $(a, b] = \{x \in \mathbb{R} : a < x \le b\}$ (vii) $(-\infty, b) = \{x \in \mathbb{R} : x < b\}$

(iv) $[a, b) = \{x \in \mathbb{R} : a \le x < b\}$ (viii) $(-\infty, b] = \{x \in \mathbb{R} : x \le b\}$

where $a < b$. The sets in (i)–(iv) are bounded; a is the infimum of these sets and b is the supremum. The sets in (v)–(viii) are unbounded; however, the sets in (v) and (vi) are bounded below, while the sets in (vii) and (viii) are bounded above. In these two cases a is the infimum and b is the supremum, respectively.

We will say that the intervals in (i), (v), and (vii) are *open intervals*, and call the sets in (ii), (vi), and (viii) *closed intervals*. Later in this section and the next chapter we will see why we have employed the adjectives open and closed to describe these intervals. Keep in mind that we do not presently consider intervals of the form $[a, \infty]$ as infinity is not a real number value. The symbol ∞ indicates that the interval extends in a never ending fashion in the positive direction while the symbol $-\infty$ indicates that the interval extends indefinitely in the negative direction.

With the idea of intervals in hand, we prove a property which is possibly the best mathematical equivalent to the statement that $\mathbb{R}$ has no gaps.

Theorem 1.4.2 (Nested Interval Property). *For each $n \in \mathbb{N}$, let $I_n = [a_n, b_n]$ be a closed, bounded interval and suppose that the collection $\{I_n : n \in \mathbb{N}\}$ has the property that $I_n \subseteq I_{n-1}$ (we call this a collection of nested intervals). Then $\bigcap_{n=1}^{\infty} I_n$ is not empty.*

Proof. For each $n \in \mathbb{N}$, let I_n be as in the statement of the theorem. To show that the desired intersection is not empty, we must produce a real number x that is in I_n for every $n \in \mathbb{N}$. To do this, consider the set

$$A = \{a_n : n \in \mathbb{N}\}.$$

The fact that the sets I_n are nested indicates that we have the string of inequalities

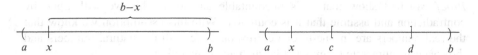

Fig. 1.3 Construction from Lemma 1.4.3

$$a_1 \leq a_2 \leq a_3 \leq \ldots \leq b_3 \leq b_2 \leq b_1.$$

From this observation, it is clear that A is bounded above. By the completeness property of $\mathbb{R}$, we then know that $\sup(A)$ exists; call it x. It is then immediately clear that $a_n \leq x$ for every $n \in \mathbb{N}$. Also, the fact that every b_n is an upper bound for A indicates that $x \leq b_n$ for every $n \in \mathbb{N}$. These two inequalities imply that $a_n \leq x \leq b_n$, i.e. $x \in I_n$, for every $n \in \mathbb{N}$. Therefore we conclude that x is in the intersection of the I_n's as desired. □

The following lemma displays a useful property of open intervals by saying that no matter how small a given interval may be, we can construct a closed interval inside it which excludes a point. In effect this essentially says that we can zoom in on a particular part of an interval to avoid dealing with bothersome points in the interval (Fig. 1.3).

Lemma 1.4.3. *Let $(a,b) \subseteq \mathbb{R}$ be an open interval and let $x \in (a,b)$. Then there is a closed interval $[c,d] \subseteq (a,b)$ with $x \notin [c,d]$.*

Proof. Let $(a,b) \subseteq \mathbb{R}$ be an open interval and let $x \in (a,b)$. The fact that this interval does not include the endpoints dictates that $a < x < b$ and hence we have $b - x > 0$. From this it follows that $\frac{1}{4}(b-x)$ and $\frac{3}{4}(b-x)$ are both positive and less than $b - x$. We define real numbers c and d such that

$$c = x + \frac{1}{4}(b-x) \quad \text{and} \quad d = x + \frac{3}{4}(b-x).$$

It follows immediately that $a < x < c < d < b$ and thus $[c,d] \subseteq (a,b)$ with $x \notin [c,d]$. □

We come again now to a discussion of the cardinality of the real number system. Most of the early examples of sets that we encounter, e.g. the natural numbers, the integers, and the rational numbers, are infinite sets, but only countably infinite. It is somewhat surprising then that the real number system is uncountable; recall our discussion from the end of the previous section. This fact immediately implies the equally surprising result that the irrational numbers are also an uncountable set even though we have so few concrete examples of irrational numbers.

The proof of the result here is by contradiction which is typical when trying to show a set is uncountable since we have a more tangible means of identifying countable sets.

Theorem 1.4.4. *The real and irrational numbers are uncountable.*

Proof. We first show that $\mathbb{R}$ is uncountable and to do this we will argue by contradiction and assume that it is countable. With this assumption we know that the real numbers are in one-to-one correspondence with the natural numbers and thus we can enumerate them; let A be such an enumeration,

$$A = \{x_1, x_2, x_3, \ldots\},$$

where we claim that every real number appears exactly once in A. For a contradiction we will work to construct a real number which is not represented in A and to do this we will appeal to the Nested Interval Property of $\mathbb{R}$.

To this end, we construct a sequence of closed, bounded, nested intervals via the following algorithm. Choose I_1 to be a closed bounded interval in $\mathbb{R}$ not containing x_1. For those who prefer to *see* an actual interval, we can take $[x_1 + 1, x_1 + 10]$ as our set though keep in mind that 1 and 10 have been chosen at random. Continuing, if $x_2 \notin I_1$, then choose $I_2 = I_1$. If $x_2 \in I_1$, use Lemma 1.4.3 to produce a closed, bounded interval I_2 with $I_2 \subseteq I_1$ and $x_2 \notin I_2$. We then carry out this process inductively and, for each $n \in \mathbb{N}$, we produce a closed, bounded interval I_n such that $I_n \subseteq I_{n-1}$ and $x_n \notin I_n$.

Here we make two observations. First, since $x_n \notin I_n$, we know that $x_n \notin \cap_{j=1}^{\infty} I_j$. The fact that this holds for every $n \in \mathbb{N}$ implies that no point of A is in this intersection. Restated, there is no real number in this intersection as A contains *all* the real numbers. On the other hand, by the Nested Interval Property there must be a real number in this intersection. This is a contradiction and thus $\mathbb{R}$ is uncountable.

To see that $\mathbb{I}$ is uncountable, we play a similar game and assume that it is countable. This immediately implies that $\mathbb{Q} \cup \mathbb{I}$ is countable, but $\mathbb{R} = \mathbb{Q} \cup \mathbb{I}$. This is a contradiction as we have just shown that $\mathbb{R}$ is uncountable. Therefore it must be the case that $\mathbb{I}$ is also uncountable. $\square$

Open, Closed, and Compact Sets

The open intervals defined earlier in this section provide the motivation for a large class of subsets of $\mathbb{R}$. The following definition characterizes the bounded open intervals in terms of their length rather than their endpoints, after which we use this to define general open sets.

Definition 1.4.5. Let $c \in \mathbb{R}$ and let $\varepsilon > 0$. We define the *ε-neighborhood* about c, denoted $B_\varepsilon(c)$, to be the set

$$B_\varepsilon(c) = \{x \in \mathbb{R} : |x - c| < \varepsilon\} = (c - \varepsilon, c + \varepsilon).$$

We call c the *center* of the neighborhood.

Fig. 1.4 Construction of ε-neighborhood

Example 1.4.6. The 1-neighborhood about 0 is the open interval $(-1, 1)$. Similarly, for $\varepsilon > 0$, the ε-neighborhood about 0 is the interval $(-\varepsilon, \varepsilon)$. We can replace 0 with any other real number.

The idea being conveyed here is that a real number x is in the ε-neighborhood about c if x is not more than ε units from c, in either the positive or negative direction. In practice we will typically think of ε as a small value, and hence these neighborhoods are a concrete representation of the real numbers which are "close" to c.

Next we define a general open set. We mentioned before that this is intended to be a generalization of open intervals. To understand this definition then, it makes sense to first determine a suitable defining characteristic for open intervals. Let $(a, b) \subseteq \mathbb{R}$. The crucial observation here is that for a point $c \in (a, b)$, there is a buffer between c and *all* points outside (a, b). Said another way, every real number which is close to c is in (a, b). We can interpret this in terms of ε-neighborhoods provided we can decide what the term "close" means in this context. With the idea of a buffer against the points outside of (a, b) in mind, take

$$\varepsilon = \frac{1}{2} \min\{b - c, c - a\},$$

half the distance from c to the closest endpoint of (a, b). Then we have

$$B_\varepsilon(c) = (c - \varepsilon, c + \varepsilon) \subseteq (a, b).$$

Thus we have constructed a neighborhood around c that stays *entirely* inside (a, b). Here the value ε is defining closeness to c in relation to the interval (a, b) and all the points within ε units of c are also in (a, b); see Fig. 1.4. This leads to the following definition.

Definition 1.4.7. Let $O \subseteq \mathbb{R}$. We say that O is an *open set* if for every $c \in O$, there is an $\varepsilon > 0$ such that $B_\varepsilon(c) \subseteq O$.

Example 1.4.8. By the discussion preceding the definition we have that every open interval of the form (a, b) is an open set. It should also be clear that $\mathbb{R}$ itself is an open set since, given $c \in \mathbb{R}$, we have $B_\varepsilon(c) \subseteq \mathbb{R}$ for every $\varepsilon > 0$. It is also vacuously true that the empty set is an open set.

The exercises will ask you to show that the sets of the form (a, ∞) and $(-\infty, b)$ are open.

Fig. 1.5 Closed interval not
an open set

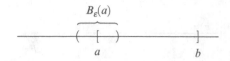

For examples of sets that are not open, consider sets of the form $[a, b]$, $[a, b)$, $(a, b]$, $[a, \infty)$, and $(-\infty, b]$. The problem here is with the endpoints and the requirement that an ε-neighborhood extend to the right *and* left of its center. For the set $[a, b]$, consider the lower endpoint a. No matter how small we choose $\varepsilon > 0$, it is impossible to conclude that

$$B_\varepsilon(a) = (a - \varepsilon, a + \varepsilon) \subseteq [a, b]$$

since any real number $x \in (a - \varepsilon, a)$ is not in $[a, b]$; see Fig. 1.5. Thus the definition of open fails since a is an element inside the set. This is an important point—the definition has to hold *for every* $c \in O$, so to show that it fails, it suffices to show that it fails for *a single* point in O.

For a more exotic example, consider the rational numbers. If $r \in \mathbb{Q}$, is it possible to construct an interval about r which consists only of rational numbers? Thinking back to our density statement for the irrational numbers, if $\varepsilon > 0$, we know that there must be an irrational t with $r - \varepsilon < t < r + \varepsilon$. Thus there is no $\varepsilon > 0$ for which $B_\varepsilon(r) \subseteq \mathbb{Q}$ and hence $\mathbb{Q}$ is not open.

With these few examples in mind, the following theorem provides a means of constructing other open sets.

Theorem 1.4.9. *(a) The union of a countable collection of open sets is an open set.*
(b) The intersection of a finite collection of open sets is an open set.

Proof. First suppose that $\{O_1, O_2, O_3, \ldots\}$ is a countable collection of open sets and set $O = \cup_{n=1}^\infty O_n$. To show that O is open, let $c \in O$. We must now produce an $\varepsilon > 0$ so that $B_\varepsilon(c) \subseteq O$. By the definition of union, we know that $O_n \subseteq O$ for every $n \in \mathbb{N}$ and $c \in O_N$ for some $N \in \mathbb{N}$. Furthermore, since O_N is open, there is an $\varepsilon > 0$ such that $B_\varepsilon(c) \subseteq O_N \subseteq O$. Thus O is open.

For the second statement, suppose $\{O_1, O_2, \ldots, O_N\}$ is a finite collection of open sets and let $O = \cap_{n=1}^N O_n$. As before, let c be an arbitrary point of O; the statement is trivially true if O is empty. To construct an ε-neighborhood here, we must be a bit more careful. In order for a neighborhood to be in the intersection, we must find an $\varepsilon > 0$ so that $B_\varepsilon(c) \subseteq O_n$ for $n = 1, 2, \ldots, N$. The fact that each O_n is open means that there is an ε_n such that $B_{\varepsilon_n}(c) \subseteq O_n$, but this provides no relationship between $B_{\varepsilon_1}(c)$ and O_2 or any other combination. Thinking creatively, we know that c is in each O_n, as are all the points close to c where the term "close" here is relative to the particular open set under consideration. It stands to reason that if we take the smallest measure of closeness, then we should be able to guarantee that these points are in *each* O_n. We set $\varepsilon = \min\{\varepsilon_1, \varepsilon_2, \ldots, \varepsilon_N\}$. Notice that $\varepsilon > 0$ since it is defined as a minimum and every $\varepsilon_n > 0$, and thus we have

$$B_\varepsilon(c) \subseteq B_{\varepsilon_n}(c) \subseteq O_n$$

for $n = 1, 2, \ldots, N$ by Exercise 1.4.6. Therefore $B_\varepsilon(c) \subseteq O$ as desired and hence O is open. $\square$

Thinking about the second statement above, the exercises will ask you to explain where the hypothesis "finite collection of open sets" is used in the proof. You will also be asked to provide an example to show that the statement can fail without this hypothesis.

Of the various set structures we will study, the open sets are by far the simplest. The evidence for this claim is provided by the fact that a strengthened version of the converse to the statement in (a) is true: every open set can be written as a countable union of *open intervals* (Exercise 1.4.8). Thus understanding open sets boils down to understanding open intervals.

We turn now to closed sets. It is natural to think of open and closed as antonyms, but it is not the case in this situation. A set that is not closed will not necessarily be open, nor vice versa. Our relationship will be via complements. We will revisit closed sets in Chap. 2 where we will encounter a different characterization which is usually taken as the definition of a closed set, but we find the following to be simpler to deal with at the outset. The second characterization will also provide insight as to why we use the word closed to describe such sets.

Definition 1.4.10. Let $F \subseteq \mathbb{R}$. We say that F is a *closed set* if F^c is an open set.

Example 1.4.11. The definition immediately implies that $\mathbb{R}$ and $\emptyset$ are both closed since their respective complements, $\emptyset$ and $\mathbb{R}$, are both open.

Similarly, the set $[a, b]$ is closed since its complement $(-\infty, a) \cup (b, \infty)$ is open by Exercise 1.4.7 and Theorem 1.4.9.

The same logic can be used to show that *any* finite set is closed. For a simple example, take $F = \{1, 2\}$. Then $F^c = (-\infty, 1) \cup (1, 2) \cup (2, \infty)$ which is a union of open intervals and hence open. The exercises will ask you to extend this to general finite sets.

As a last example, notice that $\mathbb{I}$ is not closed since its complement $\mathbb{Q}$ is not open.

Theorem 1.4.9 has an immediate analog for closed sets. The proof here is based on De Morgan's Laws for countable unions and intersections.

Theorem 1.4.12. *(a) The union of a finite collection of closed sets is a closed set. (b) The intersection of a countable collection of closed sets is a closed set.*

The final set structure we consider here is that of compact sets. These are natural extensions of closed, bounded intervals and will have particular relevance for our study of continuous functions; see Exercise 1.3.4 for the definition of a bounded set.

Definition 1.4.13. Let $K \subseteq \mathbb{R}$. We say K is *compact* if K is closed and bounded.

Example 1.4.14. Examples of compact sets include finite sets and closed, bounded intervals. For sets that are not compact we have $\mathbb{R}$, any unbounded set, and any open set (the empty set is not considered to be a bounded set).

While we did say that compact sets are meant to be a generalization of closed, bounded intervals, do not be fooled into thinking that compact sets necessarily have a simple appearance. For example, consider the set

$$A = \{1, 1/2, 1/3, \ldots\}.$$

The set is certainly bounded. To determine if it is compact, we need to need to check that it is closed. Taking the complement we find

$$A^c = (-\infty, 0] \cup (1, \infty) \cup \left(\bigcup_{n=1}^{\infty} \left(\frac{1}{n+1}, \frac{1}{n} \right) \right).$$

This set is not open since there is no ε-neighborhood about 0 which lies entirely inside the set. To convince yourself of this, think about the Archimedean Property. On the other hand, the set

$$K = \{0, 1, 1/2, 1/3, \ldots\}$$

is compact since it is bounded and its complement

$$K^c = (-\infty, 0) \cup (1, \infty) \cup \left(\bigcup_{n=1}^{\infty} \left(\frac{1}{n+1}, \frac{1}{n} \right) \right)$$

is open.

Example 1.4.15. Here we give an example of one of the most famous subsets of $\mathbb{R}$; the structure of this set is remarkable and we will investigate several of its properties here and in later sections. The set is named for Georg Cantor who was interested in the study of infinite sets and is responsible for much of what we know about the cardinalities of such sets. This first encounter provides the construction of the set and demonstrates that uncountable sets can appear rather small at first glance.

We will construct the set in a series of stages. To begin, let C_0 be the unit interval $[0, 1]$. For the set C_1, we remove the open interval that comprises the middle third of C_0,

$$C_1 = \left[0, \frac{1}{3} \right] \cup \left[\frac{2}{3}, 1 \right],$$

a union of two closed intervals. Next, we remove the open middle third of the two intervals which make up C_1,

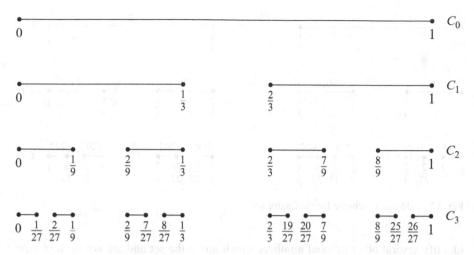

Fig. 1.6 Construction of the Cantor set

$$C_2 = \left[0, \frac{1}{9}\right] \cup \left[\frac{2}{9}, \frac{1}{3}\right] \cup \left[\frac{2}{3}, \frac{7}{9}\right] \cup \left[\frac{8}{9}, 1\right],$$

a union of four closed intervals. This process then continues indefinitely; Fig. 1.6 shows the first few of these sets. Notice that for $n \in \mathbb{N}$, C_n is a union of 2^n closed intervals of length $1/3^n$. It is then immediate that each C_n is closed since these are finite unions. We then define the *Cantor set* C by

$$C = \bigcap_{n=1}^{\infty} C_n;$$

put simply, C is what is left over after we have removed all the open middle thirds. It follows immediately that C is closed since intersections of closed sets are always closed. Moreover, $C \subseteq [0, 1]$ showing that C is bounded and therefore compact. We will revisit this set on several occasions to investigate further properties, but for now we will content ourselves with understanding the cardinality of C.

It should be apparent that 0 and 1 are in the Cantor set as these points are in C_n for every $n \in \mathbb{N}$. Furthermore, so is 1/3 and 2/3, as is any point which is an endpoint of one of the removed open intervals. This provides a means of constructing points in C, all of the form $m/3^n$, $m, n \in \mathbb{N}$. Notice that the set $\{\frac{m}{3^n} : m, n \in \mathbb{N}\}$ is a countable set. And although many of the points in this set are in C, there are also points that are not, 4/9 for instance. However, we do know that C is necessarily infinite from its construction and thus we know that C is at least countably infinite.

The question then is whether there are any more points in the Cantor set other than the endpoints of the removed intervals. As alluded to earlier, the set is actually uncountable, and hence there must be many more points. In Example 3.3.11 we will

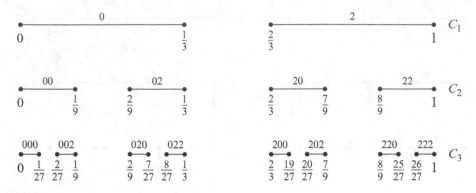

Fig. 1.7 Addressing scheme for the Cantor set

identify several other rational numbers which are in the set and are not of the form $m/3^n$; the rational number 1/4 is an example of such an element. Our goal now is to show that set is uncountable, which we will do not by constructing points in C, but by constructing a bijective function from C to an uncountable set.

To define such a function, we first devise an *addressing scheme* for C; this process is akin to defining a decimal expansion except that we work base 3 instead of base 10. If you are familiar with ternary expansions, then you will recognize our process. The idea is to assign a sequence of digits to each member of the Cantor set where the nth digit in the sequence is based on the point's location in C_n. Let $x \in C$ be fixed. We begin in C_1. If x is in the left-hand subinterval $[0, 1/3]$, then we assign 0 as the first digit in x's expansion. If x is in the right-hand subinterval $[2/3, 1]$, then we assign 2 as the first digit.

For the second digit, we have to remember where x lives in C_1, so for the moment let us call that interval I, either $[0, 1/3]$ or $[2/3, 1]$. Now, in C_2, I has been split into two subintervals. If x is in the left-hand subinterval of I, we assign 0 as the second digit, or 2 if x is in the right-hand subinterval. We then continue this process at each stage of the construction of C as detailed in Fig. 1.7.

For example, notice that 0 is assigned to the string consisting only of 0s since 0 is in a left-hand subinterval at every stage. Likewise, 1 is assigned to the string of all 2s. The endpoint 2/9 is assigned the string $02\overline{0}$ where the overline represents that 0 is repeated indefinitely. As mentioned earlier, the point 1/4 is in the Cantor set and it is an exercise to check that its expansion is $\overline{02}$; we will see why this is true in Sect. 3.3.

Now, since each point of the Cantor set is in each C_n, then we are assured that each point of the Cantor set can be assigned to such a string of 0s and 2s. At this point, define D to be the collection of all possible strings of 0s and 2s. It is important that each string in D has countably many digits, one for each natural number. We then claim that the function $f : C \rightarrow D$ which maps an element in C to its expansion is bijective; it is certainly well defined since each member of C can be assigned to such a string. It should also be clear that the function is 1-to-1 since any string of 0s and 2s can define at most one element of the Cantor set. To see that it is onto, notice that any string of 0s and 2s necessarily *defines* a real number according to the addressing above. Moreover, this real number must be an element of C as such an expansion dictates where the real number falls in each stage of the construction of C.

With this bijection established, we will be able to conclude that C is uncountable if we can show that D is uncountable. To do this, we will assume that D is countable just as we did when showing $\mathbb{R}$ is uncountable. With this assumption, we fix an enumeration

$$D = \{x_1, x_2, x_3, \ldots\}$$

and claim that every element of D is represented exactly once in this set. We now construct an element in D which is not represented in this list. First keep in mind that each element of D is an infinite string of 0s and 2s. We represent this with the array

$$x_1 = a_{11}\, a_{12}\, a_{13} \ldots a_{1n} \ldots$$
$$x_2 = a_{21}\, a_{22}\, a_{23} \ldots a_{2n} \ldots$$
$$x_3 = a_{31}\, a_{32}\, a_{33} \ldots a_{3n} \ldots$$
$$\vdots \qquad \vdots$$
$$x_n = a_{n1}\, a_{n2}\, a_{n3} \ldots a_{nn} \ldots$$
$$\vdots \qquad \vdots$$

where each of the a_{ij}'s is either 0 or 2; the digits in bold will be used to define a string of 0s and 2s which is distinct from every member in the list above. For each $n \in \mathbb{N}$, we define

$$b_n = \begin{cases} 0, & a_{nn} = 2; \\ 2, & a_{nn} = 0. \end{cases}$$

The string $b_1 b_2 b_3 \ldots$ is then an element of D. On the other hand, for $n \in \mathbb{N}$, this string differs from x_n in the nth digit and thus is not in our representation for D. This is a contradiction and we therefore conclude that D, and hence C, are both uncountable.

Exercises

Exercise 1.4.1. The following exercise asks you to challenge the statement of the Nested Interval Property and show that it fails if we alter the hypotheses.

(a) Prove that $\displaystyle\bigcap_{n=1}^{\infty} \left(0, \frac{1}{n}\right) = \emptyset$.

(b) Why does this not dispute the Nested Interval Property?

(c) Provide an example which shows that the Nested Interval Property fails if we
use unbounded, closed, nested intervals.

Exercise 1.4.2. (a) Show that the interval $(0, 1)$ is uncountable.

(b) Show that the interval (a, b) is uncountable.

Exercise 1.4.3. (a) Show that the interval $[0, 1]$ is uncountable.

(b) Show that the interval $[a, b]$ is uncountable.

Exercise 1.4.4. Suppose $A \subseteq \mathbb{R}$ is a countable set. Show that A is not open.

Exercise 1.4.5. (a) Suppose $p, q, n \in \mathbb{N}$. If $\frac{p}{q} \in (0, \frac{1}{n})$, show that $\frac{1}{q} \in (0, \frac{1}{n})$.

(b) Suppose $p, q, n \in \mathbb{N}$. If $\frac{p}{q} \in (1 - \frac{1}{n}, 1)$, show that $\frac{1}{q} \in (0, \frac{1}{n})$.

(c) Suppose $p, q, n \in \mathbb{N}$. If $\frac{p}{q} \in (1, 1 + \frac{1}{n})$, show that $\frac{1}{q} \in (0, \frac{1}{n})$.

Exercise 1.4.6. Let $c \in \mathbb{R}$ and suppose $0 < \delta < \varepsilon$. Show that $B_\delta(c) \subseteq B_\varepsilon(c)$.

Exercise 1.4.7. (a) Show that sets of the form (a, ∞) and $(-\infty, b)$ are open sets.

(b) Show that sets of the form $(a, b]$, $[a, b)$, $[a, \infty)$, and $(-\infty, b]$ are not open.

(c) Show that $\mathbb{I}$ is not open.

Exercise 1.4.8. Show that every open set in $\mathbb{R}$ can be written as a countable union
of open intervals.

Exercise 1.4.9. Explain where the hypothesis "finite collection of open sets" is
used in the proof of the second statement in Theorem 1.4.9 and provide an example
showing that the theorem can fail if this hypothesis is altered.

Exercise 1.4.10. (a) Show that sets of the form $[a, \infty)$ and $(-\infty, b]$ are closed sets.

(b) Show that sets of the form $(a, b]$, $[a, b)$, (a, ∞), and $(-\infty, b)$ are not closed.

(c) Show that any finite set is closed.

(d) Show that $\mathbb{Q}$ is not closed.

Exercise 1.4.11. Provide a proof for Theorem 1.4.12.

Exercise 1.4.12. Let $F, K, L \subseteq \mathbb{R}$ with F closed, and K and L compact. Further
assume that $K \cap L$ and $K \cap F$ are not empty.

(a) Show that $K \cup L$ is compact.

(b) Show that $K \cap L$ is compact.

(c) Show that $K \cap F$ is compact.

(d) Is $K \cup F$ compact? Prove or provide a counterexample.

Exercise 1.4.13. Let $A, B \subseteq \mathbb{R}$.

(a) If A is open and B is a closed set, show that $A \backslash B$ is open.

(b) If A is closed and B is open, show that $A \backslash B$ is closed.

(c) If A is compact and B is open, show that $A \backslash B$ is compact (you may assume
$A \setminus B$ is not empty).

1.5 Normed Linear Spaces

As a reminder, a *vector space* is a set X whose elements are called *vectors*, a field of scalars $\mathbb{F}$ (numbers, typically real or complex), and two operations: an *addition* defined on X, and a *scalar multiplication* of the vectors in X by the scalars in $\mathbb{F}$. In addition to these four objects, there are also several operational axioms which dictate how the elements in X and $\mathbb{F}$ interact with respect to the defined operations of addition and scalar multiplication. As a matter of terminology, we sometimes refer to vector spaces as *linear spaces*. Since this text deals exclusively with the real number system, we will only be interested in *real vector spaces* (the scalars are real numbers) and we provide a formal definition for the sake of reference. Also remember that the definition of an operation on a set X requires the rule to return an element *in* the set X. This is not always an obvious feature and something that will often need to be verified.

Definition 1.5.1. Let X be a set. We define the *operation of scalar multiplication* on X over $\mathbb{R}$ to be a rule which associates a real number c and a vector $x \in X$ with a vector $cx \in X$.

A *real vector space* is then a set X and two operations, a binary operation on X denoted $+$, and a scalar multiplication on X over $\mathbb{R}$. Furthermore, the following hold:

Addition Axioms

(a) *Commutativity*: $x + y = y + x$ for every $x, y \in X$;
(b) *Associativity* $(x + y) + z = x + (y + z)$ for every $x, y, z \in X$;
(c) *Additive identity* or *zero vector*: there is an element, denoted $\mathbf{0}$, in X such that
   $x + \mathbf{0} = x$ for every $x \in X$;
(d) *Additive inverse*: for each $x \in X$, there is element, denoted $-x$, in X such that
   $x + (-x) = \mathbf{0}$;

Scalar Multiplication Axioms

(e) *Distributivity*: $c(x + y) = cx + cy$ for every $x, y \in X$ and $c \in \mathbb{R}$;
(f) *Distributivity*: $(c + d)x = cx + dx$ for every $x \in X$, and $c, d \in \mathbb{R}$;
(g) *Associativity*: $(cd)x = c(dx) = d(cx)$ for every $x \in X$, and $c, d \in \mathbb{R}$;
(h) *Scalar identity*: $1x = x$ for every $x \in X$.

The following examples should be familiar from linear algebra.

Example 1.5.2. Let $X = \mathbb{R}^2$, the set of all ordered pairs of real numbers. We define addition and scalar multiplication by

$$(x_1, x_2) + (y_1, y_2) = (x_1 + y_1, x_2 + y_2) \qquad \text{and} \qquad c(x, y) = (cx, xy).$$

It is easy to check that this addition is commutative and associative. There is a zero vector, namely, $\mathbf{0} = (0, 0)$ and each element in the vector space has an additive inverse, $-(x, y) = (-x, -y)$. Furthermore, the scalar multiplication is distributive and associative, and the scalar 1 acts as a (scalar) multiplicative identity. More generally, for $n \in \mathbb{N}$ we can consider the set of all n-tuples of real numbers, denoted $\mathbb{R}^n$. By defining addition and scalar multiplication in a similar fashion, one can verify that this set also forms a real vector space.

Example 1.5.3. Let $X = P_n$ denote the collection of all polynomials with real coefficients of degree n or less. If we define the addition and scalar multiplication by the standard pointwise operations for functions, then this set becomes a real vector space. Checking that the operations behave appropriately is essentially the same as working with $\mathbb{R}^n$.

We will encounter many more real vector spaces along the way, but for the remainder of this section, we focus on various structures we can impose on such collections. The first such structure can be applied to spaces of greater generality, and while we state the definition for such spaces, we will only concern ourselves with examples which are real vector spaces.

Definition 1.5.4. A *metric space* is a pair (X, d) where X is a set and $d : X \times X \to [0, \infty)$ is a function satisfying

(a) $d(x, y) = 0$ if and only if $x = y$ in X;
(b) $d(x, y) = d(y, x)$ for all $x, y \in X$;
(c) $d(x, z) \leq d(x, y) + d(y, z)$ for all $x, y, z \in X$.

The function d is referred to as a *metric* on X.

The first property here is a non-degeneracy requirement, the second is a symmetric property, and the third is referred to as the triangle inequality; the name here is derived from the triangle inequality satisfied by the absolute value function acting on $\mathbb{R}$. This function is in fact the concrete notion which provides the motivation for the definition above and is our first example below. Notice also that the function d takes on no negative values. In essence, imposing a metric on a space is simply a means of defining distance between elements in the set X.

Example 1.5.5. Let $X = \mathbb{R}$ and define $d : \mathbb{R} \times \mathbb{R} \to [0, \infty)$ by $d(x, y) = |x - y|$. It should be clear that this rule satisfies properties (a) and (b) of a metric; property (c) follows from the triangle inequality property of the absolute value.

Typically, spaces can be endowed with multiple metrics and, often, these lead to varying conclusions which can be made about behavior in the space. For an extreme example of this, consider the following.

Example 1.5.6. Let $X = \mathbb{R}$ and define $d : \mathbb{R} \times \mathbb{R} \to [0, \infty)$ by

$$d(x, y) = \begin{cases} 1, \ x \neq y; \\ 0, \ x = y. \end{cases}$$

This is called a *discrete metric* and defines distance on $\mathbb{R}$ in such a way that all distinct real numbers are exactly 1 unit apart. An approach of this manner is quite contrary to our typical realization of $\mathbb{R}$ as a continuum with the metric defined in the previous example. For instance, imagine trying to sketch $\mathbb{R}$ such that all the real numbers were exactly one unit apart.

Example 1.5.7. Let $X = \mathbb{R}^2$ and define $d : \mathbb{R}^2 \times \mathbb{R}^2 \to [0, \infty)$ by

$$d((x_1, x_2), (y_1, y_2)) = |x_1 - y_1| + |x_2 - y_2|.$$

This rule is often referred to as the *taxicab metric* on $\mathbb{R}^2$ and is probably the most obvious generalization of Example 1.5.5. This name comes from the idea of a well-planned city as a grid with streets running only horizontally and vertically. In such a utopia, a taxi traveling from a point (x_1, x_2) to a point (y_1, y_2) would likely travel horizontally a distance of $|x_1 - y_1|$ units and then vertically a distance $|x_2 - y_2|$ units to reach its destination.

Alternatively, if we think of $|x|$ as $\sqrt{x^2}$, we could also generalize Example 1.5.5 by defining

$$d((x_1, x_2), (y_1, y_2)) = \sqrt{(x_1 - y_1)^2 + (x_2 - y_2)^2}.$$

This is called the *Euclidean metric* on $\mathbb{R}^2$. We leave it as an exercise to show that these rules are both metrics on $\mathbb{R}^2$. Note that property (c) for the Euclidean metric will be more easily proven after we have encountered the Cauchy–Schwarz inequality.

The idea of a metric is to generalize distance on the real number line to more complicated spaces. This functionality will allow us to consider convergence of sequences in arbitrary vector spaces and to discuss continuity of functions on these spaces. However, a metric isn't required to cooperate with the vector space structure, and since we will always be considering these types of spaces, it seems natural and not too restrictive to consider such specifications.

Definition 1.5.8. Let X be a real vector space. A *norm* on X is a function $\| \cdot \| : X \to [0, \infty)$ such that

(a) $\|x\| = 0$ if and only if $x = \mathbf{0}$ in X;
(b) $\|cx\| = |c|\|x\|$ for every $c \in \mathbb{R}$ and $x \in X$;
(c) $\|x + y\| \leq \|x\| + \|y\|$ for all $x, y \in X$.

We call the pair $(X, \| \cdot \|)$ a *(real) normed linear space.*

Some of these properties resemble those for a metric. The first is requiring a non-degenerate function, the second, called *homogeneity*, indicates that the norm respects the scalar multiplication, and the third, the triangle inequality, encapsulates the interaction of the vector addition with the norm.

Example 1.5.9. Let $X = \mathbb{R}$ and define $\| \cdot \| : \mathbb{R} \to [0, \infty)$ by $\|x\| = |x|$. The properties above follow quickly as the absolute value function is the prototypical norm.

The next theorem ties the ideas of norm and metric.

Proposition 1.5.10. *If $(X, \|\cdot\|)$ is a normed linear space, then the rule $d : X \times X \to [0, \infty)$ defined by*

$$d(x, y) = \|x - y\|$$

gives a metric on X.

Proof. The fact that a metric defined in this manner satisfies parts (a) and (b) of the definition of a metric follows immediately from norm properties (a) and (b). For the triangle inequality, let $x, y, z \in X$. Then

$$d(x, z) = \|x - z\| = \|(x - y) + (y - z)\| \le \|x - y\| + \|y - z\| = d(x, y) + d(y, z).$$

$\square$

Example 1.5.11. Let $X = \mathbb{R}^2$. We leave it as an exercise to check that the functions

$$\|x\| = |x_1| + |x_2|$$

and

$$\|x\| = \sqrt{(x_1)^2 + (x_2)^2},$$

where $x = (x_1, x_2)$, are in fact norms on $\mathbb{R}^2$ and generate the metrics described in Example 1.5.7.

There is one more structure possibility to discuss. While the real number system has been the motivation for both metric spaces and normed linear spaces, the following idea is a generalization of $\mathbb{R}^2$ and enables us to consider angles between vectors, though generally we will only be concerned with discussing when vectors are orthogonal (i.e., perpendicular).

Definition 1.5.12. Let X be a real vector space. An *inner product* on X is a function $\langle \cdot, \cdot \rangle : X \times X \to \mathbb{R}$ such that

(a) $\langle x, y \rangle = \langle y, x \rangle$ for all $x, y \in X$;
(b) $\langle x, x \rangle \ge 0$ for all $x \in X$ with equality if and only if $x = \mathbf{0}$ in X;
(c) $\langle cx, y \rangle = c \langle x, y \rangle$ for all $c \in \mathbb{R}$ and $x, y \in X$;
(d) $\langle x + y, z \rangle = \langle x, z \rangle + \langle y, z \rangle$ for all $x, y, z \in X$.

The pair $(X, \langle \cdot, \cdot \rangle)$ is called an *inner product space*.

Example 1.5.13. The prototype for all inner product spaces is the *Euclidean inner product* or *dot product* on $\mathbb{R}^2$ and is given by

$$\langle (x_1, x_2), (y_1, y_2) \rangle = x_1 y_1 + x_2 y_2.$$

Verification that the rule satisfies the definition of an inner product is rudimentary and left to the reader.

It may seem unclear as to how the inner product fits in with the previous two structures. It should first be noted that the inner product is the most restrictive and this will provide the most structure. The remainder of this section aims to clarify these two points.

Proposition 1.5.14 (Cauchy–Schwarz Inequality). *Let $(X, \langle \cdot, \cdot \rangle)$ be an inner product space. Then*

$$|\langle x, y \rangle|^2 \leq \langle x, x \rangle \langle y, y \rangle$$

holds for every $x, y \in X$.

Proof. First notice that the inequality is trivially true if either x or y is the zero vector by Exercise 1.5.5. When both x and y are nonzero, consider the vector $x + cy$ where $c \in \mathbb{R}$. If we take the inner product of this vector with itself, the properties of the inner product along with those given in the aforementioned exercise show

$$\langle x + cy, x + cy \rangle = \langle x, x \rangle + 2c \langle x, y \rangle + c^2 \langle y, y \rangle.$$

The fact that x and y are fixed vectors in X allows us the freedom to view this last expression as a quadratic function of the parameter c. Moreover, the fact that we have taken the inner product of a vector with itself allows us to conclude that this quadratic expression is nonnegative. Recall that a nonnegative quadratic function has either one real root or no real roots, a situation which occurs exactly when the discriminant of the quadratic is nonpositive. The discriminant of this particular expression and the nonpositivity are expressed as

$$4\langle x, y \rangle^2 - 4\langle x, x \rangle \langle y, y \rangle \leq 0.$$

Rearranging this inequality immediately yields the desired conclusion. □

Proposition 1.5.15. *Let $(X, \langle \cdot, \cdot \rangle)$ be an inner product space. Then $\|x\| = \sqrt{(\langle x, x \rangle)}$ defines a norm on X.*

The proof of this fact relies on the Cauchy–Schwarz inequality and is requested as Exercise 1.5.7. It is then easy to check, and should be checked, that the inner product defined in Example 1.5.13 generates the second norm for $\mathbb{R}^2$ given in Example 1.5.11.

Definition 1.5.16. Let $(X, \langle \cdot, \cdot \rangle)$ be an inner product space. We say that two vectors x and y in X are *orthogonal* if $\langle x, y \rangle = 0$. In this case we use the symbol $x \perp y$.

In $\mathbb{R}^2$, we tend to think of the vector (a, b) as a directed line segment with initial point at the origin and terminal point at (a, b). When two vectors dot product to zero, then these two vectors form a right angle at the origin and the above definition generalizes this to more abstract spaces. The last theorem of this section illustrates one reason why this is beneficial.

Proposition 1.5.17 (Pythagorean Theorem). *Let $(X, \langle \cdot, \cdot \rangle)$ be an inner product space. If x and y are orthogonal in X and the norm on X is defined as in Proposition 1.5.15, then*

$$\|x + y\|^2 = \|x\|^2 + \|y\|^2.$$

To close this section, it is worthwhile to notice that we have worked in order of increasing rigidity. A metric can be defined on almost any collection of objects, but since we will be dealing quite specifically with vector spaces, the additional structure of a norm is welcome. Exercise 1.5.2 will ask you to consider whether this relationship can be reversed. What will be of greater importance to us is the fact that not every normed linear space is an inner product space. In some sense this added level of structure is too much. To be specific here, while every inner product space is a normed linear space, not every norm can be obtained via an inner product as in Proposition 1.5.15 and Exercise 1.5.9 will give a simple technique for understanding when this can be done. For a simple example, we presented two norms for $\mathbb{R}^2$ in Example 1.5.11, but only one of these arises as an inner product. Section 2.5 incorporates the idea of completeness into this setting.

Exercises

Exercise 1.5.1. Let $(X, \mathbb{R}, +, \cdot)$ be a real vector space. Prove each of the following statements.

(a) The additive identity in X is unique.
(b) For each $x \in X$, the additive inverse $-x$ is unique.
(c) $0x = \mathbf{0}$ for each $x \in X$.
(d) $-x = (-1)x$ for each $x \in X$.
(e) $c\mathbf{0} = \mathbf{0}$ for each $c \in R$.

Exercise 1.5.2. Complete the proof of Proposition 1.5.10 by showing that the defined rule satisfies metric properties (a) and (b). Is the converse of this statement true? In other words, given a metric on a vector space X, does a similar relationship allow us to define a norm on X?

Exercise 1.5.3. Let $(X, \| \cdot \|)$ be a normed linear space. Show that the inequality $\left| \|x\| - \|y\| \right| \le \|x - y\|$ holds for all $x, y \in X$.

Exercise 1.5.4. Let $(X, \| \cdot \|)$ be a normed linear space and assume $x \in X$ with $x \ne \mathbf{0}$. If $r > 0$ in $\mathbb{R}$, show that there is a $c \in \mathbb{R}$ with $\|cx\| = r$.

Exercise 1.5.5. Let $(X, \langle \cdot, \cdot \rangle)$ be an inner product space. Show that each of the following properties hold.

(a) $\langle \mathbf{0}, x \rangle = \langle x, \mathbf{0} \rangle = 0$ for every $x \in X$ and where $\mathbf{0}$ denotes the zero vector;

(b) $\langle x, cy \rangle = c\langle x, y \rangle$ for all $c \in \mathbb{R}$ and $x, y \in X$;

(c) $\langle x, y + z \rangle = \langle x, y \rangle + \langle x, z \rangle$ for all $x, y, z \in X$.

(d) If $\langle x, y \rangle = 0$ for every $x \in X$, then $y = \mathbf{0}$.

(e) If $\langle x, y \rangle = \langle x, z \rangle$ for each $x \in X$, then $y = z$.

Exercise 1.5.6. (a) Verify that the taxi-cab metric defined in Example 1.5.7 defines a metric on $\mathbb{R}^2$.

(b) Show that the Euclidean inner product given in Example 1.5.13 defines an inner product on $\mathbb{R}^2$. Use this fact to show that the induced norm and metric are in fact those given in Examples 1.5.7 and 1.5.11, respectively.

Exercise 1.5.7. Supply a proof for Proposition 1.5.15.

Exercise 1.5.8. Use Proposition 1.5.15 to verify the Pythagorean Theorem as stated in Proposition 1.5.17.

Exercise 1.5.9. (a) Let $(X, \langle \cdot, \cdot \rangle)$ be an inner product space. Show that the norm (given by Proposition 1.5.15) satisfies the *parallelogram equality*, that is, for $x, y \in X$, show that

$$\|x + y\|^2 + \|x - y\|^2 = 2\|x\|^2 + 2\|y\|^2.$$

(b) Considering the Euclidean inner product in $\mathbb{R}^2$, give a geometric interpretation of this equality.

(c) Let $X = \mathbb{R}^2$ with norm given by $\|x\| = |x_1| + |x_2|$ for $x = (x_1, x_2)$. Find two vectors $x, y \in \mathbb{R}^2$ which do not satisfy the parallelogram equality. What can you conclude about this norm in relation to inner products on $\mathbb{R}^2$ and Proposition 1.5.15?

The converse of statement (a) also holds. Suppose $(X, \| \cdot \|)$ is a normed linear space for which the parallelogram equality holds for each pair of vectors $x, y \in X$. Then the function

$$\langle x, y \rangle = \frac{1}{2} \left(\|x + y\|^2 - \|x\|^2 - \|y\|^2 \right)$$

defines an inner product on X and is related to the existing norm via Proposition 1.5.15. Showing that the relation $\langle x, y + z \rangle = \langle x, y \rangle + \langle x, z \rangle$ holds is the most difficult step in this proof and the reader is pointed to p. 316 in [16] for a reference.

Exercise 1.5.10 (Hausdorff Property). Let (X, d) be a metric space. For $x \in X$ and $\varepsilon > 0$, define the *ε-neighborhood* of x by $B_\varepsilon(x) = \{y \in X : d(x, y) < \varepsilon\}$. If x_1 and x_2 are distinct elements in X, show that there exist $\varepsilon_1, \varepsilon_2 > 0$ such that $B_{\varepsilon_1}(x_1) \cap B_{\varepsilon_2}(x_2) = \emptyset$.

Exercise 1.5.11. Let X be a vector space with two norms $\| \cdot \|_1$ and $\| \cdot \|_2$. We say $\| \cdot \|_1$ is *equivalent* to $\| \cdot \|_2$ if there exist constants $m, M > 0$ such that

$$m\|x\|_1 \leq \|x\|_2 \leq M\|x\|_1$$

for all $x \in X$. Show that this relation defines an equivalence relation on the set of all norms on the vector space X.

Chapter 2
Sequences in $\mathbb{R}$

In this chapter we investigate the most basic concept of analysis, the limit of a sequence. The concepts of convergence and divergence will be the main focus of our discussion but we will also touch again on the notions of completeness and set structures in $\mathbb{R}$. It is also worth noting that each of the constructs comprising the remainder of the text, series, functional limits and continuity, the derivative, and the integral will rely heavily on our excursion here.

2.1 Sequences and Convergence

Like many mathematical definitions, a sequence can be defined rather intuitively. For instance, we could simply say that a sequence is a list of real numbers. This is likely how you have thought about sequences up to this point in your mathematical experience, however, as is often the case with definitions of this nature, there is a lack of precision in the wording which allows ambiguity to creep into the picture. Let's take a moment to think about these potential issues. For one, how long is this list? We will require that a sequence be a countably infinite list of real numbers, but our wording above would permit finite strings of numbers. Does the order of the terms in the sequence matter? Of course they do, but this is not reflected in our definition. These two problems are easily eliminated by restating our definition: A sequence is an ordered, countably infinite list of real numbers. This does not roll off the tongue quite as nicely and simply adding more adjectives is a rather imprecise use of language. As a formal definition, consider the following.

Definition 2.1.1. A *sequence* of real numbers is a function $f : \mathbb{N} \to \mathbb{R}$.

This phrasing immediately gives some concrete meaning to our informal notion stated above. Specifically, if $f : \mathbb{N} \to \mathbb{R}$ is our sequence, then we can represent f as the collection of range values $(f(1), f(2), f(3), \ldots)$ of f, i.e. we have represented f as an ordered, countably infinite list of real numbers. This encompasses all of the intuition that we previously hoped for but we have only used well-defined

M.A. Pons, *Real Analysis for the Undergraduate: With an Invitation to Functional Analysis*, 55
DOI 10.1007/978-1-4614-9638-0_2, © Springer Science+Business Media New York 2014

mathematical terms to produce the result. In particular, we have an ordered list and the order of the *terms* (particular numbers appearing in the list) is important, e.g. $f(1)$ appears before $f(2)$, and so on. Repeated terms are allowed, but are not necessary. And, most importantly, the list has countably many terms.

The next step is to understand the several types of notation used to represent sequences. As it is impossible to write out an infinite string, we need a means of representing sequences which leaves no room for guesswork on the part of the reader. As a simple example, consider the list $(1, 1, 1, 1, \ldots)$. What is the fifth term? What is the ninth term? Are you sure that both of these are 1? Typically this is what the "..." should represent, but how do you know for certain that the pattern you've deciphered is in fact the correct one? Hence the ambiguity with our notion of lists. On the other hand, consider the function $f : \mathbb{N} \to \mathbb{R}$ given by $f(n) = 1$ for all $n \in \mathbb{N}$. What is the fifth term in this sequence? The function gives us a concrete means of identifying terms whereas the list leaves room for error. As such, we will prefer some form of function, or formulaic, notation for sequences.

Often we will abandon the formal use of f for our function and replace this with the simpler notation $f(n) = a_n$ representing the nth term of the sequence. We then represent the sequence by the symbol (a_n). For an explicit situation, consider the function $f(n) = 1/n$. We could represent this as $(1, 1/2, 1/3, \ldots, 1/n, \ldots)$ or $(1/n)_{n=1}$ or $(a_n)_{n=1}$ where $a_n = 1/n$. Also, we will often want to begin our sequence at a position other than $n = 1$. When this is necessary we will use the notation $(a_n)_{n=k}$ if beginning with a natural number $k > 1$ or $k = 0$. This can be done in a more formal manner by expanding our definition of sequence to functions $f : K \to \mathbb{R}$ where $K \subset \mathbb{Z}$ has the form $\{n \in \mathbb{Z} : n \geq m\}$ for some fixed $m \in \mathbb{Z}$. As a last comment on notation, we use parenthesis in the notation for a sequence to preserve the distinction between the sequence as a newly defined object and not merely as a set of real numbers. However, each sequence is associated with its set of range values and we would use braces when representing this set. The set of range values for the sequence defined by the function $f(n) = 1$ is $\{1\}$ while the set of range values for the sequence $f(n) = 1/n$ is $\{1, 1/2, 1/3, \ldots, 1/n, \ldots\}$.

Definition 2.1.2. Let (a_n) be a sequence in $\mathbb{R}$ and let $a \in \mathbb{R}$. We say that (a_n) *converges to* a if for every $\varepsilon > 0$, there is an $N \in \mathbb{N}$ such that $|a_n - a| < \varepsilon$ for every $n \geq N$. In this case we call a the *limit* of the sequence and represent this symbolically by $(a_n) \to a$ or $\lim_{n \to \infty} a_n = a$.

As a first observation, the notational comment following the definition is phrased to imply that a sequence can have only one limit, signified by the phrase "the limit." This fact is not explicitly asserted in the definition of convergence, it requires proof which you will provide in Exercise 2.1.1. On occasion we will simplify the notation and write $\lim a_n = a$; this should cause no ambiguity for sequential limits as interesting phenomena only occur as we proceed further and further out into the sequence.

Before considering examples, it is beneficial to analyze what has just been set before us. We will see definitions of this sort on several occasions and an ever increasing level of comfort with such statements will be necessary. Suppose we

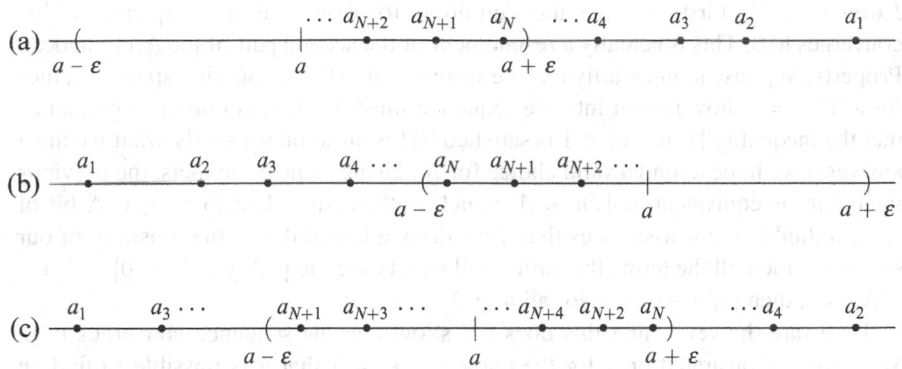

(a)

$\cdots a_{N+2} \quad a_{N+1} \quad a_N \quad \cdots a_4 \quad a_3 \quad a_2 \qquad a_1$

$a - \varepsilon \qquad\qquad\qquad\qquad a \qquad\qquad\qquad a + \varepsilon$

(b)

$a_1 \qquad a_2 \quad a_3 \qquad a_4 \cdots \quad a_N \quad a_{N+1} \quad a_{N+2} \cdots$

$a - \varepsilon \qquad\qquad\qquad a \qquad\qquad\qquad\qquad a + \varepsilon$

(c)

$a_1 \qquad a_3 \cdots \qquad a_{N+1} \quad a_{N+3} \cdots \qquad \cdots a_{N+4} \quad a_{N+2} \quad a_N \qquad \cdots a_4 \qquad a_2$

$a - \varepsilon \qquad\qquad\qquad a \qquad\quad a + \varepsilon$

Fig. 2.1 Examples of convergent sequences: $(a_n) \to a$

are given a sequence (a_n). To show that it converges, we must first make a guess as to what its limit should be. In many of the exercises, you will be given the limit candidate; however, this is often not the case and we will also investigate elementary means of computing limits in the later sections. Suppose also then that we have a limit candidate, a.

The ε in the definition above is acting as an error bound, meaning that our goal is to only consider sequence terms which are within ε units of a. Moreover, we must do this according to a very specific rule. The N represents a position in the sequence, the 10th position, or the 104th position, and so on. The rule that we must abide by is as follows: *Given a positive number ε, we must find $N \in \mathbb{N}$ so that all terms at or beyond the Nth position in the sequence (signified by the statement $n \geq N$) are within ε units of a.*

In example (a) in Fig. 2.1, notice that the sequence terms approach a in a very specific manner while the opposite behavior is exhibited in example (b). The third example then demonstrates a more complicated case of sequence convergence. In each of the three examples it is clear that we have produced a natural number N so that each term at or beyond the Nth position satisfies $|a_n - a| < \varepsilon$.

The key to proving a convergence statement is then understanding how to make appropriate choices for N. And, as with many skills, this requires practice, a fact we will repeat over and over. People, for whatever reason, often respond positively to the idea of games and the challenge of making deliberate choices for N has been introduced as such on more than one occasion. Specifically, I choose an error bound ε and the challenge (by hook or by crook!) is then for you to come up with a choice for N which has the property that $|a_n - a| < \varepsilon$ for all $n \geq N$. Often this is done by algebraically manipulating the expression $|a_n - a| < \varepsilon$ and by using elementary estimating techniques which we will demonstrate below. Moreover, it is usually not the case that we choose N to be a specific number, but rather we will use facts such as the Archimedean Property to assert the existence of a natural number with a specific property. Our first example demonstrates this technique and we then move on to more involved examples.

Example 2.1.3. First, we use the definition to show that the sequence $(1/n)$ converges to 0. This is actually a restatement of the second part of the Archimedean Property. Suppose momentarily that we were to begin by considering specific values for ε. If $\varepsilon = 1$, how far out into the sequence must we travel in order to guarantee that the inequality $|1/n - 0| < 1$ is satisfied? This question is exactly what we must answer if we hope to find a valid choice for N. Using some basic facts, the previous inequality is equivalent to $1/n < 1$, which is then equivalent to $n > 1$. A bit of critical thinking now assures us that if we draw a line at the second position of our sequence, then all the terms thereafter will satisfy the inequality $|1/n - 0| < 1$, i.e. if $N = 2$, then $|1/n - 0| < \varepsilon$ for all $n \geq N$.

Be aware, however, that this does not show that the sequence converges to 0. We have demonstrated only for the choice of $\varepsilon = 1$ that it is possible to find an appropriate choice of N though the definition says that we must do this *for every* positive number ε. It would be an endless task if we were to consider the positive numbers one by one. Thus we will use the parameter ε to represent any positive value which will allow us to take care of all possibilities at once. In practice however, it is often instructive to first consider some specific values before moving on to the most general setting. Continuing with this example, find a suitable choice for N corresponding to $\varepsilon = 1/2, 1/10$, and $1/\sqrt{2}$.

Moving on, what happens if we simply set $\varepsilon > 0$? The task remains the same and the expression $|1/n - 0| < \varepsilon$ is equivalent to the statement that $1/n < \varepsilon$. Keep in mind also that after making our choice for N, we will only consider sequence terms which satisfy $n \geq N$, so we can think this way from the outset. This inequality can be rearranged to read $1/n \leq 1/N$. Thus if we can find an N for which $1/N < \varepsilon$, then we will have that

$$\left| \frac{1}{n} - 0 \right| = \frac{1}{n} \leq \frac{1}{N} < \varepsilon$$

for all $n \geq N$. The Archimedean Property now supplies the final part of this process. Piecing these thoughts together, we now give a formal proof.

Proof. We begin with an arbitrary $\varepsilon > 0$. Then, by the Archimedean Property we can find an $N \in \mathbb{N}$ with $1/N < \varepsilon$. Now, if we assume that $n \geq N$, it follows that $1/n \leq 1/N$ and thus

$$\left| \frac{1}{n} - 0 \right| = \frac{1}{n} \leq \frac{1}{N} < \varepsilon$$

for all $n \geq N$. Therefore we conclude that 0 is the limit of the sequence $(1/n)$.   □

Later we will encounter another notion related to sequences which reproduces the first part of the Archimedean Property. For the moment however, let's continue on with a few more examples. The strategy is basically the same as in the simple example above, but as our sequences become more complicated, the relationship that determines how N is chosen for a given value of $\varepsilon > 0$ becomes less obvious.

One basic fact to remember is that when showing convergence for a specific sequence, the relationship between ε and N is likely, though not always, dictated by the Archimedean Property as you will see in the examples below.

Example 2.1.4. For a second example, let's show that $(1/(3n^2 + 1)) \to 0$. For $\varepsilon > 0$, how do we choose N? The idea is to first try to relate $|a_n - a|$ to n and then to N while keeping in mind that in our proof we will always assume that $n \geq N$. And off we go! Notice first that

$$3n^2 + 1 \geq 3n^2 \geq n^2 \geq n \geq N$$

where we are using the order properties of $\mathbb{R}$ to generate this string of inequalities. Still assuming that $n \geq N$, the previous inequality is equivalent to

$$\left| \frac{1}{3n^2 + 1} - 0 \right| = \frac{1}{3n^2 + 1} \leq \frac{1}{3n^2} \leq \frac{1}{n^2} \leq \frac{1}{n} \leq \frac{1}{N}.$$

With this estimate in hand, it is apparent that we will obtain the desired conclusion by choosing $N \in \mathbb{N}$ with $1/N < \varepsilon$.

Proof. Let $\varepsilon > 0$. By the Archimedean Property we can find $N \in \mathbb{N}$ so that $1/N < \varepsilon$. Now, if we assume that $n \geq N$, we have that

$$\left| \frac{1}{3n^2 + 1} - 0 \right| = \frac{1}{3n^2 + 1} \leq \frac{1}{3n^2} \leq \frac{1}{n^2} \leq \frac{1}{n} \leq \frac{1}{N} < \varepsilon.$$

and therefore we conclude that $(1/(3n^2 + 1)) \to 0$. □

Example 2.1.5. As a third example, let's show that the sequence $(3/\sqrt{n+2}) \to 0$. Here we will take for granted the fact that the root function is increasing on $(0, \infty)$. With this, we see that if $n \geq N$, then

$$\sqrt{n+2} \geq \sqrt{n} \geq \sqrt{N}$$

which implies that

$$\frac{3}{\sqrt{n+2}} \leq \frac{3}{\sqrt{n}} \leq \frac{3}{\sqrt{N}}.$$

Now, it is not the case that $\sqrt{N} \geq N$ and this leaves us seemingly one step short. However, we can make use of the universal quantifier in the statement of the Archimedean Property. To be precise, $\varepsilon^2/9$ is positive and thus we may choose $N \in \mathbb{N}$ so that $1/N < \varepsilon^2/9$; an algebraic manipulation then shows that this is equivalent to $3/\sqrt{N} < \varepsilon$.

Proof. Let $\varepsilon > 0$. By the Archimedean Property we can choose $N \in \mathbb{N}$ so that $1/N < \varepsilon^2/9$. Notice that this is equivalent to $3/\sqrt{N} < \varepsilon$. If we now consider $n \geq N$, we know that $\sqrt{n+2} \geq \sqrt{n} \geq \sqrt{N}$ and hence

$$\left| \frac{3}{\sqrt{n+2}} - 0 \right| = \frac{3}{\sqrt{n+2}} \leq \frac{3}{\sqrt{n}} \leq \frac{3}{\sqrt{N}} < \varepsilon.$$

Therefore we conclude that $(3/\sqrt{n+2}) \to 0$ as desired. $\square$

Why so many examples? The idea is that observation is the key to learning a proof technique. Take a moment now to compare each of the above proofs with the definition of convergence. What do you notice? Think about the flow of the proof compared to the flow of the definition. You should find that there are three important parts in the definition and each proof above exhibits these same three components in the same order. Each proof begins with the designation that ε is a positive real number. Then, a choice for N is exhibited; however, the reason for the choice is not obvious until later. Then the implication "$n \geq N$ implies $|a_n - a| < \varepsilon$" is verified. In the second and third examples above, there are some additional thoughts involving algebraic manipulations which are present to aid the reader. They add a seamlessness to the proof, a continuity of ideas if you will. This strategy is basic to most of the convergence proofs that we will encounter and should therefore be practiced. As a final word of wisdom, keep in mind that in each of the examples above, the discussion proceeding the proof was an equally important component of our solution as it provided us the means to make the choice for N, necessary for the second part of the proof, and also provided the third component of the proof. This is how math is done! Proofs do not often come easily; ideas are scratched out before hand, often several times, and then a proof is given. Keep this in mind as you work through the exercises.

We will encounter the idea of convergence over and over in several different settings and a thorough grasp of each will be critical. It will also be equally important to understand when we have a non-convergent situation. We will use the word *divergent* to signify this. Our formal definition will simply be that a sequence diverges if it does not converge, and while this sounds simple enough, it should be noted that writing down a rigorous definition of this concept is a bit more cumbersome than our definition of convergence; we will provide a more formal discussion of such a statement in Example 2.1.11. One reason for this is that there are a variety of ways for a sequence to diverge. The remainder of this section will provide two divergence criteria and we will encounter several more in Sect. 2.2.

Definition 2.1.6 (Divergence Criterion 1). Let (a_n) be a sequence of real numbers. We say that (a_n) *diverges to* ∞ if for every $M > 0$, there is an $N \in \mathbb{N}$ such that $a_n > M$ for all $n \geq N$. In this situation we use the notation $\lim_{n \to \infty} a_n = \infty$.

The definition above has the same form as our definition of convergence and the proof style is also similar in that some scratch work is often necessary in order to be able to make an adequate choice for N corresponding to a fixed value of M. We omit these details for the two examples considered below.

Example 2.1.7. Consider the sequences (n^2) and $(n^2 + \frac{n}{3} + 2)$. For the first sequence, given $M > 0$, let's choose $N \in \mathbb{N}$ so that $N > \sqrt{M}$, which is permissible by the Archimedean Property. This is equivalent to requiring that $N^2 > M$. Using this last estimate, if $n \geq N$ we see that

$$n^2 \geq N^2 > M.$$

Therefore we may conclude that (n^2) diverges to infinity. Notice here that we made our initial choice for capital N based on this last inequality.

For the second sequence we let $M > 0$ and again choose $N \in \mathbb{N}$ so that $N > \sqrt{M}$. Now, if $n \geq N$, we can use the fact that $\frac{n}{3} + 2$ is a positive quantity to obtain the estimate

$$n^2 + \frac{n}{3} + 2 > n^2 > N^2 > M$$

showing that the sequence $(n^2 + \frac{n}{3} + 2)$ also diverges to infinity.

As another example, think about the sequence $(n^2 - \frac{n}{3} + 2)$. It looks very similar to the second sequence above, except for a sign change. How will this affect our proof? The details are requested of you in Exercise 2.1.5(b).

Definition 2.1.8. Let (a_n) be a sequence in $\mathbb{R}$. We say that (a_n) is a *Cauchy sequence* if for every $\varepsilon > 0$, there is an $N \in \mathbb{N}$ such that $|a_n - a_m| < \varepsilon$ for all $n, m \geq N$.

This definition looks very similar to our definition of convergence and the proof strategy is also similar as we shall see below. The definition of limit dictates that the sequence terms get close and stay close to the limit while this definition asserts that the terms eventually get close and stay close to each other. Considering the fact that closeness is transitive in a certain sense (this is really what the triangle inequality says!), it seems likely that every convergent sequence must have this property.

Proposition 2.1.9. *Every convergent sequence is Cauchy.*

Before giving the proof, let us again consider what will be required. For a sequence (a_n) and $\varepsilon > 0$, we need to identify an $N \in \mathbb{N}$ so that $|a_n - a_m| < \varepsilon$ whenever $n, m \geq N$. However, we also have the fact that the sequence converges to some limit a. In other words, we can force the sequence terms to be as close to a as we desire. The measure of closeness that we will use in this case is $\varepsilon/2$. Also, we can modify the expression $|a_n - a_m|$ by (what some may call the mathematician's favorite trick) adding zero and using the triangle inequality. Consider the following:

$$|a_n - a_m| = |a_n - a + a - a_m| = |(a_n - a) + (a - a_m)|$$
$$\leq |a_n - a| + |a - a_m|$$
$$= |a_n - a| + |a_m - a|.$$

With this rather slick bit of algebra it follows that the quantity on the far left will be less than ε if each of the two terms on the right is less than $\varepsilon/2$.

Proof. Let (a_n) be a sequence of real numbers which converges to some limit a. In order to show that (a_n) is Cauchy, for each $\varepsilon > 0$ we must produce $N \in \mathbb{N}$ so that $|a_n - a_m| < \varepsilon$ whenever $n, m \geq N$. Let $\varepsilon > 0$. Since the sequence converges to a, we know that we can choose $N \in \mathbb{N}$ so that $|a_n - a| < \varepsilon/2$ for all $n \geq N$. With our choice of N, suppose $n, m \geq N$. Adding 0 and using the triangle inequality, we have

$$|a_n - a_m| = |a_n - a + a - a_m| \leq |a_n - a| + |a - a_m| < \varepsilon/2 + \varepsilon/2 = \varepsilon.$$

Therefore $|a_n - a_m| < \varepsilon$ for all $n, m \geq N$ and thus we have shown that our sequence is Cauchy.   □

To close this section, we have another divergence criterion provided by the contrapositive of this last result.

Corollary 2.1.10 (Divergence Criterion 2). *If a sequence is not Cauchy, then it must diverge.*

Example 2.1.11. Consider the sequence $(1, -1, 1, -1, \ldots)$. It should be apparent that this sequence does not converge. How would we go about using the negation of the definition of convergence to provide a proof of this fact? Our definition centers around a limit candidate and thus we would need to show that there is no real number which acts as a limit for this sequence, i.e. for each real number we would have to show that the definition of convergence fails. To begin, let's show that the sequence does not converge to 1. Negating the definition of sequential limit we have:

A sequence of real numbers (a_n) does not converge to a if there exist an $\varepsilon > 0$ so that for every $N \in \mathbb{N}$ there is an $n \geq N$ such that $|a_n - a| \geq \varepsilon$.

The statement says that no matter how far out in the sequence we traverse, we can always find terms that are more than some fixed distance away from our proposed limit. The game here lies in identifying an appropriate quantity for the fixed distance. Using the fact that the terms in the even positions in our sequence are 2 units away from 1, let's choose $\varepsilon = 2$. For $N \in \mathbb{N}$, choose n to be some even number greater than N. Then $|a_n - 1| = |-1 - 1| = 2 \geq \varepsilon$. Therefore we conclude that the sequence does not converge to 1.

Does this show that the sequence diverges? No, we have simply shown that the sequence does not converge to 1. Now we must argue that something similar happens for every other real number. There are some simplifying arguments that we could make, but it's beginning to sound like a lot of work! This is why Corollary 2.1.10 is so useful—it eliminates the need for working with a potential limit. To show that the sequence is not Cauchy we must find an $\varepsilon > 0$ so that for every $N \in \mathbb{N}$ there exist $n, m \geq N$ so that $|a_n - a_m| \geq \varepsilon$ (this is the formal negation of the definition of Cauchy). The game here is much the same except that in this case we are trying to show that no matter how far out in the sequence we look, we can always find terms that are more than a fixed distance away from each other, rather than from a potential limit. Considering the behavior of this specific sequence, we choose $\varepsilon = 2$ and let $N \in \mathbb{N}$. Now fix n to be some even number greater than N and m to be some odd number greater than N. It follows then that $a_n = -1$ and $a_m = 1$ and thus $|a_n - a_m| = |-1 - 1| = 2 \geq \varepsilon$. Thus we conclude that this given sequence is not Cauchy and hence not convergent.

Exercises

Exercise 2.1.1. Show that a convergent sequence has a unique limit.

Exercise 2.1.2. Verify each of the following limits using the definition of convergence.

(a) $\lim\limits_{n \to \infty} a = a$

(b) $\lim\limits_{n \to \infty} \dfrac{1}{6n^2 + 1} = 0$

(c) $\lim\limits_{n \to \infty} \dfrac{3n + 1}{2n + 5} = \dfrac{3}{2}$

(d) $\lim\limits_{n \to \infty} \dfrac{2}{\sqrt{n + 3}} = 0$

(e) $\lim\limits_{n \to \infty} \dfrac{n^2 + 6}{n^2} = 1$

(f) $\lim\limits_{n \to \infty} \dfrac{2n + 3}{3n + 1} = \dfrac{2}{3}$

Exercise 2.1.3. Let (a_n) be a sequence of positive numbers.

(a) If (a_n) converges to 0, show that

$$\lim_{n \to \infty} \frac{1}{a_n} = \infty.$$

(b) If (a_n) diverges to ∞, show that

$$\lim_{n \to \infty} \frac{1}{a_n} = 0.$$

Exercise 2.1.4. Let (a_n) and (b_n) be sequences with $a_n \leq b_n$ for all $n \in \mathbb{N}$. If $\lim a_n = \infty$, show that $\lim b_n = \infty$.

Exercise 2.1.5. (a) Suppose (a_n) and (b_n) are sequences of positive numbers with $\lim_{n \to \infty} a_n = a$ for $a > 0$, and $\lim_{n \to \infty} b_n = \infty$. Show that

$$\lim_{n \to \infty} a_n b_n = \infty.$$

(b) Use part (a) to show that the sequence $(n^2 - \frac{n}{3} + 2)$ from Example 2.1.7 diverges to infinity.

(c) Show that the sequence $(n^4/(n^2 + 1))$ diverges to infinity.
(d) Produce a definition similar to that of Definition 2.1.6 for the case that a sequence diverges to $-\infty$.
(e) Use your definition to show that the sequences $(-n^3)$ and $(n - n^3)$ diverge to $-\infty$.

Exercise 2.1.6. (a) Suppose that (a_n) and (b_n) are sequences with $(a_n) \to a$. Show that if $(a_n - b_n) \to 0$, then $(b_n) \to a$.
(b) Show that it is possible for two sequences (a_n) and (b_n) to both diverge even if $(a_n - b_n) \to 0$.

Exercise 2.1.7. Consider the statement of the Nested Interval Property (Theorem 1.4.2) with the additional hypothesis that $(b_n - a_n) \to 0$. Show that in this case the intersection $\cap_{n=1}^{\infty} I_n$ contains exactly one point.

Exercise 2.1.8. Let $r \in \mathbb{Q}$ and $t \in \mathbb{I}$. For each of the following, construct a sequence with the specified property and verify the convergence using only facts that we have proved.

(a) Construct a nonconstant sequence of rational numbers converging to r.
(b) Construct a sequence of irrational numbers converging to r.
(c) Construct a sequence of rational numbers converging to t.
(d) Construct a nonconstant sequence of irrational numbers converging to t.

2.2 Properties of Convergent Sequences

Convergent sequences have an abundance of useful properties. In this section we examine properties possessed by all convergent sequences as a means of identifying limits more easily and discuss several other divergence criteria.

Definition 2.2.1. Let (a_n) be a sequence in $\mathbb{R}$. We say (a_n) is *bounded* if there is an $M > 0$ so that $|a_n| \leq M$ for every $n \in \mathbb{N}$.

For example, the sequences $(1, -1, 1, -1, \ldots)$, $(1/n)$, and $(\sin(n))$ are bounded while the sequences (n) and $(1, 1, 2, 1/2, 3, 1/3, 4, 1/4, \ldots)$ are not bounded. A sequence that is not bounded is typically said to be *unbounded*.

Proposition 2.2.2. *If (a_n) is a convergent sequence, then (a_n) is bounded.*

Proof. Let (a_n) be a sequence with limit a. We begin by first finding a bound for the tail of the sequence and then extend this to a bound for the entire sequence. Since the sequence converges, choose $N \in \mathbb{N}$ such that $|a_n - a| < 1$ for all $n \geq N$. The choice of $\varepsilon = 1$ here is arbitrary. The triangle inequality now implies that

$$|a_n| = |a_n - a + a| \leq |a_n - a| + |a| \leq 1 + |a|$$

for all $n \geq N$ and thus $1 + |a|$ is a bound for all terms at or beyond position N. To complete the proof, consider the terms at the front end of the sequence. There are at most $N - 1$ distinct terms and we can simply choose the largest in absolute value to act as a bound for this set. Combining this idea with the previous bound for the tail, we set $M = \max\{|a_1|, |a_2|, \ldots, |a_{N-1}|, 1 + |a|\}$. It then follows immediately that $|a_n| \leq M$ for all $n \in \mathbb{N}$.   □

Corollary 2.2.3 (Divergence Criterion 3). *A sequence that is not bounded must diverge.*

How does this relate to our first divergence criterion? An example of an unbounded sequence which does not diverge to infinity is given by

$$(1, 1, 2, 1/2, 3, 1/3, 4, 1/4, \ldots).$$

On the other hand, any sequence which does diverge to infinity is unbounded. Thus this last result is a generalization of sorts of its predecessor. The concept of bounded can be broken down into two categories, bounded above and below. These ideas are explored in Exercise 2.2.2. We now examine the algebraic properties of convergent sequences.

Theorem 2.2.4 (Algebraic Limit Theorem). *Let (a_n) and (b_n) be sequences converging to a and b, respectively. The following algebraic properties then hold:*

(a) (ka_n) converges to ka for every $k \in \mathbb{R}$;
(b) $(a_n + b_n)$ converges to $a + b$;
(c) $(a_n b_n)$ converges to ab;
(d) if $b_n \neq 0$ for every $n \in \mathbb{N}$ and $b \neq 0$, then (a_n/b_n) converges to a/b.

The proof below will verify the statements (b) and (c) while the proof of the remaining parts will be considered in the exercises. In order to verify the statement in part (b), for $\varepsilon > 0$, we need to find $N \in \mathbb{N}$ so that $|(a_n + b_n) - (a + b)| < \varepsilon$ for all $n \geq N$. Though the proof here is abstract in nature, the strategy is very much the same as we encountered when showing convergence for a specific sequence. We know what is required, but we need to do some scratch work before giving a formal proof in order to understand how to make a suitable choice for N. In this situation, we know that $(a_n) \to a$, $(b_n) \to b$ and, by the triangle inequality,

$$|(a_n + b_n) - (a + b)| = |(a_n - a) + (b_n - b)| \leq |a_n - a| + |b_n - b|.$$

Notice that this inequality relates the quantity we wish to make small to the quantities we know we can control. To be specific, in order to guarantee that the desired quantity is less than ε, it suffices to consider terms from both of the given sequences which are within $\varepsilon/2$ units of their respective limits. The question now is whether or not we can make both of these things happen simultaneously, and the technique we employ will be quite common as we proceed.

By the definition of convergence, there exists $N_1 \in \mathbb{N}$ such that $|a_n - a| < \varepsilon/2$ for all $n \geq N_1$, and $N_2 \in \mathbb{N}$ so that $|b_n - b| < \varepsilon/2$ for all $n \geq N_2$. We must now choose an N which will guarantee that $|(a_n + b_n) - (a + b)| < \varepsilon$ for all $n \geq N$. The key to making this choice is being aware that we need the assumption $n \geq N$ to imply that $n \geq N_1$ and $n \geq N_2$. Thus we set $N = \max\{N_1, N_2\}$. To verify that our logic is in order, if $n \geq N$, then $n \geq N \geq N_1$ and $n \geq N \geq N_2$ which forces both quantities $|a_n - a|$ and $|b_n - b|$ to be less than $\varepsilon/2$, which in turn implies that $|(a_n + b_n) - (a + b)| < \varepsilon$. At this point, we have essentially given a proof, but a formal write-up is always necessary.

Proof. Let (a_n) and (b_n) be sequences converging to a and b, respectively, and let $\varepsilon > 0$. By the definition of convergence, choose $N_1 \in \mathbb{N}$ such that $|a_n - a| < \varepsilon/2$ for all $n \geq N_1$, and $N_2 \in \mathbb{N}$ so that $|b_n - b| < \varepsilon/2$ for all $n \geq N_2$. Also, set $N = \max\{N_1, N_2\}$. If we now consider $n \geq N$, it follows that $n \geq N_1$ and $n \geq N_2$. With this fact and the triangle inequality, we see that

$$|(a_n + b_n) - (a + b)| = |(a_n - a) + (b_n - b)| \leq |a_n - a| + |b_n - b| < \varepsilon/2 + \varepsilon/2 = \varepsilon$$

whenever $n \geq N$. Therefore we conclude that $(a_n + b_n) \to a + b$. $\square$

The proof of (c) is more involved due to the fact that multiplication is a more complicated operation than addition, but the basic approach is the same. For $\varepsilon > 0$, we must find an $N \in \mathbb{N}$ such that $|a_n b_n - ab| < \varepsilon$ for all $n \geq N$, and our first goal is to relate the quantity $|a_n b_n - ab|$ to $|a_n - a|$ and $|b_n - b|$. This is where the complication arises. It is unfortunately *not true* that $(a_n b_n - ab) = (a_n - a)(b_n - b)$ and hence we will need a bit of cleverness to find a viable relationship. Observe what happens if we add and subtract the term $a_n b$,

$$|a_n b_n - ab| = |a_n b_n - a_n b + a_n b - ab| \leq |a_n b_n - a_n b| + |a_n b - ab|$$
$$= |a_n||b_n - b| + |b||a_n - a|.$$

The introduction of this term allows us to factor, resulting in an expression consisting of terms which we can control. Thus, for $\varepsilon > 0$, we can guarantee that $|a_n b_n - ab| < \varepsilon$ if we can force the terms $|a_n||b_n - b|$ and $|b||a_n - a|$ to be less than $\varepsilon/2$.

Considering the first term, the fact that (a_n) converges guarantees us that $|a_n|$ is bounded by Proposition 2.2.2, i.e. there is an $M > 0$ such that $|a_n| \leq M$ for all $n \in \mathbb{N}$. If we then choose $N_1 \in \mathbb{N}$ such that $|b_n - b| < \varepsilon/2M$ for all $n \geq N_1$, we have

$$|a_n||b_n - b| \leq M|b_n - b| < M\left(\frac{\varepsilon}{2M}\right) = \varepsilon/2$$

whenever $n \geq N_1$.

For the second of these terms, we use a similar technique, but we also have to be cautious as it may be the case that $b = 0$; in short, we must avoid division by 0.

The proof can be split into cases $b = 0$ and $b \neq 0$, but we can avoid this since $|b| < |b| + 1$ and $|b| + 1 > 0$. With this in mind, choose $N_2 \in \mathbb{N}$ such that $|a_n - a| < \varepsilon/2(|b| + 1)$ for all $n \geq N_2$. Then,

$$|b||a_n - a| < (|b| + 1)|a_n - a| < (|b| + 1)\left(\frac{\varepsilon}{2(|b| + 1)}\right) = \varepsilon/2$$

for all $n \geq N_2$. At this point we have all the pieces needed to give a formal proof. We will again choose N to be the maximum of N_1 and N_2 and then verify that our work provides the desired conclusion.

Proof. Let $(a_n) \to a$ and $(b_n) \to b$, and let $\varepsilon > 0$. By Proposition 2.2.2, there is an $M > 0$ such that $|a_n| \leq M$ for all $n \in \mathbb{N}$. By the definition of convergence, we can choose $N_1 \in \mathbb{N}$ such that $|b_n - b| < \varepsilon/2M$ for all $n \geq N_1$. We also choose $N_2 \in \mathbb{N}$ such that $|a_n - a| < \varepsilon/2(|b| + 1)$ for all $n \geq N_2$. Now set $N = \max\{N_1, N_2\}$. If $n \geq N$, our assumptions and the triangle inequality show that

$$
\begin{aligned}
|a_n b_n - ab| = |a_n b_n - a_n b + a_n b - ab| &\leq |a_n b_n - a_n b| + |a_n b - ab| \\
&= |a_n||b_n - b| + |b||a_n - a| \\
&< M|b_n - b| + (|b| + 1)|a_n - a| \\
&< M\left(\frac{\varepsilon}{2M}\right) + (|b| + 1)\left(\frac{\varepsilon}{2(|b| + 1)}\right) \\
&= \varepsilon/2 + \varepsilon/2 = \varepsilon.
\end{aligned}
$$

Thus $(a_n b_n) \to ab$. $\qquad\square$

The theorem above will be a powerful tool in theoretical situations but can also be used to calculate limits when the given function can be broken down algebraically into simpler pieces. For more complicated sequences, identifying a candidate for N is often difficult and the theorem provides a way around this.

Example 2.2.5. Consider the sequence $(3/(3n^2 + 4))$. We can factor this and write

$$\frac{3}{3n^2 + 4} = \frac{1}{n^2} \cdot \frac{3}{(3 + (4/n^2))}.$$

For the second factor here, we know that the sequence $(1/n^2)$ converges to zero and thus the denominator $(3 + (4/n^2))$ converges to 3 by parts (a) and (b) of the Algebraic Limit Theorem. This forces the entire second term to converge to 1 by part (d). Applying part (c), we see that the original sequence then converges to zero since the first factor above converges to zero.

The previous theorem allows us to understand the algebra of sequences, that is, the theorem demonstrates that convergent sequences respect the field operations in $\mathbb{R}$. Our next result will show that similar properties hold with respect to the order on $\mathbb{R}$.

Theorem 2.2.6 (Order Limit Theorem). *Let* (a_n) *and* (b_n) *be sequences converging to a and b, respectively.*

(a) If $a_n \geq 0$ *for all* $n \in \mathbb{N}$, *then* $a \geq 0$. *On the other hand, if* $a_n \leq 0$ *for all* $n \in \mathbb{N}$, *then* $a \leq 0$.
(b) If $a_n \leq b_n$ *for every* $n \in \mathbb{N}$, *then* $a \leq b$.

Proof. For the first statement in (a), we suppose there is a sequence $(a_n) \to a$ with $a_n \geq 0$ for all $n \in \mathbb{N}$. To show that $a \geq 0$, we assume to the contrary that $a < 0$. With this additional hypothesis, we will show that there must be a term in the sequence which is negative, thus providing our contradiction. To use the given convergence, we should progress far enough out in our sequence to force the terms into the negative realm, but what ε will accomplish this goal? Consider the distance between a and 0. This is represented most easily by $|a|$. It should also be apparent that any term which satisfies $|a_n - a| < |a|$ will be closer to a than a is to 0, that is, any such term must be negative.

With this reasoning, let $\varepsilon = |a|$. Since the sequence converges to a, it must then be the case that there is an $N \in \mathbb{N}$ with $|a_n - a| < \varepsilon$ for all $n \geq N$. In particular, using our choice for ε, we have that $|a_N - a| < |a|$ which implies that $-|a| < a_N - a < |a|$. Again our goal is to show that $a_N < 0$. Working with the right-hand inequality we obtain $a_N < a + |a|$. Also, since $a < 0$ by hypothesis, we have

$$a_N < a + |a| = a + (-a) = 0.$$

Thus we have shown that $a_N < 0$, contradicting our hypothesis that $a_n \geq 0$ for all $n \in \mathbb{N}$. Therefore it must be the case that $a \geq 0$.

For the second statement in (a), suppose we are given a sequence $(a_n) \to a$ with $a_n \leq 0$ for all $n \in \mathbb{N}$. The sequence $(-a_n)$ then satisfies $-a_n \geq 0$ for all n and by Theorem 2.2.4(a), the sequence converges to $-a$. By the previous case, we have that $-a \geq 0$ which immediately implies that $a \leq 0$.

Part (b) now follows immediately from (a) by considering the sequence $(a_n - b_n)$. The fact that $a_n \leq b_n$ implies that $a_n - b_n \leq 0$ and Exercise 2.2.4(b) indicates that $(a_n - b_n) \to a - b$. Applying part (a), we see that $a - b \leq 0$ which allows us to conclude that $a \leq b$. $\qquad\square$

Our next result is in the same vein as the Order Limit Theorem, but it has a subtlety which we will exploit several times in the coming chapters.

Theorem 2.2.7 (Squeeze Theorem). *Let* $(a_n), (b_n),$ *and* (c_n) *be sequences with* $a_n \leq b_n \leq c_n$ *for every* $n \in \mathbb{N}$. *If* $(a_n) \to a$ *and* $(c_n) \to a$, *then* $(b_n) \to a$.

Before talking about the proof, consider this statement in light of the Order Limit Theorem. Naively, we could take the given inequality, $a_n \leq b_n \leq c_n$, and apply the limit process to arrive at the statement $a \leq \lim b_n \leq a$. From this it follows immediately that $\lim b_n = a$. Is this a valid argument? In actuality it is not as our hypotheses do not specify convergence of the sequence (b_n), which would be necessary in order to apply the order statements previously presented. This is the beauty of the Squeeze Theorem—it asserts that the sequence (b_n) converges *and* specifies the value of its limit.

Seeking a rigorous argument, we discuss some important points and leave the presentation of a formal proof as Exercise 2.2.13. First, observe that

$$|b_n - a| = |b_n - a_n + a_n - a| \leq |b_n - a_n| + |a_n - a|. \tag{2.1}$$

Thus we can show that $(b_n) \to a$ if we can demonstrate control over both terms on the right. Next, notice that the inequality $a_n \leq c_n$ combined with the fact that both sequences converge to a shows that the sequence $(c_n - a_n)$ consists only of positive terms and converges to 0. The compound inequality $a_n \leq b_n \leq c_n$ can also be rewritten as $0 \leq b_n - a_n \leq c_n - a_n$. Thus we can control the size of the quantity $|b_n - a_n|$ by controlling $|c_n - a_n|$, since we know that $|c_n - a_n| = c_n - a_n$ and that this sequence converges to 0. This is half the battle. The second term on the right hand side of Eq. (2.1) can then be made small by appealing to the hypothesis that $(a_n) \to a$. The task of formalizing a rigorous argument is now left to you.

Subsequences

Definition 2.2.8. Let (a_n) be a sequence in $\mathbb{R}$ and let $n_1 < n_2 < n_3 < \ldots$ be a sequence of natural numbers. The function $f : \mathbb{N} \to \mathbb{R}$ defined by $f(j) = a_{n_j}$ is called a *subsequence* of the given sequence (a_n). We use the notation $(a_{n_j})_{j=1}$ or simply (a_{n_j}) to denote the subsequence.

The multiple subscripts can be confusing at first attempt, but your intuition concerning the common usage of the prefix "sub" is likely correct. A subsequence is simply a sequence chosen from the terms of a given sequence in which we have not changed the order in which the terms appear. The function notation simply allows us to define this in a rigorous fashion and the subscripts enable us to keep track of the position of a term both in the original sequence and in the subsequence. In particular, the term a_{n_j} appears as the jth term of the subsequence and as the n_jth term of the original sequence. The fact that the sequence $n_1 < n_2 < n_3 < \ldots$ is chosen as a strictly increasing sequence of natural numbers guarantees that the order of the terms is preserved.

Example 2.2.9. Just as sets can have many subsets, the same is true of sequences and subsequences. Consider, for example, the sequence

$$(1/n) = (1, 1/2, 1/3, 1/4, \ldots, 1/n, \ldots).$$

The most obvious subsequences are those in which we take terms appearing in the original sequence in related positions, e.g. the even terms, the odd terms, every third term, and so on. To write these out explicitly is a simple matter. For the three subsequences just mentioned we have

$$(a_{n_j})_{j=1} = (1/2, 1/4, 1/6, 1/8, \ldots),$$

$$(a_{n_k})_{k=1} = (1, 1/3, 1/5, 1/7, \ldots),$$

$$(a_{n_l})_{l=1} = (1/3, 1/6, 1/9, 1/12, \ldots).$$

It will also be necessary on occasion to explicitly identify the function $j \mapsto n_j$ which defines the relationship between position in the subsequence to position in the original. The three maps needed for this example are $n_j = 2j$, $n_k = 2k - 1$, and $n_l = 3l$.

Example 2.2.10. As a more particular class of examples, let $(a_n)_{n=1}$ be a sequence and let $k \in \mathbb{N}$. We define the k–*tail* of (a_n) by the sequence $(a_n)_{n=k}$; in essence, we simply remove the first $k - 1$ terms of the sequence. In this situation, $n_1 = k$, $n_2 = k + 1$, and in general, $n_j = k + (j - 1)$ producing the subsequence

$$(a_{n_j})_{j=1} = (a_{n_1}, a_{n_2}, a_{n_3}, \ldots) = (a_k, a_{k+1}, a_{k+2}, \ldots) = (a_n)_{n=k}.$$

Exercise 2.2.18 explores this type of subsequence in more detail.

The following proposition demonstrates that the convergence behavior of a sequence carries over to every subsequence.

Proposition 2.2.11. *If (a_n) is a sequence converging to a, then every subsequence of (a_n) also converges to a.*

Before giving the proof, let us again sketch out a few ideas. We know that (a_n) converges to a and thus for every $\varepsilon > 0$, there is an $N \in \mathbb{N}$ with $|a_n - a| < \varepsilon$ for all $n \geq N$. To show that (a_{n_j}) converges to a, our focus will be on identifying $J \in \mathbb{N}$ such that $|a_{n_j} - a| < \varepsilon$ whenever $j \geq J$. In short, we need to choose J large enough so that all the terms in the subsequence appearing after the J position appear in the original sequence past the N position in order to guarantee the desired degree of closeness to a. The fact that a subsequence is defined by a strictly increasing sequence of natural numbers implies that if we choose $J \in \mathbb{N}$ such that $n_J \geq N$, which we can do since such a sequence is unbounded in $\mathbb{N}$, it follows that $n_j \geq n_J \geq N$ whenever $j \geq J$. In other words, the position of the term a_{n_j} in the original sequence is at or beyond position N.

Proof. Let (a_n) be a sequence with limit a, and let (a_{n_j}) be any subsequence. Let $\varepsilon > 0$ and choose $N \in \mathbb{N}$ so that $|a_n - a| < \varepsilon$ for all $n \geq N$. Next choose $J \in \mathbb{N}$ such that $n_J \geq N$. Then if $j \geq J$, it is apparent that $n_j \geq n_J \geq N$ and thus $|a_{n_j} - a| < \varepsilon$. Therefore we conclude that (a_{n_j}) converges to a. $\qquad\square$

As with our previous divergence criteria, the contrapositive of Proposition 2.2.11 provides us with yet another means of detecting divergence, but we first have a definition to help with terminology.

Definition 2.2.12. Let (a_n) be a sequence and let $a \in \mathbb{R}$. We call a a *subsequential limit of* (a_n) if there is a subsequence (a_{n_j}) of (a_n) which converges to a.

Example 2.2.13. Consider again the sequence $(1, -1, 1, -1, \ldots)$. This sequence has exactly two subsequential limits since the subsequence consisting of all the odd terms converges to 1 while the subsequence consisting of all the even terms converges to -1. To be explicit, the subsequence defined by $n_j = 2j - 1$ gives

$$(a_{n_j}) = (a_1, a_3, a_5, \ldots) = (1, 1, 1, \ldots)$$

and the rule $n_j = 2j$ yields the subsequence

$$(a_{n_j}) = (a_2, a_4, a_6, \ldots) = (-1, -1, -1, \ldots).$$

Corollary 2.2.14 (Divergence Criterion 4). *If a sequence* (a_n) *has two distinct subsequential limits, then the sequence* (a_n) *diverges.*

Example 2.2.15. Consider the sequence $(1, 1, 1, 1/2, 1, 1/3, 1, 1/4, \ldots)$. You should see two subsequences embedded here, one which converges to 1 and one which converges to 0. Take a moment to write out the defining functions for these two subsequences, after which we can conclude that this sequence diverges. What would happen if we changed the sequence to $(0, 1, 0, 1/2, 0, 1/3, 0, 1/4, \ldots)$?

Examining the various divergence criteria, it seems that a sequence can diverge in several distinct ways. While this is true to some degree, we have purposefully overplayed some of these scenarios. There are actually only two distinct cases. An unbounded sequence must have a subsequence that diverges to ∞ or $-\infty$ (Exercise 2.2.2), while a bounded sequence that diverges must have two subsequences with distinct limits. This statement encompasses divergence Criteria 1, 3, and 4, and is restated for the sake of convenience below. Divergence Criterion 2 can potentially be applied to any divergence situation. Its true role in the behavior of convergent sequences will be examined in the next section.

Proposition 2.2.16 (Divergence Criterion for Sequential Limits). *A sequence* (a_n) *diverges provided one of the following occur:*

(a) the sequence is unbounded, i.e. the sequence has a subsequence which diverges to either ∞ or $-\infty$;
(b) the sequence has two subsequences with distinct limits.

Exercises

Exercise 2.2.1. Let (a_n) be a Cauchy sequence. Show that (a_n) is bounded.

Exercise 2.2.2. We say a sequence is *bounded below* if there exist $m \in \mathbb{R}$ such that $m \leq a_n$ for all $n \in \mathbb{N}$ and *bounded above* if there exist $M \in \mathbb{R}$ such that $a_n \leq M$ for all $n \in \mathbb{N}$.

(a) Show that a sequence is bounded if and only if it is bounded above and below.
(b) Show that a sequence which is not bounded above has a subsequence which diverges to ∞.
(c) Show that a sequence which is not bounded below has a subsequence which diverges to $-\infty$.

Exercise 2.2.3. Verify each of the following limits.

(a) $\lim\limits_{n\to\infty} \dfrac{4n + 7}{2n - 9} = 2$
(d) $\lim\limits_{n\to\infty} \dfrac{n^3 + 6n - 1}{n^4 + n + 2} = 0$

(b) $\lim\limits_{n\to\infty} \sqrt{\dfrac{1}{6n^2 + 1}} = 0$
(e) $\lim\limits_{n\to\infty} \dfrac{n^2 + -3n + 6}{n^2} = 1$

(c) $\lim\limits_{n\to\infty} \dfrac{\sqrt{9n^2 + n}}{n + 1} = 3$
(f) $\lim\limits_{n\to\infty} \sqrt{9n^2 + n} - 3n = \dfrac{1}{6}$

Exercise 2.2.4. (a) Prove part (a) of Theorem 2.2.4.
(b) Suppose that $(a_n) \to a$ and $(b_n) \to b$. Show that $(a_n - b_n) \to a - b$.
(c) In the proof of part (c) of Theorem 2.2.4, we relied on the fact that a convergent sequence is bounded in order to show that the term $|a_n||b_n - b| < \varepsilon/2$. Explain why this was necessary. In other words, why couldn't we simply choose $N_1 \in \mathbb{N}$ such that $|b_n - b| < \varepsilon/2|a_n|$ for all $n \geq N$?
(d) Suppose (b_n) is a sequence satisfying $b_n \neq 0$ for all $n \in \mathbb{N}$ and $b \neq 0$. Show then that there is a number $K > 0$ such that $|b_n| > K$ for all $n \in \mathbb{N}$.
(e) Use part (d) and techniques similar to those used to prove Theorem 2.2.4(c) to prove Theorem 2.2.4(d).
(f) Extend part (d) of Theorem 2.2.4 by showing that the statement still holds if we remove the requirement that $b_n \neq 0$ for all $n \in \mathbb{N}$. First try extending part (d) to this more general situation.

Exercise 2.2.5. Let (a_n) be a sequence with $(a_n) \to a$. Show that $(a_n^k) \to a^k$ for every $k \in \mathbb{N}$.

Exercise 2.2.6. Let (a_n) be a sequence with $(a_n) \to a$ and $a_n \geq 0$ for every $n \in \mathbb{N}$. Show that $(\sqrt{a_n}) \to \sqrt{a}$. To do this, consider two cases: $a = 0$ and $a > 0$.

Exercise 2.2.7. Let (a_n) be a sequence with $(a_n) \to a$.

(a) Show that $(|a_n|)$ converges to $|a|$. It may be helpful to consider Exercise 1.3.2.
(b) On the other hand, provide an example of a sequence (a_n) such that $(|a_n|) \to 1$, but (a_n) does not converge to 1 or -1.
(c) However, if $(|a_n|) \to 0$, show that $(a_n) \to 0$.

Exercise 2.2.8. Let (a_n) and (b_n) be sequences.

(a) If (a_n) converges but (b_n) diverges, show that $(a_n + b_n)$ must diverge.
(b) If (a_n) and (b_n) both diverge, must it be the case that $(a_n + b_n)$ diverges? Explain in detail.
(c) If (a_n) diverges and $c \in \mathbb{R}$ with $c \neq 0$, show that (ca_n) diverges.

Exercise 2.2.9. Let (a_n) and (b_n) be sequences.

(a) If (a_n) is a bounded sequence and (b_n) converges to zero, show that the product sequence $(a_n b_n)$ converges to zero. Note: You cannot use the Algebraic Limit Theorem to prove this. Why not?
(b) Does a similar result hold if (b_n) converges to a nonzero limit?
(c) Does the result hold if (a_n) is not bounded and $(b_n) \to 0$.

Exercise 2.2.10. Let (a_n) and (b_n) be sequences which diverge to infinity.

(a) If $c > 0$, show that (ca_n) diverges to infinity.
(b) Show that $(a_n + b_n)$ diverges to infinity.
(c) Show that $(a_n b_n)$ diverges to infinity.
(d) What conclusions can we draw about the behavior of (a_n/b_n)?

Exercise 2.2.11. Provide an example for each of the following statements.

(a) A sequence (a_n) which diverges and

$$\lim_{n \to \infty} \frac{a_{n+1}}{a_n} = 1.$$

(b) A sequence (a_n) which converges to 0 and a sequence (b_n) which diverges to infinity with

$$\lim_{n \to \infty} a_n b_n = 1.$$

Exercise 2.2.12. (a) Extend Theorem 2.2.6(a) by showing that the result holds when we replace 0 by any other real number c.
(b) Do the results of Theorem 2.2.6 hold if we replace $\leq$ and $\geq$ with $<$ and $>$? Explain.

Exercise 2.2.13. Using the discussion from the text, supply a formal proof of Theorem 2.2.7.

Exercise 2.2.14. Let $A \subseteq \mathbb{R}$ be nonempty and bounded. Show that there exist sequences (x_n) and (y_n) in A which converge to $\sup(A)$ and $\inf(A)$, respectively.

Exercise 2.2.15. Suppose (a_n) is a sequence with positive terms. If $\lim na_n$ exists, show that $(a_n) \to 0$.

Exercise 2.2.16. For each of the following, give an example of a sequence with the specified property.

(a) A sequence with a subsequence that diverges to ∞ and a subsequence which diverges to $-\infty$.
(b) A sequence for which 0, 1, -1 are subsequential limits. Write out the function that defines the sequence and the three subsequences.
(c) A sequence for which each natural number is a subsequential limit.
(d) A sequence for which every real number is a subsequential limit.

Exercise 2.2.17. Suppose that (a_n) is a sequence with the property that the subsequences (a_{n_j}) and (a_{n_k}), where $n_j = 2j$ and $n_k = 2k - 1$, both converge to a. Show that (a_n) also converges to a.

Exercise 2.2.18. Let $(a_n)_{n=1}$ be a sequence. Show that the following are equivalent.

(a) The sequence (a_n) converges.
(b) There is a $k \in \mathbb{N}$ so that the k–tail of (a_n) converges.
(c) The k–tail of (a_n) converges for every $k \in \mathbb{N}$.

2.3   Completeness in ℝ Revisited

Completeness is essentially the defining property of the real number system and enriches the system with an abundance of properties not shared by its subsystems—the natural numbers, the integers, and the rational numbers. Here we present three theorems which are crucial to the study of sequences all of which follow from the fact that ℝ is complete. After discussing these theorems we will investigate their true connection to our axiom of completeness.

Definition 2.3.1. A sequence (a_n) is said to be *increasing* if $a_n \leq a_{n+1}$ for every $n \in \mathbb{N}$ and *decreasing* if $a_n \geq a_{n+1}$ for every $n \in \mathbb{N}$. A *monotone sequence* is one that is either increasing or decreasing.

Example 2.3.2. The sequences $(n/(n + 1))$ and (n) are both increasing and the sequences $(1/n)$ and $(n - n^2)$ are both decreasing; we could also say that all four sequences are monotone sequences; however, it's often best to be as specific about the behavior as possible. For a non-example, the sequence $((-1)^n)$ is not monotone since it is neither increasing nor decreasing.

   Consider the first two sequences above. The first converges to 1 while the second diverges to infinity even though they have the same monotonic behavior. The next theorem provides the rationale for this distinction in behavior.

Theorem 2.3.3 (Monotone Convergence Theorem). *Every bounded and monotone sequence must converge. In particular, if a sequence is bounded above and increasing then it converges and likewise, if it is bounded below and decreasing then it converges.*

Fig. 2.2 Monotone Convergence Theorem

Proof. Without loss of generality, let us assume that (a_n) is an increasing sequence which is bounded above and consider the set of range values of the sequence, $A = \{a_n : n \in \mathbb{N}\}$. Since the sequence in question is bounded above, the set A must also be bounded above; it is also nonempty. Therefore the supremum exists and we set $a = \sup(A)$.

We claim now that $(a_n) \to a$. To verify this, let $\varepsilon > 0$. By Lemma 1.2.10, there is an element of A, call it a_N, such that $a - \varepsilon < a_N$. If we now consider $n \in \mathbb{N}$ with $n \geq N$, the fact that our sequence is increasing together with the fact that a is the supremum of A guarantees us that

$$a - \varepsilon < a_N \leq a_n \leq a < a + \varepsilon$$

(see Fig. 2.2); we can rephrase this by saying that $|a_n - a| < \varepsilon$ whenever $n \geq N$ and therefore conclude that $(a_n) \to a$.   □

Notice that the proof provides a bit more information than is explicitly stated in the theorem—the sequence must converge and the supremum of the set of terms is the limit. We now encourage the reader to supply a proof for the alternate situation involving decreasing sequences. In the exercises you will also show that any increasing sequence that is not bounded above must diverge to infinity and the analogous statement for decreasing sequences which are not bounded below.

Example 2.3.4. The Monotone Convergence Theorem is a very practical tool, particularly for sequences defined recursively. Take the sequence (a_n) defined by the recursive relationship $a_1 = 3$ and $a_{n+1} = (2a_n + 1)/3$. In order to apply the theorem, we must first show that the sequence is monotone and bounded. To do this we write out the first few terms in an effort to observe behavior and then work to prove our claims. Computing, we see that the first three terms are 3, 7/3, and 17/9. It appears that the sequence is decreasing with an upper bound of 3 (obvious since the terms are decreasing). Coming up with a lower bound is a more complicated process, but since all the operations involve positive values, it seems likely that 0 will serve as a lower bound. We must now work to show that the inequality

$$0 \leq a_{n+1} \leq a_n \leq 3$$

holds for each $n \in \mathbb{N}$. Notice we have summed up both the decreasing nature of the sequence and the proposed bounds with this single compound inequality. Since the statement must be shown for each natural number, we proceed by mathematical induction.

For our base case, we take $n = 1$ and it is certainly true that

$$0 \le 7/3 \le 3 \le 3.$$

For the inductive hypothesis, suppose that for some $j \in \mathbb{N}$ we know that the inequality

$$0 \le a_{j+1} \le a_j \le 3$$

holds true. We now show that the corresponding inequality holds for $j + 1$. The strategy here will be to produce the desired inequality by algebraically manipulating the known inequality using the algebraic processes given in the recursive definition of the sequence, with care given to the order of the operations. Moving forward, we multiply across the inequality

$$0 \le a_{j+1} \le a_j \le 3$$

by 2, add 1, and then divide by 3 to obtain the inequality

$$1/3 \le (2a_{j+1} + 1)/3 \le (2a_j + 1)/3 \le 7/3.$$

The inner most expressions take a simpler form as a_{j+2} and a_{j+1}, respectively,

$$1/3 \le a_{j+2} \le a_{j+1} \le 7/3.$$

We had hoped for an upper bound of 3 and a lower bound of 0 and the previous inequality certainly implies that

$$0 \le a_{j+2} \le a_{j+1} \le 3.$$

By induction, we conclude that our claim is true for all $n \in \mathbb{N}$. By the Monotone Convergence Theorem, the sequence converges to some value L.

To find the limit, we apply the limit process to the both sides of the recursive equation $a_{n+1} = (2a_n + 1)/3$. By the Algebraic Limit Theorem, the right-hand side converges to $(2L + 1)/3$. The sequence (a_{n+1}) is a subsequence of the original sequence and therefore also converges to L. Hence we arrive at the equation $L = (2L + 1)/3$, which we can solve to find that $L = 1$.

Example 2.3.5. Let $x > 1$ and consider the sequence (x^n). In order to understand the behavior of such a sequence, let's examine the case $x = 2$ for a moment. The resulting sequence is $(2, 4, 8, 16, \ldots, 2^n, \ldots)$. It appears as if we have an increasing sequence which diverges to infinity. Returning to our general situation, we will show that this behavior is common to all sequences of the form above. First, to show that the sequence is increasing, we use the order properties of $\mathbb{R}$. Let $n \in \mathbb{N}$. Multiplying the inequality $x > 1$ by x^n yields $x^n x > x^n$ which implies that $x^{n+1} > x^n$. We thus conclude that the sequence is increasing since this relationship holds for all $n \in \mathbb{N}$.

If the sequence is unbounded, then by Exercise 2.3.6 we can conclude that it diverges to infinity, providing the desired conclusion. At the present, we are not capable of dealing with powers in an efficient manner as we have not discussed logarithms. To alleviate this issue, we will use a proof by contradiction. Assume to the contrary then that the sequence (x^n) is bounded. By the Monotone Convergence Theorem, it must be the case that $(x^n) \to L$ for some $L \in \mathbb{R}$. Moreover, we also know that $L = \sup\{x^n : n \in \mathbb{N}\}$. To obtain our contradiction, we will work to produce a term in the sequence greater than L. To accomplish this, we will use the convergence of the sequence to force the terms close enough to L so that when we move to a subsequent term, the increasing behavior of the sequence forces a term to bypass L. The key in all of this is the fact that the sequence is strictly increasing and the distance between successive terms is completely determined by the distance between x and 1. Let $\varepsilon = |x - 1|$ and choose $N \in \mathbb{N}$ such that $|x^n - L| < \varepsilon$ for all $n \geq N$. In particular, the term x^N satisfies $|x^N - L| < |x - 1|$. We claim that $x^{N+1} > L$. Using the fact that $x > 1$ and the increasing nature of the sequence shows that

$$|x^{N+1} - x^N| = x^N |x - 1| > |x - 1| > |x^N - L|.$$

The inequality indicates that the distance between x^{N+1} and x^N is greater than the distance between x^N and L. Combining this fact with the increasing behavior of the sequence and the fact that L is the supremum of the sequence terms allows us to conclude that $x^N \leq L < x^{N+1}$. However this is a contradiction since $x^n \leq L$ for all $n \in \mathbb{N}$. Therefore it must be the case that the sequence (x^n) is unbounded and hence diverges to ∞.

Example 2.3.6. Exercise 2.3.7 shows that for a real number $0 < x < 1$, the sequence $(x^n) \to 0$. With this we can show that the Cantor set (Example 1.4.15) contains no intervals. In our construction of C, we first constructed sets C_n which had the property that each is a union of 2^n closed intervals of length $1/3^n$. If the Cantor set contained an interval, say of length $\varepsilon > 0$, then this interval would have to be contained in C_n for every $n \in \mathbb{N}$; in particular, for any fixed $n \in \mathbb{N}$, this interval would have to be entirely contained in *one* of the 2^n intervals of length $1/3^n$ which comprise C_n. However, since $(1/3^n) \to 0$, we can choose $N \in \mathbb{N}$ such that $0 < 1/3^N < \varepsilon$. From this we conclude that the Cantor Set contains no intervals.

The second major theorem of this section concerns subsequences of bounded sequences and is one of the most notable theorems in the study of real analysis. The proof here is somewhat more involved than those we have encountered up to this point and we will discuss some of the ideas before presenting the proof.

Theorem 2.3.7 (Bolzano–Weierstrass Theorem). *Every bounded sequence has a convergent subsequence.*

The proof technique we use is not as obvious as some of the others we've encountered and the basic outline is as follows. Using the fact that the sequence is bounded, we can identify an $M > 0$ so that $|a_n| \leq M$ for all $n \in \mathbb{N}$. We can

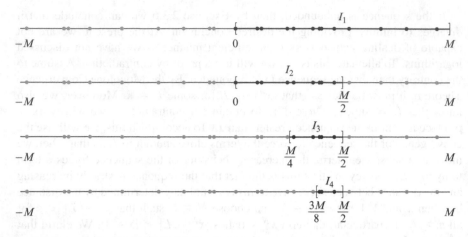

Fig. 2.3 Construction of $\{I_j\}$ within $[-M, M]$

rephrase this in terms of set inclusion, $a_n \in [-M, M]$ for all n. The idea then is to construct a collection of closed, bounded, nested intervals $\{I_j : j \in \mathbb{N}\}$ based on the behavior of our sequence and such that the length of these intervals decreases to zero. Exercise 2.1.7 then guarantees us that the intersection of these sets contains a unique element a. We then construct a subsequence (a_{n_j}) with the term a_{n_j} coming from the interval I_j and claim that this subsequence must converge to a. One of the most beautiful aspects of this theorem is that it is constructive in its approach, meaning that it constructs both the subsequence and the limit in an explicit manner, rather than simply using a reduction technique and applying an abstract result. The computer scientist will recognize this process as a binary search algorithm. We also mention that the use of the word countable in the proof will mean a countably infinite collection.

Proof. Let (a_n) be a bounded sequence of real numbers. Using this hypothesis, we can find $M > 0$ such that $a_n \in [-M, M]$ for all $n \in \mathbb{N}$. To construct a sequence of closed, bounded, nested intervals with lengths decreasing to zero, we will successively cut the interval $[-M, M]$ into halves. For I_1, we bisect $[-M, M]$ into the closed subintervals $[-M, 0]$ and $[0, M]$. The fact that sequences must have countably many terms guarantees us that at least one of these two subintervals must contain countably many terms of the sequence (a_n). Choose I_1 to be one such subinterval (if both $[-M, 0]$ and $[0, M]$ contain infinitely many terms of the sequence, then the choice is completely arbitrary). Note that the length of I_1 is M.

To select I_2, we bisect I_1 into closed subintervals of length $M/2 = M/2^1$. Again, at least one of these subintervals must contain countably many terms of the sequence (a_n) and we choose I_2 to be one such closed interval. Notice also that $I_2 \subseteq I_1$. Continuing inductively, we can construct a collection of closed, bounded, nested intervals $\{I_j\}$ (Fig. 2.3) with the property that each interval contains countably many terms of the sequence (a_n) and the length of I_j is $M/2^{j-1}$. From Exercise 2.3.7, we see that the sequence $M/2^{j-1}$ converges to zero as $j \to \infty$. By Exercise 2.1.7, it follows that the intersection $\cap_{j=1}^{\infty} I_j$ contains a unique element which we shall denote by a. The convergence of the sequence $M/2^{j-1}$ will also be key in the final component of the proof.

At this point we are ready to define a subsequence; this is also done in an inductive manner. Choose $n_1 \in \mathbb{N}$ so that $a_{n_1} \in I_1$. Next, choose $n_2 \in \mathbb{N}$ so that $n_2 > n_1$ and $a_{n_2} \in I_2$, which is permissible since I_2 contains countably many terms of the sequence (a_n). Continuing, for each $j \in \mathbb{N}$, we choose $n_j \in \mathbb{N}$ so that $n_j > n_{j-1}$ and $a_{n_j} \in I_j$.

We claim now that the subsequence (a_{n_j}) converges to a. Let $\varepsilon > 0$ and choose $J \in \mathbb{N}$ such that $M/2^J < \varepsilon$. Then if $j \geq J + 1$, we have that $j - 1 \geq J$ and $M/2^{j-1} \leq M/2^J < \varepsilon$. In other words, for $j \geq J + 1$, the length of the interval I_j is less than ε. Finally, if we consider a term a_{n_j} with $j \geq J + 1$, then both a_{n_j} and a are in I_j and, using the restriction on the length of this interval, it is clear that

$$|a_{n_j} - a| \leq \frac{M}{2^{j-1}} \leq \frac{M}{2^J} < \varepsilon.$$

Thus we conclude that $(a_{n_j}) \to a$ as desired.   □

Notice that boundedness is a hypothesis in both of the previous two statements. We then are able to draw conclusions regarding the convergence behavior of the sequence in question. The connection here is provided by Theorem 2.2.2; both of these theorems are attempts to prove the converse of the statement that every convergent sequence is bounded. It is clear that there are bounded sequences which diverge, however, these two theorems show that there is more to be said. On the one hand, with an additional hypothesis we can conclude convergence, and on the other, with boundedness alone the sequence may diverge, but some part of it must converge.

The third main theorem of this section returns to the idea of a Cauchy sequence. We explored this in a previous section as a means of detecting divergence but it turns out that it is also capable of indicating convergence. The beauty of the theorem below is that it provides us a means of determining whether or not a sequence converges without the explicit knowledge of a limit. The two theorems presented above share this property to some degree.

Theorem 2.3.8 (Cauchy Criterion). *A sequence converges if and only if it is Cauchy.*

Proof. We showed one direction of the statement in Proposition 2.1.9 and now we need to verify that every Cauchy sequence is convergent. Let (a_n) be a Cauchy sequence. It was shown in Exercise 2.2.1 that every Cauchy sequence is bounded. Thus we can apply the Bolzano–Weierstrass Theorem to generate a subsequence (a_{n_j}) which converges. Denote the limit of this subsequence by a.

To show that the original sequence (a_n) also converges to a, let $\varepsilon > 0$. Since the original sequence is Cauchy, there exists an $N \in \mathbb{N}$ such that $|a_n - a_m| < \varepsilon/2$

whenever $n, m \geq N$. Furthermore, there exists a $J \in \mathbb{N}$ such that $|a_{n_j} - a| < \varepsilon/2$ for all $j \geq J$. At this point we have two ideas to work with, one which concerns the original sequence and one which concerns only the subsequence. To connect these ideas, we must take an extra bit of care. Choose $K \in \mathbb{N}$ such that $K \geq J$ and $n_K \geq N$. This second choice is permissible since the sequence (n_j) is an increasing sequence of natural numbers and will therefore eventually exceed the value that we have specified for N. These two conditions on the value of K guarantee us that $|a_n - a_{n_K}| < \varepsilon/2$ for all $n \geq N$ since $n_K \geq N$, and $|a_{n_K} - a| < \varepsilon/2$ since $K \geq J$. Using these two conditions with the triangle inequality, for $n \geq N$,

$$|a_n - a| = |a_n - a_{n_K} + a_{n_K} - a| \leq |a_n - a_{n_K}| + |a_{n_K} - a| < \varepsilon/2 + \varepsilon/2 = \varepsilon.$$

Hence we conclude that $(a_n) \to a$. $\square$

Each of these three theorems and the Nested Interval Property are direct consequences of the completeness axiom of $\mathbb{R}$ and each other. Indeed, consider the proofs of these four statements. The Nested Interval Property depended on the axiom of completeness and the structure of intervals while the Monotone Convergence Theorem came directly from the axiom and the behavior of monotone sequences. The Bolzano–Weierstrass Theorem was a result of the Nested Interval Property, which in turn supplied the machinery for the Cauchy Criterion. Exploring these implications indicates that while each result seems to be doing something very different with respect to the real number system, they are connected to each other. The connection is brought to light in a more obvious manner by considering the bigger picture, that is, each result is asserting the existence of a real number satisfying a specific set of hypotheses. And though the hypotheses of these five statements are drastically different, each result can be thought of in a visual sense as asserting that the real number line is a continuum.

It is also typical when considering mathematical implications to understand when statements are logically equivalent. As it turns out, all five of the statements discussed in the previous paragraph are equivalent. In other words, the axiom of completeness is not the only way to move from the ordered field $\mathbb{Q}$ to the complete, ordered field $\mathbb{R}$. We could have taken any of these other statements as our completeness axiom. For example, it is common to take the statement that all Cauchy sequences converge as the completeness axiom. The excellent text [30] takes this approach. Our choice of using the existence of suprema was simply the easiest to motivate early on. However, we will see in Sect. 2.5 that our choices change when considering completeness in an arbitrary metric space.

To close this section, we present a proof that the Monotone Convergence Theorem implies our axiom of completeness. The proof also depends on the Order Limit Theorem for sequences; however, the proof of that fact does not depend on the completeness property in $\mathbb{R}$. The reader is then encouraged to consider using the other statements under consideration to prove the axiom of completeness and/or any of the other statements.

Proof. We assume that every bounded, monotone sequence in $\mathbb{R}$ converges. We then assume that $A \subseteq \mathbb{R}$ is nonempty and bounded above. Our task is to then produce a point $s \in \mathbb{R}$ which satisfies the definition of supremum. In order to apply the Monotone Convergence Theorem, we begin by constructing two sequences. Let $a_1 \in A$ and let y_1 be an upper bound for A. If $a_1 = y_1$, then this number will satisfy the definition of supremum and we are done. Indeed, the point is certainly an upper bound, and since the point is in A, it is also less than or equal to any upper bound.

If this is not the case, then $M = y_1 - a_1 > 0$ and we consider the interval $[a_1, y_1]$. Let b_1 be the midpoint of this interval. If b_1 is an upper bound for A, we define $y_2 = b_1$ and set $a_2 = a_1$. If b_1 is not an upper bound for A, then we set $y_2 = y_1$ and choose a point $a_2 \in [b_1, y_1]$. If $a_2 = y_2$, then this point is the supremum of A and we are done. If not, then the interval $[a_2, y_2]$ has a positive length less than or equal to $M/2$.

Continuing in this manner, we construct two sequences (a_n) and (y_n). If at some point we have $a_n = y_n$, then the process terminates and the supremum of the set is this common value. If the process continues without terminating, the sequence (a_n) is an increasing sequence of elements of A and is bounded above, while (y_n) is a decreasing sequence of upper bounds for A and is hence bounded below. Moveover, $|y_n - a_n| \le M/2^{n-1} \to 0$ as $n \to \infty$. By the Monotone Convergence Theorem, (a_n) and (y_n) both converge, and since $|y_n - a_n| \to 0$, they have the same limit; call it s. We claim now that s is the supremum of A.

For each $n \in \mathbb{N}$, we know that $a \le y_n$ since each y_n is an upper bound for A. The Order Limit Theorem implies that $a \le s$, showing that s is an upper bound. Finally, assume that x is some other upper bound for A. To show that $s \le x$, we assume to the contrary that $x < s$. Since $(a_n) \to s$, we can find an $n \in \mathbb{N}$ such that $|a_n - s| < |s - x|$ for all $n \ge N$. In particular, the term a_N satisfies $|a_N - s| < |s - x|$, i.e. the distance between a_N and s is less than the distance between s and x. Our hypothesis concerning the order of these values now implies that $x < a_N \le s$. But this is a contradiction since x is an upper bound for A. Thus $s \le x$ and s is the supremum of A.   □

The proof above is technical in nature and should remind you of the technique we used to prove the Bolzano–Weierstrass Theorem. It should also be easy to see that with some simple modifications, we could use a very similar argument to show that the Monotone Convergence Theorem implies the Nested Interval Property or that the Nested Interval Property implies the axiom of completeness. While we won't pursue this collection of statements further, we encourage the reader to attempt proving several of the other equivalences mentioned before the proof.

Exercises

Exercise 2.3.1. Show that the sequence defined recursively by $y_1 = 1$ and $y_{n+1} = (3y_n + 4)/4$ converges and find its limit.

Exercise 2.3.2. Show that the sequence defined recursively by $y_1 = 8$ and $y_{n+1} = (3y_n + 4)/4$ converges and find its limit.

Exercise 2.3.3. Show that the sequence defined recursively by $y_1 = 1$ and $y_{n+1} = 4 - 1/y_n$ converges and find its limit.

Exercise 2.3.4. Consider the sequence defined by

$$x_1 = \sin 1, \quad x_2 = \max\{\sin 1, \sin 2\}, \quad x_3 = \max\{\sin 1, \sin 2, \sin 3\}$$

and in general,

$$x_k = \max\{\sin 1, \sin 2, \ldots, \sin k\}.$$

Show that the sequence converges (you do not need to find the limit).

Exercise 2.3.5. Consider the sequence defined by the pattern

$$\sqrt{2}, \sqrt{2\sqrt{2}}, \sqrt{2\sqrt{2\sqrt{2}}}, \ldots.$$

(a) Find a recursive definition for this sequence.
(b) Show that the sequence converges.
(c) Find the limit of this sequence.

Exercise 2.3.6. (a) Let (a_n) be an increasing sequence which is not bounded above. Show that the sequence diverges to ∞.
(b) Let (a_n) be an decreasing sequence which is not bounded below. Show that the sequence diverges to $-\infty$.

Exercise 2.3.7. (a) Let $0 < x < 1$. Show that the sequence $(x^n) \to 0$.
(b) Let $-1 < x < 0$. Show that the sequence $(x^n) \to 0$.

Exercise 2.3.8. A sequence (a_n) is called *additive* if $a_{m+n} = a_m + a_n$ for all $m, n \in \mathbb{N}$.

(a) Show that the sequence $(2n)_{n=1}^{\infty}$ is additive.
(b) Suppose that (a_n) is an additive sequence and show that the sequence $\left(\dfrac{a_n}{n}\right)$ converges. What is the limit?

Exercise 2.3.9. A sequence (a_n) is called *subadditive* if $a_{m+n} \leq a_m + a_n$ for all $m, n \in \mathbb{N}$.

(a) Show that the sequence $(2n + 1)_{n=1}^{\infty}$ is subadditive.
(b) Suppose that (a_n) is a subadditive sequence of positive real numbers. Show that the sequence $\left(\dfrac{a_n}{n}\right)$ has a convergent subsequence.
(c) Suppose that (a_n) is a subadditive sequence of positive real numbers. Extend the previous exercise by showing that the sequence $\left(\dfrac{a_n}{n}\right)$ is convergent.

Exercise 2.3.10. The following sequence of exercises explore the notion of the limit superior and inferior of a bounded sequence. Assume that (a_n) is a bounded, not necessarily convergent, sequence of real numbers.

(a) Define a new sequence (y_n) by $y_n = \sup\{a_k : k \geq n\}$. For the sake of clarity,

$$y_1 = \sup\{a_1, a_2, a_3, \ldots\}$$

and

$$y_2 = \sup\{a_2, a_3, a_4, \ldots\}$$

and so on. Show that the sequence (y_n) converges.
(b) The *limit superior* of the sequence (a_n) is defined to be the limit of the sequence (y_n) constructed in part (a) and is denoted by

$$\limsup a_n = \lim y_n.$$

Provide a definition for the *limit inferior*, denoted $\liminf a_n$. Does this quantity always exist? Briefly explain.
(c) Let (a_n) be the sequence $(1, -1, 1, -1, 1, -1, \ldots)$. Find $\limsup a_n$ and $\liminf a_n$.
(d) Show that $\liminf a_n \leq \limsup a_n$ for any bounded sequence (a_n).
(e) Suppose that $\liminf a_n = \limsup a_n$. Show that $\lim a_n$ exists and has the same value.
(f) Suppose that $\lim a_n$ exists. Show that $\liminf a_n$ and $\limsup a_n$ both exist and have the same value.
(g) Show that $\liminf(-a_n) = -\limsup a_n$.

Exercise 2.3.11. (a) Let (a_n) be a bounded sequence of real numbers. Show that there is a subsequence of (a_n) which converges to $\limsup a_n$.
(b) Can a similar statement be made for the $\liminf a_n$? Supply the details.
(c) Use either (a) or (b) to supply a short proof for the Bolzano–Weierstrass Theorem.
(d) Let (a_n) be a bounded, non-convergent sequence of real numbers. Show that (a_n) has at least two subsequences which converge to different values.

(e) Assume that (a_n) is a bounded sequence with the property that every convergent subsequence of (a_n) converges to the same limit $a \in \mathbb{R}$. Show then that (a_n) must converge to a. (It's probably easiest to use parts (a) and (b); however, another proof can be given by assuming that (a_n) doesn't converge and using the Bolzano–Weierstrass Theorem.)

Exercise 2.3.12. Give an example of each of the following. If it is not possible to give an example, supply a proof explaining why.

(a) A convergent sequence which is not Cauchy.
(b) A Cauchy sequence which is not increasing.
(c) A bounded sequence which is not Cauchy.
(d) A divergent sequence with a subsequence that is Cauchy.

Exercise 2.3.13. Let (a_n) and (b_n) be Cauchy sequences.

(a) Use the definition of Cauchy to show that $(a_n + b_n)$ is Cauchy.
(b) Show that $(a_n b_n)$ is Cauchy.

2.4 Set Structures in $\mathbb{R}$ via Sequences

In Sect. 1.4 we discussed various set structures in $\mathbb{R}$ with open sets forming the basis of our discussion. Here we return to closed and compact sets and explore their characterizations via sequential criteria.

Definition 2.4.1. Let $A \subseteq \mathbb{R}$. A point $c \in \mathbb{R}$ is called a *limit point of* A if there is a sequence (a_n) in A with $a_n \neq c$ for every $n \in \mathbb{N}$ and such that $(a_n) \to c$. A point $c \in A$ which is not a limit point is called an *isolated point*.

It is important to keep in mind that limit points may be outside the set A but isolated points must be in the set in question.

Example 2.4.2. Consider the set $A = \{1, 1/2, 1/3, 1/4, \ldots\}$. The point 0 is not in A but it is a limit point of A since the sequence $(1/n)$ is contained in A, no point of the sequence is equal to 0, and $(1/n) \to 0$. Is 1 a limit point? Keep in mind that to satisfy the definition we must build a sequence converging to 1, but we cannot use the point 1. It's obvious then that 1 is not a limit point since all other points of A are at least $\frac{1}{2}$ a unit away from 1 and hence there is no way for a sequence in A to converge to 1. This shows that 1 is an isolated point of A. Similar reasoning will show that no other point of A, or of $\mathbb{R}$ for that matter, is a limit point. From this we conclude that every point of A is isolated.

For another example, think about the set $A = (0, 1)$. If $a \in (0, 1)$, then we can find a sequence in A with the properties of Definition 2.4.1. In particular, the sequence $(a - a/n)_{n=2}$ is in A, no point of the sequence is equal to a and the sequence converges to a. Thus every point of A is a limit point of A. Are there any others? What about 0 and 1? For 0, the sequence $(1/n)$ will suffice and, for 1, we can choose the sequence $(n/(n+1))$. Now consider a real number $a > 1$. Since all the points of A are less than one, it is clear by the Order Limit Theorem that there is no way to construct a sequence in A which converges to a. A similar argument shows that no real number $a < 0$ is a limit point of A. Does A have any isolated points?

In the last two examples, the set in question had limit points which are not elements of the set. The set $[0, 1]$ does not have this property. Reasoning just as with the open interval above, we can show that every point of the set is a limit point, but this time 0 and 1 are also in the set. Hence $[0, 1]$ contains all its limit points. The next theorem states that this is a way to characterize closed sets. On a related note, *every* element of $[0, 1]$ is also a limit point, but this is not necessary for general closed sets; the set $\{1, 2\}$ has no limit points, only isolated points, but is still closed.

Theorem 2.4.3. *A set $F \subseteq \mathbb{R}$ is closed if and only if it contains all its limit points.*

Proof. First suppose that F is closed, which means that F^c is open by definition. To show that this set contains all its limit points, let $c \in \mathbb{R}$ be a limit point of F, that is, suppose there is a sequence (a_n) in F with $a_n \neq c$ for all n and $(a_n) \to c$. Now we must show that $c \in F$. To make use of our assumption that F^c is open, we will assume towards a contradiction that $c \in F^c$. Hence there is an $\varepsilon > 0$ such that $B_\varepsilon(c) \subseteq F^c$. However, since $(a_n) \to c$, there is an $N \in \mathbb{N}$ such that $|a_N - c| < \varepsilon$, meaning that $a_N \in B_\varepsilon(c) \subseteq F^c$. But a_N is in F, a contradiction. Thus $c \in F$.

For the converse, suppose that F contains all its limit points. To show that F is closed, we must show that F^c is open. Here we will also work by contradiction and we assume that F^c is not open. Hence there is a point $c \in F^c$ such that for every $\varepsilon > 0$, $B_\varepsilon(c)$ is not a subset of F^c. We claim now that the point c must be a limit point of F, which will provide a contradiction since $c \notin F$.

To verify the claim, we construct a sequence in F which converges to c. To define this sequence, choose a_n to be a point in F which is in $B_{1/n}(c)$; this is possible since for every $\varepsilon > 0$, we know that $B_\varepsilon(c)$ is *not* a subset of F^c. It follows immediately that (a_n) is in F and $a_n \neq c$ for all n. Furthermore,

$$c - \frac{1}{n} < a_n < c + \frac{1}{n}$$

and hence converges to c by the Squeeze Theorem. This shows that c is a limit point of F and provides our contradiction. $\square$

Thinking about terminology for a moment, when we use the word closed it is typically with respect to a set and operations. For example, we say that $\mathbb{R}$ is closed with respect to the operations of addition and multiplication. The theorem above demonstrates that a closed set is closed with respect to the operation of limits. Continuing this line of inquiry, we now define the closure of a set.

Definition 2.4.4. Let $A \subseteq \mathbb{R}$. The *closure of A*, denoted $\overline{A}$, is defined to be the set which consists of A together with all its limit points. Letting L denote the set of limit points of A, we can write $\overline{A} = A \cup L$.

Example 2.4.5. Let $A = \{1, 1/2, 1/3, \ldots\}$. From our discussion in the previous example we see that $L = \{0\}$ and $\overline{A} = \{0, 1, 1/2, 1/3, \ldots\}$. From that example we also see that $\overline{(0, 1)} = \overline{[0, 1]} = [0, 1]$ with $L = [0, 1]$ for both $(0, 1)$ and $[0, 1]$.

Exercise 2.1.8 shows that $\overline{\mathbb{Q}} = \overline{\mathbb{I}} = \mathbb{R}$ with $L = \mathbb{R}$ for both sets. Considering $\mathbb{N}$, we have $L = \emptyset$ and hence $\overline{\mathbb{N}} = \mathbb{N}$.

Notice that in all of these examples, both L and $\overline{A}$ are closed. The following theorem formalizes this.

Proposition 2.4.6. *Let $A \subseteq \mathbb{R}$. The set $\overline{A}$ is a closed set. Furthermore, A is closed if and only if $A = \overline{A}$.*

Proof. We will show that $\overline{A}$ is closed and leave the second statement as an exercise. Our goal is to show that $\overline{A}^c$ is an open set and hence we choose an arbitrary point $c \in \overline{A}^c$. In order to generate an ε–neighborhood for c, notice that the definition of closure implies $c \in (A \cup L)^c = A^c \cap L^c$, i.e. c is not in A and c is not a limit point of A. Using the negation of Exercise 2.4.1, the statement that $c \notin L$ means that there is an $\varepsilon > 0$, such that for every $a \in A$, either $a = c$ or $a \notin B_\varepsilon(c)$. Now, we know that $c \notin A$, and thus $a \neq c$, so it must be the case that no point of A is in $B_\varepsilon(c)$ for this particular value of ε.

We can conclude that $B_\varepsilon(c) \subseteq \overline{A}^c$ if we can also show that no point of L is in this ε-neighborhood. To this end, suppose that $d \in L$ and $d \in B_\varepsilon(c)$. The fact that the ε-neighborhood is open then implies that there is a $\delta > 0$ such that $B_\delta(d) \subseteq B_\varepsilon(c)$. However, since d is a limit point of A, Exercise 2.4.1 implies the existence of a point $a \in A$ with $a \neq d$ and $a \in B_\delta(d) \subseteq B_\varepsilon(c)$. But this is impossible since we know that $B_\varepsilon(c)$ contains no point of A. Thus $B_\varepsilon(c)$ also contains no limit points of A and we have $B_\varepsilon(c) \subseteq \overline{A}^c$. Therefore $\overline{A}^c$ is open and $\overline{A}$ is closed. $\square$

Recall from the previous example that the closure of both the rationals and irrationals is $\mathbb{R}$. This is a restatement of the fact that both these sets are dense in $\mathbb{R}$.

Proposition 2.4.7. *Let S be a subset of $\mathbb{R}$. Then S is dense in $\mathbb{R}$ if and only if $\overline{S} = \mathbb{R}$.*

Proof. First suppose that S is dense in $\mathbb{R}$. In order to show that $\overline{S} = \mathbb{R}$, we must show that $\mathbb{R} \subseteq \overline{S}$; the reverse inclusion is trivially true. In order to verify the desired inclusion, it suffices to show that every point of $\mathbb{R}$ is a limit point of S. To this end, let $t \in \mathbb{R}$. Applying the density condition, for each $n \in \mathbb{N}$, choose $s_n \in S$ such that

$$t - \frac{1}{n} < s_n < t.$$

It is clear that $s_n \neq t$ for all n and the Squeeze Theorem guarantees us that $(s_n) \to t$. Thus every point of $\mathbb{R}$ is a limit point of S.

For the converse suppose that $\overline{S} = \mathbb{R}$. To show that S is dense in $\mathbb{R}$, let $r, t \in \mathbb{R}$ with $r < t$. To produce a point $s \in S$ with $r < s < t$, let $r_1 = (r + t)/2$, the midpoint of r and t. If $r_1 \in S$, then we are done. If $r_1 \notin S$, the fact that $\overline{S} = \mathbb{R}$ implies that there is a sequence (s_n) in S with $s_n \neq r_1$ for all n and $(s_n) \to r_1$. If we then choose $\varepsilon = \frac{1}{2}(r_1 - r) = \frac{1}{2}(t - r_1)$, the definition of convergence implies that there is an $N \in \mathbb{N}$ such that $|r_1 - s_N| < \varepsilon$. This immediately implies that

$$r < r_1 - \varepsilon < s_N < r_1 + \varepsilon < t$$

as desired. □

We turn our attention now to compact sets which were originally defined as closed and bounded subsets of $\mathbb{R}$. With the sequential characterization of closed sets in mind, we know that any compact set must contain its limit points. In order to understand the bounded condition one only need to consider the Bolzano Weierstrass Theorem which states that any bounded sequence must have a convergent subsequence. Thus if we take a sequence (a_n) in a compact set K, then we know that (a_n) has a convergent subsequence. The fact that K is closed then implies that this subsequential limit must be in K. This motivates the following definition which turns out to be equivalent to compactness.

Definition 2.4.8. Let $K \subseteq \mathbb{R}$. We say that K is *sequentially compact* if every sequence in K has a subsequence which converges to a point in K.

Example 2.4.9. From the comments before the definition, we know that any compact set will be sequentially compact which provides an abundance of examples. For an example of a set which is not sequentially compact, take the open interval $(0, 1)$. We must now identify a sequence (x_n) in $(0, 1)$ for which no subsequence converges to a point in $(0, 1)$. The easiest way to do this is to think about limit points. Certainly 0 is a limit point, and the sequence $(1/n)$ is in $(0, 1)$, but no subsequence converges to a point in $(0, 1)$. Indeed, every subsequence converges to 0 which is not in this interval.

The following theorem relates the notion of compactness and sequential compactness in $\mathbb{R}$ though the proof is postponed to the end of the section where we include a third characterization of compact subsets of $\mathbb{R}$.

Theorem 2.4.10 (Heine–Borel Theorem). *A set $K \subseteq \mathbb{R}$ is compact if and only if it is sequentially compact.*

Compactness has been characterized in terms of closed sets and sequences, but it turns out that it is also possible to do so in terms of open sets (our first characterization in terms of closed sets was really about open sets too).

Definition 2.4.11. Let $K \subseteq \mathbb{R}$. We say a collection of open sets $\{O_\lambda : \lambda \in \Lambda\}$ is an *open cover* for K if $K \subseteq \bigcup_{\lambda \in \Lambda} O_\lambda$; the collection Λ may be an uncountable set. For an open cover $\{O_\lambda : \lambda \in \Lambda\}$ of K, we say that $\{O_{\lambda_1}, O_{\lambda_2}, \ldots, O_{\lambda_N}\}$ is a *finite subcover* if $K \subseteq \bigcup_{n=1}^{N} O_{\lambda_n}$. In this case we say that the open cover *admits* a finite subcover for K.

Example 2.4.12. The definition above is somewhat more general than sequential compactness and can be harder to digest at first glance. An open cover is a set consisting of sets and the union of these sets contains the set in question. The collection $\{(-n, n) : n \in \mathbb{N}\}$ is an open cover for $\mathbb{R}$ since

$$\mathbb{R} \subseteq \bigcup_{n=1}^{\infty} (-n, n).$$

For the interval $(0, 1)$, we can take $\{(0, 1)\}$ as an open cover, though this is rather boring. Other open covers include $\{(1/n, 1) : n \geq 2\}$, $\{(0, 1 - 1/n) : n \geq 2\}$, and $\{(1/n, 1 - 1/n) : n \geq 3\}$.

What we are truly interested in are sets whose open covers *always* admit at least one finite subcover. For an example where this does not happen, we again consider $(0, 1)$ and the open cover $\{(1/n, 1) : n \geq 2\}$. To see that there is no finite subcover, suppose that $N \in \mathbb{N}$ is fixed and represents the largest natural number index in a particular finite subcover. But then the union of these finitely many open intervals is $(1/N, 1)$ which does not cover all of $(0, 1)$. Hence there is no finite subcover for this particular open cover.

Theorem 2.4.13 (Heine–Borel Theorem). *Let* $K \subseteq \mathbb{R}$. *The following are equivalent.*

(a) K *is compact.*
(b) K *is sequentially compact.*
(c) *Every open cover for* K *admits a finite subcover.*

When proving a string of equivalent statements as we have above it is common place to show the string of implications (a)$\Rightarrow$(b)$\Rightarrow$(c)$\Rightarrow$(a). Our proof will hinge on our original definition for compactness and we will show that both (b) and (c) are logically equivalent to (a).

Proof. $(a) \Rightarrow (b)$. The proof for this direction follows immediately from the comments which motivated the definition of sequential compactness though there is one tiny detail to consider which arises from the requirement that "$a_n \neq c$ for all $n \in \mathbb{N}$" in the definition of a limit point.

$(b) \Rightarrow (a)$. Here we assume that K is sequentially compact and we must show that K is closed and bounded. First suppose to the contrary that K is not bounded. For a contradiction we will construct a sequence such that every subsequence is unbounded and hence not convergent. For $n \in \mathbb{N}$, choose $x_n \in K$ such that $|x_n| > n$. Now let (x_{n_k}) be a subsequence of (x_n) and let $M > 0$. Then there is an $N \in \mathbb{N}$ such that $N > M$. Furthermore, since (n_k) is an increasing sequence of natural numbers, there is a $k \in \mathbb{N}$ such that $n_k > N > M$ which implies that $|x_{n_k}| > n_k > N > M$. Hence (x_{n_k}) is unbounded and cannot possibly converge. But this contradicts the fact that K is sequentially compact and thus it must be the case that K is bounded.

To show that K is also closed, we use the sequential characterization of closed sets. Let $x \in \mathbb{R}$ be a limit point of K. Then we know there is a sequence (x_n) in K with $x_n \neq x$ for all n and $(x_n) \to x$. The sequential compactness then implies that there is a subsequence (x_{n_k}) which converges to a point $y \in K$. By

Proposition 2.2.11 we conclude that $x = y$ and thus $x \in K$ showing that K is closed.

$(a) \Rightarrow (c)$. We again assume that K is closed and bounded, and let $\mathcal{O} = \{O_\lambda : \lambda \in \Lambda\}$ be an open cover for K. Our goal is to show that there is a finite subcollection of $\mathcal{O}$ whose union contains K. First notice that $\inf(K)$ and $\sup(K)$ both exist and are elements of K by Exercise 2.4.8; call these a and b, respectively. It follows that

$$K \subseteq [a, b];$$

modifying our previous goal, we will now seek to show that there is a finite subcollection of $\mathcal{O}$ whose union contains $[a, b]$, from which our desired conclusion will immediately follow. To this end, we define a set

$$E = \left\{ x \in [a, b] : K \cap [a, x] \subseteq \bigcup_{i=1}^{N} O_{\lambda_i} \text{ for some } \{\lambda_1, \lambda_2, \ldots, \lambda_N\} \subseteq \Lambda \right\}.$$

The set E is certainly not empty; it should be clear that $a \in E$ since a is in some $O_\lambda \in \mathcal{O}$. Notice also that E is bounded above since $E \subseteq [a, b]$; set $c = \sup(E)$. As a first observation, we show that $c \in K$. For a contradiction, suppose $c \notin K$. Then, since K^c is open, there is a $\delta > 0$ such that $B_\delta(c) \subseteq K^c$. By definition of c and our assumption that $c \notin K$, we have $a < c < b$ and hence it must be the case that $B_\delta(c) \subseteq (a, b)$ since $a, b \in K$. However, by Lemma 1.2.10, there is a $y \in E$ with $c - \delta < y$. Using the fact that $B_\delta(c) \subseteq K^c$, it follows that $K \cap [a, y] = K \cap [a, c + \delta/2]$. But this implies that $c + \delta/2 \in E$ which contradicts the fact that c is the supremum of E. Thus it must be the case that $c \in K$.

Next we will show that $b = c$. Using the fact that $c \in K$, we immediately find that there is an open set $O_\alpha \in \mathcal{O}$ with $c \in O_\alpha$. With this open set specified, we know that there is an $\varepsilon > 0$ such that $(c - \varepsilon, c + \varepsilon) \subseteq O_\alpha$, and appealing to Lemma 1.2.10 again, there must be a point $d \in E$ with $d \in (c - \varepsilon, c)$. This is the crux. The fact that $d \in E$ implies that there is a finite set $\{\lambda_1, \lambda_2, \ldots, \lambda_N\} \subseteq \Lambda$ such that

$$K \cap [a, d] \subseteq \bigcup_{i=1}^{N} O_{\lambda_i}.$$

From this and the fact that $d \in (c - \varepsilon, c) \subseteq O_\alpha$, it is clear that

$$K \cap [a, c] \subseteq O_\alpha \cup \left(\bigcup_{i=1}^{N} O_{\lambda_i} \right);$$

this last union is also a finite subcover. However, O_α contains all points in the interval $(c, c + \varepsilon)$, which shows that

$$K \cap [a, c + \varepsilon] \subseteq O_\alpha \cup \left(\bigcup_{i=1}^{N} O_{\lambda_i} \right).$$

If $c < b$, then there are points of $[a, b]$ in the interval $(c, c + \varepsilon)$ and the above set inclusion then contradicts the fact that $c = \sup(E)$. Thus it must be the case that $c = b$ (c cannot be greater than b) and the above finite subcover satisfies

$$K = K \cap [a, b] = K \cap [a, c] \subseteq O_\alpha \cup \left(\bigcup_{i=1}^{N} O_{\lambda_i} \right),$$

as desired.

$(c) \Rightarrow (a)$. The proof here is also left as an exercise with a hint provided. $\square$

Exercises

Exercise 2.4.1. Let $A \subseteq \mathbb{R}$ with $c \in \mathbb{R}$. Show that c is a limit point of A if and only if for every $\varepsilon > 0$, there is a point $a \in A$ such that $a \neq c$ and $a \in B_\varepsilon(c)$.

Exercise 2.4.2. Let $O \subseteq \mathbb{R}$ be an open set with $a \in O$. If (a_n) converges to a, show that at most finitely many terms of this sequence are not in O.

Exercise 2.4.3. Let $O \subseteq \mathbb{R}$ be an open set. Prove that O has no isolated points.

Exercise 2.4.4. (a) Give an example of a bounded set A for which $\sup(A)$ is not a
limit point of A.
(b) Give an example of an open set which contains all the rational numbers, but
which is not all of $\mathbb{R}$.

Exercise 2.4.5. Let $A \subseteq \mathbb{R}$ and let L be the set of limit points of A. Show that L is a closed set.

Exercise 2.4.6. Let $A \subseteq \mathbb{R}$. Show that A is closed if and only if $A = \overline{A}$.

Exercise 2.4.7. Let $A \subseteq \mathbb{R}$ be nonempty and bounded above. Use the definition of closure to show that $\sup(A) \in \overline{A}$.

Exercise 2.4.8. Let $K \subseteq \mathbb{R}$ be a compact set. Show that $\sup K$ and $\inf K$ are elements of K.

Exercise 2.4.9. A set $A \subseteq \mathbb{R}$ is called a G_δ (pronounced "G-delta") set if it can be written as a countable intersection of open sets.

(a) Explain why a G_δ set is not necessarily an open set.
(b) Explain why every open set is a G_δ set.
(c) Show that the set $\{0\}$ is a G_δ set.
(d) Show that the interval $(0, 1]$ is a G_δ set.
(e) Is $[0, 1]$ a G_δ set?

Exercise 2.4.10. A set $A \subseteq \mathbb{R}$ is called an F_σ (pronounced "F-sigma") set if it can be written as a countable union of closed sets.

(a) Explain why an F_σ set is not necessarily a closed set.
(b) Explain why every closed set is a F_σ set.
(c) Show that the interval $(0, 1]$ is an F_σ set.
(d) Is $(0, 1)$ an F_σ set?
(e) Show that $\mathbb{Q}$ is an F_σ set.

Exercise 2.4.11. (a) Let $A \subseteq \mathbb{R}$. Show that A is a G_δ set if and only if A^c is an F_σ set.

(b) Show that $\mathbb{I}$ is a G_δ set.
(c) Is the set $(-\infty, 0) \cap (0, \infty)$ a G_δ set, an F_σ set, or both?

Exercise 2.4.12. Follow the outline below to complete the proof of the Heine–Borel Theorem.

(a) Verify that $(a) \Rightarrow (b)$.
(b) Verify that $(c) \Rightarrow (a)$. One possible approach is to argue by contrapositive: if K is not closed or not bounded, then there is an open cover which does not admit a finite subcover. If K is not closed, construct an open cover which does not admit a finite subcover, and likewise if K is not bounded.

2.5 Complete Spaces

As with our study of $\mathbb{R}$, the foundation of analysis in any setting is the investigation of convergent sequences. The following definitions are natural extensions of the notions we have encountered thus far.

Definition 2.5.1. Let (X, d) be a metric space. A sequence (x_n) in X *converges to* $x \in X$ if for every $\varepsilon > 0$, there exists an $N \in \mathbb{N}$ such that $d(x_n, x) < \varepsilon$ for every $n \geq N$. Likewise the sequence (x_n) is called a *Cauchy sequence* if for every $\varepsilon > 0$, there exists an $N \in \mathbb{N}$ such that $d(x_n, x_m) < \varepsilon$ for every $n, m \geq N$.

The definition of convergence here is closely related to the convergence of a sequence of real numbers. If (x_n) is a sequence in a metric space (X, d), then (x_n) converges to $x \in X$ if the sequence of real numbers $d(x_n, x)$ converges to 0 as $n \to \infty$. When working in a normed linear space, the metric of choice is the metric induced by the norm as stated in Proposition 1.5.10. In this case, the statement above is interpreted by saying that (x_n) converges to x in X if $\|x_n - x\| \to 0$ as $n \to \infty$. Similar comparisons can be made for Cauchy sequences in a metric space and Cauchy sequences of real numbers.

The first result of this section exhibits a familiar connection between these two ideas. It is also a good idea to compare the proof in this more generatl setting to the proof given for sequences in $\mathbb{R}$ in Proposition 2.1.9.

Theorem 2.5.2. *In any metric space, a convergent sequence is Cauchy.*

Proof. Let (X, d) be a metric space and suppose (x_n) is a sequence in X converging to $x \in X$. For $\varepsilon > 0$, choose $N \in \mathbb{N}$ such that $d(x_n, x) < \varepsilon/2$. For $n, m \geq N$ observe that

$$d(x_n, x_m) \leq d(x_n, x) + d(x, x_m) < \varepsilon/2 + \varepsilon/2 = \varepsilon,$$

and thus we conclude that (x_n) is a Cauchy sequence.   □

In $\mathbb{R}$, the converse of this theorem also holds, the Cauchy Criterion, and provides a means of determining convergence in situations when no limit candidate can be exhibited explicitly. Unfortunately, in a general metric space it is not always true that Cauchy sequences converge. In order to single out the exceptional metric spaces with this property, we have the following definition.

Definition 2.5.3. A metric space (X, d) is called *complete* if every Cauchy sequence in X converges to a point in X.

Recall that we asserted the completeness of $\mathbb{R}$ via the existence of suprema. However, we then explored various other equivalent statements: the Nested Interval Property, the Monotone Convergence Theorem, the Bolzano–Weierstrass Theorem, and the Cauchy Criterion. Several of these statements depend on the fact that there is an order on $\mathbb{R}$. It should be clear, consider $\mathbb{R}^2$ for example, that general metric spaces do not always afford an order, and thus we are restricted in our approach to completeness in these spaces. While there are other ways to classify completeness in general metric spaces, the use of Cauchy sequences is the most standard. Complete metric spaces which have the additional structure of a norm or an inner product are named for two of the most prominent analysts of the early twentieth century, Stefan Banach and David Hilbert.

Definition 2.5.4. A complete metric space in which the metric is induced by a norm as in Proposition 1.5.10 is called a *complete normed linear space* or, more commonly, a *Banach space*. A complete normed linear space whose norm is induced by an inner product as in Proposition 1.5.15 is called a *Hilbert space*.

We now present two examples which exhibit a standard technique for demonstrating completeness.

Example 2.5.5. The space $\mathbb{R}^2$ with the Euclidean metric is a Hilbert space. We know that this space is an inner product space, so we simply need to show that the metric induced by the inner product is complete. Before beginning, if $x = (x_1, x_2)$ and $y = (y_1, y_2)$ are elements of $\mathbb{R}^2$, recall that metric distance between x and y is given by

$$d(x, y) = \|x - y\| = \sqrt{\langle x - y, x - y \rangle} = \sqrt{(x_1 - y_1)^2 + (x_2 - y_2)^2}.$$

To show that the space is complete, let (x_n) be a Cauchy sequence in $\mathbb{R}^2$. It is then our task to show that this sequence converges to some $x \in \mathbb{R}^2$. We begin by identifying x. For notation, we will write x_n as the ordered pair (x_{n1}, x_{n2}). Since (x_n) is Cauchy, for $\varepsilon > 0$ we can find $N \in \mathbb{N}$ such that

$$d(x_n, x_m) = \sqrt{(x_{n1} - x_{m1})^2 + (x_{n2} - x_{m2})^2} < \varepsilon$$

for all $n, m \geq N$. Using a simple estimate we see that

$$|x_{n1} - x_{m1}| \leq \sqrt{(x_{n1} - x_{m1})^2 + (x_{n2} - x_{m2})^2} < \varepsilon$$

for $n, m \geq N$ and thus the sequence $(x_{n1})_{n=1}^{\infty}$ is a Cauchy sequence in $\mathbb{R}$. The Cauchy Criterion now guarantees the existence of $x_1 \in \mathbb{R}$ such that $(x_{n1}) \to x_1$ as $n \to \infty$. Similarly we can find $x_2 \in \mathbb{R}$ such that $(x_{n2}) \to x_2$ as $n \to \infty$. Let $x = (x_1, x_2) \in \mathbb{R}^2$.

To complete the proof, we show that (x_n) converges to x in the Euclidean metric. By the Algebraic Limit Theorem, $(x_{n1} - x_1)^2 + (x_{n2} - x_2)^2 \to 0$ as $n \to \infty$. Thus

$$d(x_n, x) = \sqrt{(x_{n1} - x_1)^2 + (x_{n2} - x_2)^2} \to 0,$$

completing the proof.

Our second example deals with a new space which will reappear in a different guise later on.

Example 2.5.6. Let ℓ^{∞} denote the space of all bounded sequences of real numbers, i.e.

$$\ell^{\infty} = \{(a_n)_{n=1} = (a_1, a_2, a_3 \ldots) : \sup_n |a_n| < \infty\}.$$

If we define addition and scalar multiplication by

$$(a_n) + (b_n) = (a_n + b_n) \quad \text{and} \quad c(a_n) = (ca_n),$$

then this collection of sequences forms a real vector space. Furthermore, if (a_n) is a bounded sequence, the rule

$$\|(a_n)\|_{\infty} = \sup_n |a_n|$$

defines a norm on ℓ^{∞}, referred to commonly as the *sup-norm*. We leave this as an exercise and prove the completeness.

Proposition 2.5.7. *The space ℓ^{∞} is complete with respect to the sup-norm.*

Proof. To begin, assume we have a Cauchy sequence in ℓ^∞. The proof will be broken down into several steps; the first, and most lengthy, will be to identify a limit candidate. Since the elements in this set are themselves sequences, we will need to take some care with notation. Let $\left((a_{nk})_{k=1}\right)_{n=1}$ denote this Cauchy sequence. Writing this out with an array (we borrow heavily from the standard matrix notation where the first index represents the row of the entry and the second represents the column), we have

$$
\begin{aligned}
(a_{1k}) &= a_{11}, a_{12}, a_{13}, \ldots, a_{1k}, \ldots \\
(a_{2k}) &= a_{21}, a_{22}, a_{23}, \ldots, a_{2k}, \ldots \\
(a_{3k}) &= a_{31}, a_{32}, a_{33}, \ldots, a_{3k}, \ldots \\
&\ \vdots \quad \vdots \qquad\quad \vdots \\
(a_{nk}) &= a_{n1}, a_{n2}, a_{n3}, \ldots, a_{nk}, \ldots \\
&\ \vdots \quad \vdots \qquad\qquad \vdots \qquad \vdots
\end{aligned}
$$

For the sake of clarity, applying the sup-norm to one of these sequences is expressed by

$$
\|(a_{nk})_{k=1}\|_\infty = \sup_k |a_{nk}|;
$$

the point is to be aware of the fact that the norm is taken by considering the supremum as the second subscript varies, i.e. along the rows in the array above.

Now, since this sequence is Cauchy, given $\varepsilon > 0$, we can find $N \in \mathbb{N}$ such that

$$
\|(a_{nk}) - (a_{mk})\|_\infty = \sup_k\{|a_{nk} - a_{mk}|\} < \varepsilon
$$

whenever $n, m \geq N$. Further, notice that for each $k \in \mathbb{N}$ it is true that

$$
|a_{nk} - a_{mk}| \leq \|(a_{nk}) - (a_{mk})\|_\infty.
$$

Thus $|a_{nk} - a_{mk}| < \varepsilon$ for all $n, m \geq N$, and hence the sequence $(a_{nk})_{n=1}$ is a Cauchy sequence in $\mathbb{R}$ for each fixed $k \in \mathbb{N}$; in other words, for each k there is an $a_k \in \mathbb{R}$ such that $(a_{nk})_{n=1}$ converges to a_k as $n \to \infty$. Adding this new information to our array gives

$$
\begin{aligned}
(a_{1k}) &= a_{11}, a_{12}, a_{13}, \ldots, a_{1k}, \ldots \\
(a_{2k}) &= a_{21}, a_{22}, a_{23}, \ldots, a_{2k}, \ldots \\
(a_{3k}) &= a_{31}, a_{32}, a_{33}, \ldots, a_{3k}, \ldots \\
&\ \vdots \quad \vdots \qquad\quad \vdots \\
(a_{nk}) &= a_{n1}, a_{n2}, a_{n3}, \ldots, a_{nk}, \ldots \\
&\ \ \downarrow? \quad\ \downarrow \quad \downarrow \quad \downarrow \qquad\quad \downarrow \\
(a_k) &= a_1, \quad a_2, \quad a_3, \ \ldots, \ a_k, \ \ldots
\end{aligned}
$$

We will be finished if we can show that $(a_k)_{k=1}$ is in ℓ^∞ and that the sequence $\left((a_{nk})_{k=1}\right)_{n=1}$ (the left-hand column of our array) converges to the sequence $(a_k)_{k=1}$ (the lower left-hand entry) as $n \to \infty$. Notice here that when we speak of convergence we mean convergence in the metric induced by the sup-norm.

First, we will show that (a_k) is in ℓ^∞. To see this, we appeal to the fact that Cauchy sequences are bounded, a fact you will show in Exercise 2.5.1. Thus there is an $M > 0$ such that $\|(a_{nk})_{k=1}\|_\infty \leq M$ for every $n \in \mathbb{N}$. This immediately implies that $|a_{nk}| \leq M$ for every $n, k \in \mathbb{N}$. The Order Limit Theorem now implies that $|a_k| \leq M$ for all k and hence (a_k) is a bounded sequence.

Finally, we must show that $\|(a_{nk}) - (a_k)\|_\infty \to 0$ as $n \to \infty$. Let $\varepsilon > 0$. Using the original hypothesis that our sequence is Cauchy, we can find $N \in \mathbb{N}$ such that

$$\|(a_{nk}) - (a_{mk})\|_\infty = \sup_k \{|a_{nk} - a_{mk}|\} < \varepsilon/4$$

for $n, m \geq N$. As before, this immediately implies that

$$|a_{nk} - a_{mk}| \leq \|(a_{nk}) - (a_{mk})\|_\infty < \varepsilon/4$$

for all $n, m \geq N$ and all $k \in \mathbb{N}$. The proof of the Cauchy Criterion now immediately shows that

$$|a_{nk} - a_k| < \varepsilon/2$$

for all $n \geq N$ and all $k \in \mathbb{N}$. To conclude, the previous inequality ensures us that

$$\|(a_{nk}) - (a_k)\|_\infty = \sup_k \{|a_{nk} - a_k|\} \leq \varepsilon/2 < \varepsilon$$

for all $n \geq N$. Thus the space ℓ^∞ equipped with the sup-norm is a complete normed linear space.   □

Before moving on, notice that the technique used to identify the limit candidate in both of these examples was exactly the same. In the case of $\mathbb{R}^2$ it was easy to see that the potential limit was in the space while in ℓ^∞ we had to work a little harder. The final convergence argument for these two examples differed but that was to be expected due to the nature of the norms being considered.

Definition 2.5.8. Let X be a vector space and let $W \subseteq X$. We call W a *subspace* of X if W is a vector space with respect to the operations on X.

Notice that the definition of subspace subtly requires W to be nonempty by vector space axiom (d). Also, the definition above is stated in the most general form; however, the operational axioms of vector space, (b), (c), and (g)–(j), must hold no matter what collection W we are considering since $W \subseteq X$ and X is a vector space. Thus we really only need to check that all of the correct vectors are in

W. The following proposition captures this thought and we leave the formal proof to the reader.

Proposition 2.5.9. *Let X be a vector space and let $W \subseteq X$. If W is not empty, then W is a subspace of X if and only if W is closed with respect to the addition and scalar multiplication of X.*

One immediate consequence here is that the additive identity in a subspace W is the same as the additive identity in the larger space X. This often provides a quick means of showing a set is not a subspace.

Example 2.5.10. Let $X = \mathbb{R}^2$. If $x \in \mathbb{R}^2$, the set $W = \{cx : c \in \mathbb{R}\}$ is a subspace of W. However, the set $W = \{(x_1, x_2) : 2x_1 + 3x_2 = 1\}$ is not a subspace since it does not contain the zero vector.

For the polynomial spaces P_n (see Example 1.5.3), it should be apparent that P_n is a subspace of P_{n+1} for all $n \in \mathbb{N}$.

To close this section, we consider several subspaces of ℓ^∞ and explore an example of a space which is not complete.

Example 2.5.11. Let c denote the collection of all sequences of real numbers which converge and let c_0 denote the collection of all sequence of real numbers which converge to zero. In Exercise 2.5.9 you are asked to show that $c_o \subseteq c \subseteq \ell^\infty$. When equipped with the sup-norm of the previous example, these two spaces are complete normed linear spaces.

Example 2.5.12. Define p^∞ to be the collection of sequences of real numbers with at most finitely many nonzero terms. Any sequence of this form must be in ℓ^∞ since the supremum of a finite number of points always exists; hence, we conclude that $p^\infty \subseteq \ell^\infty$. To confirm that this collection is a subspace of ℓ^∞ we need to verify that it is closed with respect to the addition and scalar multiplication; it is certainly not empty. To this end suppose that (a_k) is a sequence with n nonzero terms and (b_k) is a sequence with m nonzero terms. Then the sum $(a_k) + (b_k)$ has at most $n + m$ nonzero terms, and, for $c \in \mathbb{R}$, $c(a_k)$ has at most n nonzero terms. Thus both new elements are in p^∞.

We will now show that this space is not complete with respect to the sup-norm. To do this, we must produce a sequence of elements in p^∞ which is Cauchy but has no limit in p^∞. Take the sequence of sequences $\left((a_{nk})_{k=1} \right)_{n=1}$ defined by

$$a_{nk} = \begin{cases} 1/k, & k \leq n; \\ 0, & k > n. \end{cases}$$

Writing this out in an array form to better see what each sequence looks like, we have

$$(a_{1k}) = 1, \quad 0, \quad 0, \quad 0, \ldots, 0, 0, \ldots$$
$$(a_{2k}) = 1, 1/2, \quad 0, \quad 0, \ldots, 0, 0, \ldots$$
$$(a_{3k}) = 1, 1/2, 1/3, 0, \ldots, 0, 0, \ldots$$

$$\vdots \quad \vdots \qquad \vdots$$

$$(a_{nk}) = 1, 1/2, 1/3, \ldots, 1/n, 0, \ldots$$

$$\vdots \quad \vdots \qquad \vdots \qquad \vdots$$

To see that this sequence is Cauchy, let $\varepsilon > 0$ and choose $N \in \mathbb{N}$ such that $1/N < \varepsilon$. Then, for a fixed $k \in \mathbb{N}$, (we assume $n > m$ to simplify the calculation)

$$|a_{nk} - a_{mk}| = \begin{cases} 0, & k \le m < n; \\ 1/k, & m < k \le n; \\ 0, & m < n < k. \end{cases}$$

If we now assume that $n > m \ge N$, it follows that $1/(m+1) \le 1/N < \varepsilon$ and thus

$$\|(a_{nk}) - (a_{mk})\|_\infty = \sup_k \{|a_{nk} - a_{mk}|\} = 1/(m+1) < \varepsilon;$$

hence we have a Cauchy sequence. We also note that this sequence converges in the sup-norm to the sequence $(a_k) = (1/k)_{k=1}$ since

$$\|(a_{nk}) - (a_k)\|_\infty = \sup_k \{|a_{nk} - a_k|\} = 1/(n+1).$$

This convergence is in the larger space ℓ^∞ and hence (a_k) is the *only* possible limit for our chosen Cauchy sequence. However, this limit is not in p^∞ since it has countably many nonzero terms and we therefore conclude that p^∞ is not complete.

Exercises

Exercise 2.5.1. Let $(X, \|\cdot\|)$ be a normed linear space. We say that a sequence (x_n) is *bounded* if there exists an $M > 0$ such that $\|x_n\| \le M$ for every $n \in \mathbb{N}$.

(a) Show that a convergent sequence is bounded.
(b) Show that a Cauchy sequence is bounded. Keep in mind that we are not assuming our space is complete, and hence you cannot simply reference part (a).

Exercise 2.5.2. Let (X, d) be a metric space and suppose (x_n) is a convergent sequence in X. Show that the limit is unique.

Exercise 2.5.3 (Algebraic Limit Theorem for Normed Linear Spaces). Suppose $(X, \|\cdot\|)$ is a normed linear space and let (x_n) and (y_n) be sequences in X with limits x and y, respectively.

(a) Show that $(x_n + y_n)$ converges to $x + y$.
(b) If $c \in \mathbb{R}$, show that (cx_n) converges to cx.

Exercise 2.5.4. Let $(X, \|\cdot\|)$ be a normed linear space and let (x_n) be a sequence in X converging to $x \in X$. Use Exercise 1.5.3 to show that the sequence of real numbers $(\|x_n\|)$ converges to $\|x\|$.

Exercise 2.5.5. Let (X, d) be a metric space. Show that X is complete if and only if every Cauchy sequence has a convergent subsequence (with limit x in X).

Exercise 2.5.6. Show that ℓ^∞ is a vector space and that the rule $\|\cdot\|_\infty$ given in Example 2.5.6 defines a norm.

Exercise 2.5.7. Use the parallelogram equality to show that the sup-norm does not arise as an inner product as in Proposition 1.5.15.

Exercise 2.5.8. Let X be a vector space and let $W \subseteq X$ be a subspace. Show that $0_W = 0_X$.

Exercise 2.5.9. (a) Show that $p^\infty \subseteq c_0 \subseteq c \subseteq \ell^\infty$.
(b) Show that each space in the chain above is a subspace of its superset.
(c) Show that the spaces c and c_0 are complete normed linear spaces with respect to the sup-norm.

Chapter 3
Numerical Series

With a working knowledge of sequences in hand we are ready to explore the fundamental constructs of calculus. We begin with the topic that is arguably most closely related to sequences, that is, a series of real numbers. The general idea is to provide a functional technique for summing an infinite string of values. For the most part, our focus here will be on convergence tests rather than on devising explicit means of summing a series.

3.1 Series of Real Numbers

Definition 3.1.1. Let (a_n) be a sequence of real numbers. An *infinite series of real numbers* is a formal expression of the form

$$a_1 + a_2 + a_3 + a_4 + \cdots ; \qquad (3.1)$$

for shorthand, we use the symbol $\sum_{j=1}^{\infty} a_j$ to denote the series and call the sequence (a_n) the *defining sequence* of the series. Given an infinite series, the *sequence of partial sums* associated to the series is defined by

$$s_n = a_1 + a_2 + \cdots + a_n = \sum_{j=1}^{n} a_j.$$

If $\lim s_n$ exists and converges to a value S, then we say the series $\sum_{j=1}^{\infty} a_j$ *converges to S* and write $\sum_{j=1}^{\infty} a_j = S$. As with sequences, if the series does not converge, then we say it *diverges*.

The definition above deserves a few immediate remarks. First, Eq. (3.1) is referred to as a "formal expression." This is due to the fact that it is not a legitimate expression in the sense that it is impossible to add up infinitely many terms in

M.A. Pons, *Real Analysis for the Undergraduate: With an Invitation to Functional Analysis*, 99
DOI 10.1007/978-1-4614-9638-0_3, © Springer Science+Business Media New York 2014

finite time (given that infinitely many of the terms are nonzero). As an example, if someone asked you to find the value of the sum $1/2 + 1/4 + 1/8 + \cdots = \sum_{n=1}^{\infty} 1/2^n$, and you attempted to do this by adding the successive terms together, you would never be able to complete this task, nor would your successor, and so on. In short, the task would never be completed by standard means. However, the definition is designed to give a rigorous meaning to an expression of this form; by considering the sequence of partial sums, we mean to reduce the series to a more tractable object, a sequence. Assuming we can find a closed form for a general partial sum s_n we then take the limit using the methods developed in the previous chapter.

Second, as with sequences of real numbers it is often helpful to know whether or not we are dealing with a convergent or divergent object without the explicit knowledge of a limit in the case of convergence. The case for infinite series is even more pronounced with regard to this aspect due to the fact that it is typically difficult to find a closed form for the partial sums. In fact, finding the limit of a series is extremely tricky business and we will most often be satisfied with simply knowing whether or not the series converges or diverges.

The third remark hinges on the previous comment, though the situation is slightly different than for sequences. If a series $\sum_{n=1}^{\infty} a_n$ converges, then so does any tail, however, the value of the series could change. For instance, (though it remains to be verified) the series $\sum_{n=1}^{\infty} 1/2^n$ converges to 1 while the series $\sum_{n=0}^{\infty} 1/2^n$ converges to 2; however, the convergence of either one of these implies the convergence of the other. Exercise 3.1.1 will ask you to verify this fact. For this reason, on many occasions we will suppress the notation and simply use $\sum a_n$ without indicating a starting position.

Example 3.1.2. Let's use the definition of a series to determine whether or not the sum $\sum_{n=0}^{\infty} (-1)^n = 1 - 1 + 1 - 1 + 1 - 1 + \cdots$ converges. Computing the partial sums we have $s_0 = 1$, $s_1 = 1 - 1 = 0$, $s_2 = 1 - 1 + 1 = 1$, and so on. It is apparent that the sequence (s_n) alternates between 0 and 1, a case of divergence. Thus we also conclude that the series $\sum_{n=0}^{\infty} (-1)^n$ diverges.

This is a simple example but it has a hidden usefulness! Recall the associative property of real number addition, $(a + b) + c = a + (b + c)$ for all real numbers a, b, c. This can be extended to any finite sum, i.e. given a finite string of numbers to add, the order in which the addition is carried out is unimportant. As is often the case, things deteriorate when moving from a finite setting to an infinite setting. To see this, let's "add" the previous series in the following distinct ways, first by grouping the terms into pairs starting with the first term and then by grouping the terms into pairs but skipping the first:

$$\sum_{n=0}^{\infty} (-1)^n \stackrel{?}{=} (1 - 1) + (1 - 1) + (1 - 1) + \cdots = 0 + 0 + 0 + \cdots = 0;$$

$$\sum_{n=0}^{\infty} (-1)^n \stackrel{?}{=} 1 + (-1 + 1) + (-1 + 1) + (-1 + 1) + \cdots = 1 + 0 + 0 + 0 + \cdots = 1.$$

How do you make sense of this? The short answer is, something is wrong. A better question: what is the lesson to be learned? While we have just performed some very naive mathematics, these types of situations are typically the driving force behind definitions and theorems. The bizarre behavior exhibited above indicates that the associative property is not behaving appropriately in this infinite setting and that order should be respected, hence the notion of partial sum. By reducing the series to finite sums, we are able to use the associative property.

The following examples demonstrate one divergent and one convergent series. The methods used here are ad hoc at best and we will explore more streamlined tests for convergence in the following sections.

Example 3.1.3. First let's show that $\sum_{n=1}^{\infty} 1/n$ diverges. To do this we will show that the sequence of partial sums has an unbounded subsequence and hence cannot converge. To begin, notice that

$$s_1 = 1,$$

$$s_2 = 1 + \frac{1}{2} = 1 + 1\left(\frac{1}{2}\right),$$

$$s_4 = 1 + \frac{1}{2} + \frac{1}{3} + \frac{1}{4} > 1 + \frac{1}{2} + \frac{1}{4} + \frac{1}{4} = 1 + 2\left(\frac{1}{2}\right),$$

$$s_8 = 1 + \frac{1}{2} + \frac{1}{3} + \frac{1}{4} + \frac{1}{5} + \frac{1}{6} + \frac{1}{7} + \frac{1}{8} > 1 + \frac{1}{2} + \frac{1}{4} + \frac{1}{4} + \frac{1}{8} + \frac{1}{8} + \frac{1}{8} + \frac{1}{8} = 1 + 3\left(\frac{1}{2}\right).$$

In general, we can mimic this estimating technique to see that the subsequence (s_{2^j}) satisfies the bound

$$s_{2^j} > 1 + j\left(\frac{1}{2}\right).$$

The sequence $(1 + j/2)$ diverges to ∞ and thus the subsequence (s_{2^j}) also diverges to ∞ by Exercise 2.1.4. It follows that the sequence of partial sums (s_n) diverges which implies that the series $\sum 1/n$ also diverges. There have been many proofs of this fact and it is first attributed to the Bernoulli brothers. The series is typically referred to as the *harmonic series*.

Example 3.1.4. For an example of a convergent series, consider $\sum_{n=1}^{\infty} 1/2^n$. To show that this series converges, we will use a simple algebraic relationship between the terms,

$$\frac{1}{2^n} = \frac{1}{2^{n-1}} - \frac{1}{2^n}.$$

With this, we see that the partial sum s_n takes the form

$$s_n = \frac{1}{2} + \frac{1}{4} + \frac{1}{8} + \cdots + \frac{1}{2^n}$$

$$= \left(1 - \frac{1}{2}\right) + \left(\frac{1}{2} - \frac{1}{4}\right) + \left(\frac{1}{4} - \frac{1}{8}\right) + \cdots + \left(\frac{1}{2^{n-1}} - \frac{1}{2^n}\right)$$

$$= 1 - \frac{1}{2^n}.$$

This is a closed form for the partial sum s_n and taking a limit we have

$$\sum_{n=1}^{\infty} \frac{1}{2^n} = \lim s_n = \lim \left(1 - \frac{1}{2^n}\right) = 1$$

which is permissible since we know that $(1/2^n) \to 0$ as $n \to \infty$.

For a second approach let us use the Monotone Convergence Theorem to show that the series converges. First, since the series consists of positive terms, we know that the sequence of partial sums is increasing. Furthermore, the equation $s_n = 1 - 1/2^n$ implies that the sequence of partial sums is bounded above by 1. Hence the sequence (s_n) converges to some value L. To calculate L, notice that

$$s_n = 1 - \frac{1}{2^n} = 1 + s_{n-1} - s_n$$

which implies that $2s_n = 1 + s_{n-1}$. Applying a limit to both sides of this equation, we have $2L = 1 + L$ and it follows that $L = 1$.

The algebraic relationships used in these two examples may have seemed mysterious at first, but they are used because they work. This is often intimidating for the beginner. The goal of exploring numerous examples is to expose ourselves to a wide variety of strategies for tackling problems and simple algebraic manipulations will be key as we progress; practice with such strategies is of the utmost importance.

Theorem 3.1.5 (Cauchy Criterion for Series). *Let $\sum a_n$ be a series of real numbers. Then the series converges if and only if for every $\varepsilon > 0$, there is an $N \in \mathbb{N}$ such that*

$$|a_{m+1} + a_{m+2} + \cdots + a_n| < \varepsilon$$

for all $n > m \geq N$ (the choice $n > m$ is arbitrary and works equally well if we require $m > n \geq N$).

Proof. The proof follows from the relationship

$$|s_n - s_m| = |a_{m+1} + a_{m+2} + \cdots + a_n|,$$

for $n > m$, and the Cauchy Criterion for sequences. $\square$

The Cauchy Criterion is a powerful tool for theoretical purposes and will be a valuable tool in proving the convergence tests of the next two sections. As far as its practical use, the Cauchy Criterion is used less frequently than, say, the Comparison Test (Theorem 3.2.5), but our next example gives a case when it is a most ideal means of determining convergence.

Example 3.1.6. Use the Cauchy Criterion to determine whether the series

$$\sum_{n=1}^{\infty} \frac{1}{n(n+1)}$$

converges or diverges. Before jumping in, let's think about what we need to do in order to satisfy the criterion. We first consider the difference of the partial sums

$$|s_n - s_m| = \left| \frac{1}{(m+1)(m+2)} + \frac{1}{(m+2)(m+3)} + \cdots + \frac{1}{n(n+1)} \right|.$$

The absolute value is unnecessary here since all the terms are positive. Moreover, each term can be decomposed by the formula

$$\frac{1}{j(j+1)} = \frac{1}{j} - \frac{1}{j+1}. \tag{3.2}$$

Substituting this in for each term in the previous line shows that

$$|s_n - s_m| = \left(\frac{1}{m+1} - \frac{1}{m+2} \right) + \left(\frac{1}{m+2} - \frac{1}{m+3} \right) + \cdots + \left(\frac{1}{n} - \frac{1}{n+1} \right).$$

Notice now that the last term in the first parentheses cancels with the first term in the next and so on, i.e. we are looking at what some would call a *collapsing or telescoping sum*. Performing this bit of algebra we arrive at

$$|s_n - s_m| = \frac{1}{m+1} - \frac{1}{n+1} \le \frac{1}{m+1}.$$

With this work in hand, we are ready to give our proof.

Proof. Let $\varepsilon > 0$ and choose $N \in \mathbb{N}$ such that $1/N < \varepsilon$. Then, if $n > m \ge N$, we have that $1/(m+1) \le 1/N < \varepsilon$ and, applying Eq. (3.2),

$$|s_n - s_m| = \left| \frac{1}{(m+1)(m+2)} + \frac{1}{(m+2)(m+3)} + \cdots + \frac{1}{n(n+1)} \right|$$

$$= \left(\frac{1}{m+1} - \frac{1}{m+2} \right) + \left(\frac{1}{m+2} - \frac{1}{m+3} \right) + \cdots + \left(\frac{1}{n} - \frac{1}{n+1} \right)$$

$$= \frac{1}{m+1} - \frac{1}{n+1} \le \frac{1}{m+1} \le \frac{1}{N} < \varepsilon.$$

Therefore the given series converges by the Cauchy Criterion.   □

The proof of the Cauchy Criterion followed immediately from the Cauchy Criterion for sequences and the next result is the series analog of the Monotone Convergence Theorem.

Theorem 3.1.7. *Let $\sum a_n$ be a series of nonnegative terms. Then the series converges if and only if the partial sums are bounded. If the sequence of partial sums is unbounded, then the series diverges to ∞.*

Proof. First suppose that the series $\sum a_n$ converges. The fact that (s_n) is bounded follows immediately from the definition of a series and Proposition 2.2.2.

Conversely, suppose that the sequence of partial sums (s_n) is bounded; our goal is to apply the Monotone Convergence Theorem to this sequence. The fact that the sequence (a_n) consists of nonnegative numbers implies that $s_{n+1} = s_n + a_{n+1} \geq s_n$ which indicates that (s_n) is an increasing sequence. Our hypothesis that the sequence of partial sums is bounded then implies that the sequence of partial sums, and hence the series, must converge. On the other hand, if the sequence of partial sums is not bounded, the fact that the sequence (s_n) is increasing indicates that it must diverge to ∞.   □

Theorem 3.1.8 (Algebraic Limit Theorem for Series). *Suppose $\sum a_n$ and $\sum b_n$ are convergent series. Then*

(a) the series $\sum ca_n$ converges for every $c \in \mathbb{R}$ and $\sum ca_n = c \sum a_n$;
(b) the series $\sum (a_n + b_n)$ converges and $\sum (a_n + b_n) = \sum a_n + \sum b_n$.

Proof. For (a), we assume that $\sum a_n$ converges and let $c \in \mathbb{R}$. For notation, let (s_n) be the partial sums of the series $\sum a_n$ and let (t_n) be the partial sums of the series $\sum ca_n$. By hypothesis, we know that (s_n) converges and our goal is to show that (t_n) converges. Observe that

$$t_n = ca_1 + ca_2 + ca_3 + \cdots + ca_n = c(a_1 + a_2 + a_3 + \cdots + a_n) = cs_n.$$

By the Algebraic Limit Theorem for sequences, we know that (cs_n) converges and the equality above dictates that (t_n) also converges; hence the series $\sum ca_n$ converges. For the summation statement, notice that

$$\sum ca_n = \lim t_n = \lim cs_n = c \lim s_n = c \sum a_n.$$ □

Notice that there is no series analog for sums of the form $\sum a_n b_n$. It is certainly not the case that an expression of this form would be equal to $(\sum a_n)(\sum b_n)$. This can be seen by considering the distributive property relating real number addition and multiplication and the corresponding statement involving the partial sums of the three series. Furthermore, there is no general relation between $\sum a_n b_n$ and $(\sum a_n)(\sum b_n)$, however, Cauchy devised a means of taking the product of two series. If (a_n) and (b_n) are sequences with nonnegative terms, he showed that

$$\left(\sum_{n=0}^{\infty} a_n\right)\left(\sum_{n=0}^{\infty} b_n\right) = (a_0 b_0) + (a_0 b_1 + a_1 b_0) + (a_0 b_2 + a_1 b_1 + a_2 b_0) + \cdots$$

$$= \sum_{n=0}^{\infty}\sum_{j=0}^{n} a_j b_{n-j}.$$

We leave the proof of this fact to the reader.

The results of this section are meant to demonstrate the strong tie between sequences and series. We next move on to results which are more specific to series, but the notion of a sequence should always be at the forefront of your thinking when considering a series.

Exercises

Exercise 3.1.1. Show that the following statements are equivalent.

(a) The series $\sum_{j=1}^{\infty} a_j$ converges.
(b) For every $N \in \mathbb{N}$, the series $\sum_{j=N}^{\infty} a_j$ converges.
(c) There exists an $N \in \mathbb{N}$ so that the series $\sum_{j=N}^{\infty} a_j$ converges.

Note: We often begin a series at zero and so, to be technical, the statements above could be restated for $N \in \mathbb{N} \cup \{0\}$.

Exercise 3.1.2. Let a be a nonzero real number and let $N \in \mathbb{N}$. Show that $\sum_{n=N}^{\infty} a$ diverges.

Exercise 3.1.3. Suppose $\sum a_n$ has partial sums of the form $s_n = n/(n+1)$. Find a_n.

Exercise 3.1.4. For each $n \in \mathbb{N}$, show that $1/2^n = 1/2^{n-1} - 1/2^n$.

Exercise 3.1.5. (a) Show that the series $\sum_{n=1}^{\infty} 1/3^n$ converges to $1/2$. To begin, apply Theorem 3.1.7 to show that the series converges. To find the sum, if s_n represents a partial sum, show that

$$3s_n = 1 + s_{n-1}.$$

(b) Use this same approach to show that the series $\sum_{n=1}^{\infty} 1/4^n$ converges to $1/3$.
(c) Let $j \in \mathbb{N} \setminus \{1\}$. Show that the series $\sum_{n=1}^{\infty} 1/j^n$ converges to $1/(j-1)$.

Exercise 3.1.6. Show that the series $\sum_{n=1}^{\infty} 1/n^2$ converges.

Exercise 3.1.7. Show that the series $\sum_{n=1}^{\infty} n/(n+1)! = 1$ using the definition of a partial sum.

Exercise 3.1.8. Supply a proof for the Cauchy Criterion stated in Theorem 3.1.5.

Exercise 3.1.9. Suppose $\sum a_n$ is a divergent series and let $c \in \mathbb{R}$ with $c \neq 0$. Show that $\sum c a_n$ diverges.

Exercise 3.1.10. Supply a proof Theorem 3.1.8(b).

Exercise 3.1.11. (a) Suppose $\sum a_n$ converges and $\sum b_n$ diverges. Show that the sum $\sum (a_n + b_n)$ must diverge.
(b) Show that it is possible for two series $\sum a_n$ and $\sum b_n$ to both diverge while the series $\sum (a_n + b_n)$ converges.
(c) Show that it is possible for both $\sum a_n$ and $\sum b_n$ to diverge but for $\sum a_n b_n$ to converge.

Exercise 3.1.12. For two sequences (a_n) and (b_n), prove the summation by parts formula

$$\sum_{j=m+1}^{n} a_j b_j = s_n b_n - s_m b_{m+1} + \sum_{j=m+1}^{n-1} s_j (b_j - b_{j+1}),$$

where $s_j = a_1 + a_2 + \cdots + a_j$.

3.2   Basic Convergence Tests

The focus of this section is on basic convergence tests. The tests here depend simply on how rapidly the terms in the series become small. The next section will focus on convergence tests which depend on cancelation among the terms. Our first test is the simplest and should always be considered first when encountering a question of series convergence.

Theorem 3.2.1 (Divergence Test). *Let $\sum a_n$ be a series of real numbers. If the defining sequence (a_n) does not converge to zero, then the series diverges.*

Proof. The proof here can be simplified if we work with the contrapositive of the given statement. To be explicit, we will actually show that if a series $\sum a_n$ converges, then the defining sequence (a_n) converges to zero. To this end, suppose $\sum a_n$ converges and let (s_n) denote the sequence of partial sums. Using the Cauchy Criterion, given $\varepsilon > 0$ we can find $M \in \mathbb{N}$ such that $|s_n - s_m| < \varepsilon$ for all $n > m \geq M$. Also notice that a general term in the sequence (a_n) can be written as $a_n = s_n - s_{n-1}$. At this point, set $N = M + 1$. If we now consider a natural number $n \geq N$, we see that $n > n - 1 \geq M$ and we conclude that

$$|a_n - 0| = |s_n - s_{n-1}| < \varepsilon.$$

Ergo (a_n) converges to zero. $\square$

Example 3.2.2. Bear in mind that the Divergence Test *only* indicates divergence. For example, we can conclude that both series $\sum(n+1)/(n+2)$ and $\sum(n^2+3n)/(n^2+3n+1)$ must diverge since the defining sequences do not converge to zero. However, if a defining sequence converges to 0, this *does not* imply that the series converges. An example of this behavior is found in the series $\sum 1/n$. The sequence $(1/n)$ does converge to 0, but not fast enough to guarantee convergence of the series as shown in Example 3.1.3.

Theorem 3.2.3 (Cauchy Condensation Test). *Let (a_n) be a decreasing sequence of nonnegative real numbers. Then the series $\sum_{n=1}^{\infty} a_n$ converges if and only if the series*

$$\sum_{n=0}^{\infty} 2^n a_{2^n} = a_1 + 2a_2 + 4a_4 + 8a_8 + 16a_{16} + \cdots \tag{3.3}$$

converges.

Our approach here will be to exploit the fact that we are dealing with nonnegative terms in both series in question. By Theorem 3.1.7, the convergence of such a series is completely dependent on whether or not the partial sums are bounded. We will use this reduction in both directions of the argument.

Proof. To begin, set

$$s_n = a_1 + a_2 + \cdots + a_n \quad \text{and} \quad t_k = a_1 + 2a_2 + \cdots + 2^k a_{2^k}.$$

Now assume that the series in Eq. (3.3) converges. This means that there is an $M > 0$ such that $t_k \leq M$ for all $k \in \mathbb{N}$. Next we consider a partial sum s_n of the original series. If $n \in \mathbb{N}$, then there is an integer k such that $n < 2^k \leq 2^{k+1} - 1$. Also, the fact that our series contains only nonnegative terms indicates that the sequence of partial sums in question is increasing. If we utilize this together with the fact that the defining sequence (a_n) is decreasing, we have

$$s_n \leq s_{2^{k+1}-1} = a_1 + (a_2+a_3) + (a_4+a_5+a_6+a_7) + \cdots + (a_{2^k} + \cdots + a_{2^{k+1}-1})$$

$$\leq a_1 + (a_2 + a_2) + (a_4 + a_4 + a_4 + a_4) + \cdots + (a_{2^k} + \cdots + a_{2^k})$$

$$= a_1 + 2a_2 + 4a_4 + \cdots + 2^k a_{2^k}$$

$$= t_k \leq M.$$

The fact that this estimate holds for all $n \in \mathbb{N}$ shows that (s_n) is bounded and hence $\sum a_n$ converges.

For the converse, we will work with the contrapositive. To this end, assume that the series in Eq. (3.3) diverges. It follows then that the partial sums t_k are not bounded, i.e. given $M > 0$, there is a $k \in \mathbb{N}$ such that $t_k > M$. Our goal is then to show that the original series diverges which we will accomplish by showing that the sequence of partial sums (s_n) has a subsequence which is unbounded. The subsequence we will focus on is (s_{2^k}). To show that it is unbounded, let $M > 0$ and find $k \in \mathbb{N}$ such that $t_k > 2M$. Estimating the partial sum s_{2^k}, we have

$$s_{2^k} = a_1 + a_2 + (a_3 + a_4) + (a_5 + a_6 + a_7 + a_8) + \cdots + (a_{2^{k-1}+1} + \cdots + a_{2^k})$$

$$\geq a_1 + a_2 + (a_4 + a_4) + (a_8 + a_8 + a_8 + a_8) + \cdots + (a_{2^k} + \cdots + a_{2^k})$$

$$\geq (1/2)a_1 + a_2 + 2a_4 + 4a_8 + \cdots + 2^{k-1}a_{2^k}$$

$$= (1/2)t_k > M.$$

Thus the sequence (s_n) cannot converge as it is unbounded and hence $\sum a_n$ also diverges.   □

Example 3.2.4. Use the Cauchy Condensation test to determine whether the series $\sum_{n=1}^{\infty} 1/n$ and $\sum_{n=1}^{\infty} 1/n^2$ converge or diverge. This is slightly cheating since we already know the answer to both, but the point is to see how to apply the theorem. For the harmonic series, we consider the new series

$$\sum_{n=1}^{\infty} 2^n a_{2^n} = \sum_{n=1}^{\infty} 2^n \frac{1}{2^n} = \sum_{n=1}^{\infty} 1$$

which obviously diverges to infinity. Thus the harmonic series diverges as we previously confirmed.

For the second series we have

$$\sum_{n=1}^{\infty} 2^n a_{2^n} = \sum_{n=1}^{\infty} 2^n \frac{1}{(2^n)^2} = \sum_{n=1}^{\infty} \frac{1}{2^n}$$

which converges by Example 3.1.4. Thus the original series $\sum_{n=1}^{\infty} 1/n^2$ must also converge.

Theorem 3.2.5 (Comparison Test). *Let $(a_n), (b_n)$ be sequences with $0 \leq a_n \leq b_n$ for all $n \in \mathbb{N}$.*

(a) If $\sum_{n=1}^{\infty} b_n$ converges, then $\sum_{n=1}^{\infty} a_n$ also converges.
(b) If $\sum_{n=1}^{\infty} a_n$ diverges, then $\sum_{n=1}^{\infty} b_n$ also diverges.

Proof. To prove the first statement, suppose that $\sum_{n=1}^{\infty} b_n$ converges and let $\varepsilon > 0$. By the Cauchy Criterion, there is an $N \in \mathbb{N}$ such that

$$|b_{m+1} + b_{m+2} + \cdots + b_n| < \varepsilon$$

for all $n > m \geq N$. Using the inequality relationship between the terms of (a_n) and (b_n) together with the fact that all the terms are positive shows that

$$|a_{m+1} + a_{m+2} + \cdots + a_n| = a_{m+1} + a_{m+2} + \cdots + a_n$$
$$\leq b_{m+1} + b_{m+2} + \cdots + b_n$$
$$= |b_{m+1} + b_{m+2} + \cdots + b_n|.$$

Since this inequality holds for all $n \in \mathbb{N}$, it must be the case that

$$|a_{m+1} + a_{m+2} + \cdots + a_n| \leq |b_{m+1} + b_{m+2} + \cdots + b_n| < \varepsilon$$

for all $n > m \geq N$; applying the Cauchy Criterion once more allows us to conclude that $\sum_{n=1}^{\infty} a_n$ also converges.

For the second statement, notice that it is the contrapositive of the first and therefore must also be true. □

Both the Cauchy Condensation and Comparison Tests rely on relating the series in question to a series whose convergence behavior is understood. In order to make good use of these tests it is therefore imperative that we have a class of well understood examples. We've encountered a few examples thus far, and now we investigate two broad classes.

Example 3.2.6. A series of the form

$$\sum_{j=0}^{\infty} a r^j$$

is called a *geometric series*. In considering convergence of such a series, it turns out that the value of a has little effect. If $a = 0$, then the series trivially converges, and for $a \neq 0$, the convergence is totally independent of the value of a. This may not be totally clear at present, but will become clear as we consider the role of r. Focusing now on the parameter r, there are two cases which are extremely quick to deal with, $r = 1$ and $r = -1$. Take a moment to convince yourself that both of these are divergent cases. This leaves us with two final cases, $|r| < 1$ and $|r| > 1$. Luckily we can deal with both at once. Using the factorization identity

$$(1 - x)(1 + x + x^2 + \cdots + x^n) = 1 - x^{n+1} \tag{3.4}$$

and algebra we deduce that the partial sum s_n has the form,

$$s_n = a + ar + ar^2 + \cdots + ar^n = \frac{a(1 - r^{n+1})}{1 - r}.$$

Keep in mind that the denominator in the last expression is valid since $r \neq 1$. Now we take the limit of the partial sum as $n \to \infty$. If $|r| < 1$, then $(r^{n+1}) \to 0$ and thus (s_n) converges to $a/(1 - r)$ by the Algebraic Limit Theorem. On the other hand, if $|r| > 1$, the sequence (r^{n+1}) diverges and thus the expression for (s_n) also diverges, again by the Algebraic Limit Theorem. Our conclusion is summed up by saying that a geometric series converges if and only if $|r| < 1$, in which case we can write

$$\sum_{j=0}^{\infty} ar^j = \frac{a}{1-r}.$$

As a final note on this example, the summation formula just stated is only valid if the series begins at 0 and the value changes with the starting point, provided that $a \neq 0$.

This example makes quick work of the series $\sum_{n=0}^{\infty} 1/2^n$ considered in the previous section. For this we have $a = 1$ and $r = 1/2$ and thus the sum is equal to $1/(1 - 1/2) = 2$. Let's also consider the series $\sum_{n=1}^{\infty} 1/2^n$. The formula above doesn't explicitly apply, but with a little rewriting, we can get it into form,

$$\sum_{n=1}^{\infty} \frac{1}{2^n} = \sum_{n=1}^{\infty} \frac{1}{2 \cdot 2^{n-1}} = \sum_{n=0}^{\infty} \frac{1}{2 \cdot 2^n} = \sum_{n=0}^{\infty} \frac{1}{2} \left(\frac{1}{2}\right)^n.$$

In this form we can apply the formula above with $a = 1/2$ and $r = 1/2$ and we have

$$\sum_{n=1}^{\infty} \frac{1}{2^n} = \frac{1/2}{1 - 1/2} = 1.$$

Example 3.2.7. For our next example, a series of the form

$$\sum_{n=1}^{\infty} \frac{1}{n^p}$$

is called a *p-series*. We've already encountered this for $p = 1$ (the harmonic series) and $p = 2$. The point of interest with these two examples is that while they seem similar, they exhibit different behaviors with respect to convergence. For a general value of $p > 0$ (it clearly diverges if $p \leq 0$), we use the Cauchy Condensation Test. This results in the series

$$\sum_{n=1}^{\infty} 2^n a_{2^n} = \sum_{n=1}^{\infty} 2^n \frac{1}{(2^n)^p} = \sum_{n=1}^{\infty} \left(\frac{1}{2^{p-1}}\right)^n.$$

In this form, it is clear that we are dealing with a geometric series which converges if and only if $r = 1/(2^{p-1}) < 1$. This occurs precisely when $p - 1 > 0$, or $p > 1$. Our conclusion is then that a p-series converges if and only if $p > 1$, and hence diverges if $p \leq 1$.

The Irrationality of e

There are certain irrational numbers which seem to receive more attention than the others, the most famous of these being $\sqrt{2}$, π, and e. We have already shown that $\sqrt{2}$ is irrational and have also seen how to extend this to roots of other natural numbers. Here we take up the case for e. There are many definitions for this number and you have likely encountered many of these in previous courses. Here we take a series approach; in Sect. 7.4 we will see that this definition is equivalent to its many cousins.

Definition 3.2.8. The number e is defined by the infinite series

$$e = \sum_{n=0}^{\infty} \frac{1}{n!} = 1 + 1 + \frac{1}{2} + \frac{1}{6} + \frac{1}{24} + \frac{1}{120} + \cdots$$

where $n! = 1 \cdot 2 \cdots (n-1) \cdot n$ and $0! = 1$.

One should check that the defining series actually converges as our definition would make little sense if it did not. The Ratio Test provides the quickest means to accomplish this but a comparison will also confirm this fact. The following lemma allows us to locate e with respect to the integers. In the exercises you will compute a tighter set of bounds for the value of e.

Lemma 3.2.9. *The number e satisfies the bound $2 < e < 3$.*

Proof. For the lower bound, notice that

$$e = 1 + 1 + \frac{1}{2} + \frac{1}{6} + \frac{1}{24} + \frac{1}{120} + \cdots > 2$$

by considering only the first two terms in the series.

For the upper bound, first observe that $2 \cdot 3 \cdot 4 \cdots n > 2^{n-1}$, from which it follows that

$$\frac{1}{n!} < \frac{1}{2^{n-1}}.$$

Applying this estimate to the terms in the definition of e, starting with the third term, yields

$$e = 1 + 1 + \frac{1}{2} + \frac{1}{3!} + \frac{1}{4!} + \frac{1}{5!} + \cdots$$

$$< 1 + 1 + \frac{1}{2} + \frac{1}{2^2} + \frac{1}{2^3} + \frac{1}{2^4} + \cdots$$

$$= 1 + \sum_{n=0}^{\infty} \frac{1}{2^n}$$

$$= 3$$

where the last equality is obtained by summing the geometric series appearing in the third line.   □

Theorem 3.2.10. *The number e is irrational.*

Proof. Suppose e is rational, that is, suppose there exist positive integers a and b with

$$e = \frac{a}{b}.$$

Notice that $b \neq 1$ since we know e is not an integer, Lemma 3.2.9. In particular, $b \geq 2$ in $\mathbb{N}$. To obtain a contradiction, we begin by defining a new number

$$x = b! \left(e - \sum_{n=0}^{b} \frac{1}{n!} \right).$$

We'll show that x is an integer with $0 < x < 1$ which cannot happen simultaneously.

For the first point, observe that

$$x = b! \left(e - \sum_{n=0}^{b} \frac{1}{n!} \right) = b! \left(\frac{a}{b} - \sum_{n=0}^{b} \frac{1}{n!} \right)$$

$$= a(b-1)! - b! \sum_{n=0}^{b} \frac{1}{n!}$$

$$= a(b-1)! - \sum_{n=0}^{b} \frac{b!}{n!}$$

$$= a(b-1)! - \left(\frac{b!}{0!} + \frac{b!}{1!} + \frac{b!}{2!} + \cdots + \frac{b!}{b!} \right)$$

$$= a(b-1)! - (b! + b! + (3 \cdot 4 \cdots b) + \cdots + 1).$$

This shows that x is an integer since each term in the last sum is an integer.

To see that $x > 0$, notice that

$$e = \sum_{n=0}^{\infty} \frac{1}{n!} > \sum_{n=0}^{b} \frac{1}{n!}$$

for any value of b. Thus

$$x = b! \left(e - \sum_{n=0}^{b} \frac{1}{n!} \right) > 0.$$

Finally, we need to show that $x < 1$. First observe that

$$e - \sum_{n=0}^{b} \frac{1}{n!} = \sum_{n=b+1}^{\infty} \frac{1}{n!}$$

and hence

$$x = b!\left(e - \sum_{n=0}^{b} \frac{1}{n!}\right) = b! \sum_{n=b+1}^{\infty} \frac{1}{n!} = \sum_{n=b+1}^{\infty} \frac{b!}{n!}$$

by the Algebraic Limit Theorem for series. Algebra and an estimate now indicate that

$$x = \sum_{n=b+1}^{\infty} \frac{b!}{n!} = \frac{1}{b+1} + \frac{1}{(b+1)(b+2)} + \frac{1}{(b+1)(b+2)(b+3)} + \cdots$$

$$\leq \frac{1}{b+1} + \frac{1}{(b+1)^2} + \frac{1}{(b+1)^3} + \cdots$$

$$= \sum_{n=1}^{\infty} \frac{1}{(b+1)^n}$$

$$= \sum_{n=1}^{\infty} \left(\frac{1}{b+1}\right)^n.$$

Notice that this last sum is a geometric series with $r = 1/(b+1) < 1$, and, taking care to apply the geometric series formula appropriately, we conclude that

$$x \leq \sum_{n=1}^{\infty} \left(\frac{1}{b+1}\right)^n = \frac{\frac{1}{b+1}}{1 - \frac{1}{b+1}} = \frac{1}{b} \leq \frac{1}{2} < 1$$

since $b \geq 2$. Therefore it must be the case that e is irrational. □

Exercises

Exercise 3.2.1. Suppose (a_n) is a sequence with $a_n > 0$ for all n.

(a) If $\sum_{n=1}^{\infty} a_n$ converges, show that $\sum_{n=1}^{\infty} a_n/(1 + a_n)$ also converges.

(b) If $\sum_{n=1}^{\infty} a_n$ converges, show that $\sum_{n=1}^{\infty} 1/(1 + a_n)$ diverges.

Exercise 3.2.2. Suppose $\sum a_n$ converges and $a_n \neq 0$ for all n. Show that $\sum 1/a_n$ diverges.

Exercise 3.2.3. To determine whether the series $\sum 1/(n + \sqrt{n})$ converges or diverges, consider comparisons with $\sum 1/n$, $\sum 1/\sqrt{n}$, and $\sum 1/(2n)$. Only one of these will provide a valid comparison.

Exercise 3.2.4. Determine whether each of the given series converges or diverges. If the series converges, find its sum.

(a) $\displaystyle\sum_{n=1}^{\infty} \frac{e^n}{3^{n-1}}$

(d) $\displaystyle\sum_{n=1}^{\infty} \frac{1}{1 + (2/3)^n}$

(b) $\displaystyle\sum_{n=1}^{\infty} \frac{e^n}{2^{n+1}}$

(e) $\displaystyle\sum_{n=1}^{\infty} \frac{1 + 2^n}{3^n}$

(c) $\displaystyle\sum_{n=2}^{\infty} \frac{2^{n+4}}{5^{n-3}}$

(f) $\displaystyle\sum_{n=1}^{\infty} \left(\frac{1}{2^n} + \frac{2}{n} \right)$

Exercise 3.2.5. Find the value of c such that the series $\sum_{n=2}^{\infty}(1 + c)^{-n}$ converges to 2.

Exercise 3.2.6. Prove the following generalization of the comparison test: Let (a_n) and (b_n) be sequences of real numbers and suppose there exists $N \in \mathbb{N}$ such that $0 \le a_n \le b_n$ for all $n \ge N$. Show that $\sum_{n=1}^{\infty} a_n$ converges if $\sum_{n=1}^{\infty} b_n$ converges.

Exercise 3.2.7. Suppose $\sum a_n$ is a convergent series with nonnegative terms and let (b_n) be a bounded sequence of nonnegative real numbers. Show that the series $\sum a_n b_n$ converges.

Exercise 3.2.8. Suppose $\sum a_n$ and $\sum b_n$ are convergent series with nonnegative terms.

(a) Show that the series $\sum a_n b_n$ converges.
(b) Find an example of two such series for which we also have $(\sum a_n)(\sum b_n) \ne \sum a_n b_n$.

Exercise 3.2.9. Let $(a_n), (b_n)$ be sequences with nonnegative terms which satisfy the inequality

$$\frac{a_{n+1}}{a_n} \le \frac{b_{n+1}}{b_n}$$

for all $n \in \mathbb{N}$.

(a) If $\sum_{n=1}^{\infty} b_n$ converges, show that $\sum_{n=1}^{\infty} a_n$ also converges.
(b) If $\sum_{n=1}^{\infty} a_n$ diverges, show that $\sum_{n=1}^{\infty} b_n$ also diverges.

Exercise 3.2.10. Suppose (a_n) is a sequence with nonnegative terms and such that $\lim n^2 a_n$ exists. Show that $\sum a_n$ converges.

Exercise 3.2.11 (Limit Comparison Test). Part (a) asks you to prove the Limit Comparison Test and part (b) asks you to apply the test. Parts (d) and (e) consider exceptional cases of the test.

(a) Let $\sum a_n$ and $\sum b_n$ be series of positive terms and suppose that the terms of the two series satisfy

$$\lim \frac{a_n}{b_n} = 1.$$

Show that $\sum a_n$ converges if and only if $\sum b_n$ converges.

(b) Suppose (a_n) is a sequence of positive real numbers which satisfies $\lim n a_n = c$ with $c \neq 0$. What can you conclude about the convergence of $\sum a_n$?

(c) For a more general form of the statement, let $\sum a_n$ and $\sum b_n$ be series of positive terms and suppose that the terms of the two series satisfy

$$\lim \frac{a_n}{b_n} = c$$

where $c \neq 0$. Show that $\sum a_n$ converges if and only if $\sum b_n$ converges.

(d) Let $\sum a_n$ and $\sum b_n$ be series of positive terms and suppose $\sum b_n$ converges. If

$$\lim \frac{a_n}{b_n} = 0,$$

show that $\sum a_n$ also converges.

(e) Let $\sum a_n$ and $\sum b_n$ be series of positive terms and suppose $\sum b_n$ is divergent. If

$$\lim \frac{a_n}{b_n} = \infty,$$

show that $\sum a_n$ also diverges.

Exercise 3.2.12. Use the estimating techniques employed in Lemma 3.2.9 to show that $2.7 < e < 2.8$.

Exercise 3.2.13. Let s_n be a partial sum of the series used to define e. Show that

$$0 < e - s_n < \frac{1}{n!n}.$$

Use this to calculate e with an error of less than 10^{-6}.

3.3 Absolute and Conditional Convergence

The convergence tests we've considered to this point only concern series with nonnegative terms. In this section we focus on series with both positive and negative terms in which case cancelation among the terms plays a role in the convergence behavior. We begin with a statement concerning series whose terms alternate between positive and negative values. In a sense this is the simplest type of series containing infinitely many positive and negative terms.

Theorem 3.3.1 (Alternating Series Test). *Let (a_n) be a decreasing sequence of nonnegative real numbers converging to zero. Then both alternating series*

$$\sum_{n=1}^{\infty}(-1)^{n+1}a_n \quad \text{and} \quad \sum_{n=1}^{\infty}(-1)^n a_n$$

converge.

Proof. For the proof we will show that the sequence of partial sums (s_n) for the series $\sum_{n=1}^{\infty}(-1)^{n+1}a_n$ converges by showing that both subsequences (s_{2j}) and (s_{2j-1}) converge to a common value; the fact that (s_n) converges then follows from Exercise 2.2.17. We begin with (s_{2j}) and will use the Monotone Convergence Theorem to show that it converges. First notice that

$$s_{2j} = a_1 - a_2 + a_3 - a_4 + \cdots + a_{2j-1} - a_{2j}$$
$$= (a_1 - a_2) + (a_3 - a_4) + \cdots + (a_{2j-1} - a_{2j})$$
$$\leq (a_1 - a_2) + (a_3 - a_4) + \cdots + (a_{2j-1} - a_{2j}) + (a_{2j+1} - a_{2j+2})$$
$$= s_{2(j+1)}$$

since each parenthetical term is nonnegative due to the decreasing behavior of the sequence (a_n). Hence (s_{2j}) is an increasing sequence.

To show that the sequence is also bounded, we have that

$$0 \leq s_{2j} = (a_1 - a_2) + (a_3 - a_4) + \cdots + (a_{2j-1} - a_{2j})$$

where we are again using the fact that each parenthetical term is nonnegative. For an upper bound, we associate the terms in a slightly different configuration to see that

$$s_{2j} = a_1 - (a_2 - a_3) - (a_4 - a_5) - \cdots - (a_{2j-2} - a_{2j-1}) - a_{2j} \leq a_1$$

since these new parenthetical terms are also nonnegative. Thus (s_{2j}) is bounded and increasing and hence converges; call its limit S.

To see that the subsequence (s_{2j-1}) also converges to S, notice that

$$s_{2j-1} = s_{2j} - a_{2j}.$$

By the Algebraic Limit Theorem and the fact that $(a_n) \to 0$, it follows that

$$\lim s_{2j-1} = \lim s_{2j} - \lim a_{2j} = S.$$

Therefore we conclude that $\sum_{n=1}^{\infty}(-1)^{n+1}a_n$ converges by the aforementioned exercise.

The fact that the series $\sum_{n=1}^{\infty}(-1)^n a_n$ also converges follows from the Algebraic Limit Theorem for Series since

$$\sum_{n=1}^{\infty}(-1)^n a_n = (-1)\sum_{n=1}^{\infty}(-1)^{n+1}a_n.$$

□

Example 3.3.2. To demonstrate the theorem, consider the series $\sum_{n=1}^{\infty}(-1)^{n+1}/n$; this is called the *alternating harmonic series*. Notice that the sequence $(1/n)$ is decreasing and converges to zero and thus by the Alternating Series Test, the alternating harmonic series must converge. In general, the *alternating p-series* $\sum_{n=1}^{\infty}(-1)^{n+1}/n^p$ converges when $p > 0$.

The next result attempts to reduce a given series to a series with only nonnegative terms, in which case we can use the tests of previous section.

Theorem 3.3.3 (Absolute Convergence Test). *Let $\sum a_n$ be a series of real numbers. If $\sum |a_n|$ converges, then $\sum a_n$ also converges. If $\sum |a_n|$ converges, then*

$$\left| \sum_{n=1}^{\infty} a_n \right| \le \sum_{n=1}^{\infty} |a_n|.$$

We will leave the proof of this statement as an exercise. As a hint, the Cauchy Criterion in conjunction with the triangle inequality will produce the desired implication. The estimate will then follow from the Order Limit Theorem. As a cautionary remark, do not fall into the trap of thinking that

$$\left| \sum a_n \right| = \sum |a_n|.$$

The theorem above is giving an analog of the triangle inequality for series, and, as in the finite case, it is possible for the inequality to be strict.

Example 3.3.4. Consider the series

$$\sum_{n=1}^{\infty} \frac{\cos(n)}{n^2}.$$

Though we can't explicitly compute the value of $\cos(n)$, the oscillating nature of the cosine function indicates that $\cos(n)$ assumes both positive and negative values. Using the fact that $|\cos x| \le 1$ for all real x, we obtain the estimate

$$\left| \frac{\cos(n)}{n^2} \right| = \frac{|\cos(n)|}{n^2} \le \frac{1}{n^2}.$$

By the Comparison Test, we conclude that the series

$$\sum_{n=1}^{\infty} \left| \frac{\cos(n)}{n^2} \right|$$

converges and hence the series in question also converges.

As a second example, we return to the alternating harmonic series from Example 3.3.2. Here we see that

$$\sum_{n=1}^{\infty} \left| \frac{(-1)^{n+1}}{n} \right| = \sum_{n=1}^{\infty} \frac{1}{n}$$

which diverges, while the original series converges by the Alternating Series Test. This motivates the following definition which sorts convergent series with both positive and negative terms into two classes; there is, of course, the third class of divergent series.

Definition 3.3.5. Let $\sum a_n$ be a series of real numbers. If $\sum |a_n|$ converges, we say that the series $\sum a_n$ *converges absolutely*. If $\sum |a_n|$ diverges but $\sum a_n$ converges, then we say that the series $\sum a_n$ *converges conditionally*.

With this terminology we say that the alternating harmonic series is conditionally convergent. For a general alternating p-series, we have absolute convergence if $p > 1$ and conditional convergence for $0 < p \leq 1$.

The final two tests we explore indicate when a series converges absolutely or when it diverges. These two tests have a similar flavor in that the convergence behavior is determined solely by the terms of the series with no modifications or comparisons. They are also the most general convergence tests we have. On the down side, both tests produce a gray area which we will discuss after stating and proving the theorems.

While both proofs employ the same technique, the case for the Root Test is simpler and we present it first. The proof is centered around comparing the series in question to a geometric series which will explain the three possible conclusions of each theorem.

Theorem 3.3.6 (Root Test). *Let $\sum a_n$ be a series and suppose*

$$L = \lim |a_n|^{1/n}$$

exists or is infinite.

(a) *If $0 \leq L < 1$, then the series converges absolutely.*
(b) *If $L > 1$, then the series diverges.*
(c) *If $L = 1$, then the test is inconclusive, i.e. the series may converge absolutely or conditionally, or it may diverge.*

Proof. To begin, let L be as in the statement of the theorem and assume that $0 \leq L < 1$. Now choose $R \in (L, 1)$ and set $\varepsilon = R - L$. By our definition of L, there is an $N \in \mathbb{N}$ with

$$\left| |a_n|^{1/n} - L \right| < \varepsilon$$

for all $n \geq N$. By expanding this inequality and focusing on the resulting upper bound, we see that

$$|a_n|^{1/n} < L + \varepsilon = R$$

or

$$|a_n| < R^n$$

for all $n \geq N$. The fact that $R < 1$ then assures us that the series

$$\sum_{n=N}^{\infty} R^n$$

converges. The comparison test and Exercise 3.1.1 show that $\sum |a_n|$ converges which indicates that $\sum a_n$ converges absolutely as desired.

For (b), suppose that $L > 1$. The argument here is similar to that above, however we focus instead on finding a lower bound for the terms $|a_n|$ rather than an upper bound. This time, choose $R \in (1, L)$ and set $\varepsilon = L - R$. Using the definition of L again, there is an $N \in \mathbb{N}$ such that

$$\left| |a_n|^{1/n} - L \right| < \varepsilon$$

for all $n \geq N$. Expanding this inequality, but focusing on the lower bound for this case we have that

$$R = L - \varepsilon < |a_n|^{1/n}$$

or

$$1 < R^n < |a_n|$$

for all $n \geq N$. This guarantees us that the defining sequence (a_n) does not converge to zero and the Divergence Test supplies our conclusion.

For (c), consider $\sum 1/n^2$, $\sum (-1)^{n+1}/n$, and $\sum 1/n$. Each of these will have $L = 1$ providing specific examples of each of the possible outcomes and thus explaining why the test is inconclusive in this case. The fact that $L = 1$ for each of

these series hinges on the fact that the sequence $(n^{1/n})$ converges to 1 as $n \to \infty$.
The proof of this fact is requested as Exercise 3.3.7.   □

Theorem 3.3.7 (Ratio Test). *Let $\sum a_n$ be a series with $a_n \neq 0$ for all $n \in \mathbb{N}$ and
suppose*

$$L = \lim \left| \frac{a_{n+1}}{a_n} \right|$$

exists or is infinite.

(a) *If $0 \leq L < 1$, then the series converges absolutely.*
(b) *If $L > 1$, then the series diverges.*
(c) *If $L = 1$, then the test is inconclusive, i.e. the series may converge absolutely
    or conditionally, or it may diverge.*

Proof. First suppose that $0 \leq L < 1$. As with the previous theorem, our proof will
use a comparison of the series $\sum |a_n|$ with a convergent geometric series. To begin,
choose $R \in (L, 1)$ and set $\varepsilon = R - L$. By our hypothesis, there is an $N \in \mathbb{N}$ such
that

$$\left| \left| \frac{a_{n+1}}{a_n} \right| - L \right| < \varepsilon$$

for all $n \geq N$. Expanding and rearranging this compound inequality shows that

$$\left| \frac{a_{n+1}}{a_n} \right| < L + \varepsilon = R$$

for all $n \geq N$; we will not employ the lower estimate in this case. One last algebraic
manipulation shows that

$$|a_{n+1}| \leq R|a_n|$$

for all $n \geq N$. Moreover, repeated applications of this inequality yield

$$|a_n| \leq R^{n-N}|a_N| \qquad (3.5)$$

for all $n \geq N$.

At this point, it's best to think about where we are headed. The estimate above
shows that the N-tail of the series $\sum |a_n|$ satisfies

$$|a_N| + |a_{N+1}| + |a_{N+2}| + \cdots + |a_{N+j}| + \cdots \leq |a_N| + R|a_N|$$

$$+ R^2 |a_N| + \cdots + R^j |a_N| + \cdots$$

providing a valuable comparison due to the fact that the second series above is
geometric.

To make this argument rigorous we have the following. First, the series

$$\sum_{n=0}^{\infty} |a_N| R^n = \sum_{n=N}^{\infty} |a_N| R^{n-N}$$

is geometric and converges since $|R| < 1$. Applying the comparison from Eq. (3.5) guarantees us that the series

$$\sum_{n=N}^{\infty} |a_n|$$

converges and Exercise 3.1.1 completes the proof of (a).

We leave the proof of (b) as Exercise 3.3.8. The third statement is deduced by considering examples for which $L = 1$ and which encompass all possible convergence outcomes. In particular, consider $\sum 1/n^2$, $\sum(-1)^{n+1}/n$, and $\sum 1/n$. □

Example 3.3.8. First take the series $\sum (2/n)^n$. The series looks almost geometric but not quite. It is, however, a prime candidate for the Root Test. We have

$$L = \lim |(2/n)^n|^{1/n} = \lim 2/n = 0 < 1.$$

Thus we conclude that the series converges.

Now consider the series $\sum 3^n/n!$. Applying the Ratio Test,

$$L = \lim \left| \frac{3^{n+1}/(n+1)!}{3^n/n!} \right| = 3 \lim \frac{1}{n+1} = 0 < 1.$$

Thus this series also converges.

For a more interesting example, we investigate the series

$$\frac{1}{2} + \frac{1}{3} + \frac{1}{2^2} + \frac{1}{3^2} + \frac{1}{2^3} + \frac{1}{3^3} + \cdots.$$

Considering the Ratio test first, we will show that $\lim |a_{n+1}/a_n|$ does not exist. First observe that we can express the terms in the even position as $a_{2j} = 1/3^j$ and the terms in the odd positions as $a_{2j-1} = 1/2^j$. For the sake of notation, we set $b_n = a_{n+1}/a_n$. If we then consider the subsequences of (b_n) consisting of the even and odd terms, respectively, we have

$$\lim b_{2j} = \lim \frac{a_{2j+1}}{a_{2j}} = \lim \frac{\left(\frac{1}{2}\right)^{j+1}}{\left(\frac{1}{3}\right)^j} = \lim \frac{1}{2} \left(\frac{3}{2}\right)^j = \infty$$

while

$$\lim b_{2j-1} = \lim \frac{a_{2j}}{a_{2j-1}} = \lim \frac{\left(\frac{1}{3}\right)^j}{\left(\frac{1}{2}\right)^j} = \lim \left(\frac{2}{3}\right)^j = 0.$$

Thus we conclude that

$$\lim b_n = \lim \left| \frac{a_{n+1}}{a_n} \right|$$

does not exist which means that the Ratio Test fails.

For the root test, notice that

$$\lim (a_{2j})^{1/2j} = \lim \left(\frac{1}{3^j} \right)^{1/2j} = \frac{1}{\sqrt{3}}$$

while

$$\lim (a_{2j-1})^{1/(2j-1)} = \lim \left(\frac{1}{2^j} \right)^{1/(2j-1)} = \frac{1}{\sqrt{2}}.$$

This also shows that the limit necessary to apply the Root Test does not exist. We will discuss the resolution to this situation after the next result.

The $L = 1$ situation in both the Root and the Ratio tests cannot be alleviated and, as exhibited by the examples given in each proof, is problematic for even very simple series. However, it is typical in these situations that we can use other tests to determine convergence behavior. Moreover, there are many other series tests which we have not explored, e.g. those of Abel, Dirichlet, Raabe, Kummer, and Gauss to name a few. The text [24] treats several of these in detail while the text [15] is entirely devoted to a study of series.

That being said, there is another possible outcome which the tests do not address, the case when L does not exist; the third series of Example 3.3.8 provides an example of such a series. This issue stems from the fact that we have used a limit in our hypothesis, but we can repair this by using the notion of limit superior and inferior which always exists or is infinite. We state the most general form of the Ratio Test and leave its proof for the reader. A similar statement can be made for the Root Test, though only the limit superior is needed there.

Theorem 3.3.9 (Ratio Test). *Let $\sum a_n$ be a series with $a_n \neq 0$ for all $n \in \mathbb{N}$. Let*

$$L = \limsup \left| \frac{a_{n+1}}{a_n} \right| \qquad \text{and} \qquad l = \liminf \left| \frac{a_{n+1}}{a_n} \right|$$

where either is allowed to be infinite.

(a) *If $0 \leq L < 1$, then the series converges absolutely.*
(b) *If $l > 1$, then the series diverges.*
(c) *If $l \leq 1 \leq L$, then the test is inconclusive, i.e. the series may converge absolutely or conditionally, or it may diverge.*

Example 3.3.10. Consider again the last series from Example 3.3.8. In the context of the generalized Ratio Test just stated, we conclude that $L = \infty$ while $l = 0$ and the test is still inconclusive. For the Root Test however, we conclude that $\limsup |a_n|^{1/n} = 1/\sqrt{2} < 1$ which confirms that the series converges.

It is natural to ask now whether there are series for which the Root Test is inconclusive but the Ratio Test provides a conclusive outcome. The answer to this is no. In fact, we can show that if a series satisfies conclusion (a) or (b) in the Ratio Test, then the series also satisfies the same conclusion in the Root Test. In other words, the Ratio Test can be derived from the Root Test. The proceeding example then shows that the two tests are not logically equivalent. So, while the Ratio Test is usually easier to use from the computational perspective, the Root Test is the more powerful test.

To close this chapter we return once again to the Cantor set. It may be useful to review the construction given in Sect. 1.4. When we explored the set previously we discussed the fact that although it may seem that the only elements in the set are the endpoints of the removed intervals, this is certainly not the case since we also verified that the set is uncountable. Here we provide some examples of rational numbers which are in the set. There are of course irrational numbers in the Cantor set, but these are harder to explicitly identify.

Example 3.3.11. First we will show that $\frac{1}{4}$ is in Cantor set. Let us begin by locating $\frac{1}{4}$ in the sets C_n used to construct the Cantor set. In C_1, $\frac{1}{4}$ is in the interval $[0, \frac{1}{3}]$, the left-hand subinterval of $[0, 1]$ and is closer to $\frac{1}{3}$ than it is to zero. In moving to C_2, $[0, \frac{1}{3}]$ splits into $[0, \frac{1}{9}] \cup [\frac{2}{9}, \frac{1}{3}]$ and $\frac{1}{4}$ is in the right-hand subinterval $[\frac{2}{9}, \frac{1}{3}]$; here $\frac{1}{4}$ is closer to $\frac{2}{9}$ than it is to $\frac{1}{3}$. In C_3, $\frac{1}{4}$ is in the left-hand subinterval $[\frac{2}{9}, \frac{7}{27}]$ of $[\frac{2}{9}, \frac{1}{3}]$ and is closer to $\frac{7}{27}$.

To show that $\frac{1}{4}$ is in the Cantor set, we could continue the exercise above and attempt to demonstrate that $\frac{1}{4}$ is in C_n for every $n \in \mathbb{N}$, but this approach is cumbersome at best. Rather, we will show that there is a sequence of points in the Cantor set which converges to $\frac{1}{4}$ and then use the fact that we are working with a closed set to obtain our conclusion. The back and forth behavior exhibited above is useful as it provides some intuition as to how to construct such a sequence. Specifically, we will take the closest endpoints, $\frac{1}{3}, \frac{2}{9}, \frac{7}{27}$, as our starting point and use these to construct a sequence of points in the Cantor set converging to $\frac{1}{4}$ (Fig. 3.1).

Let $a_1 = \frac{1}{3}$. Then, setting $a_2 = \frac{2}{9}$, we can write

$$a_2 = \frac{1}{3} - \frac{1}{9} = a_1 - \frac{1}{3^2}.$$

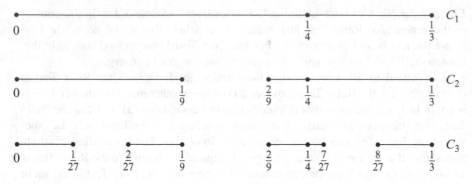

Fig. 3.1 $\frac{1}{4}$ is in the Cantor set

Now set $a_3 = \frac{7}{27}$ and notice that

$$a_3 = \frac{2}{9} + \frac{1}{27} = a_2 + \frac{1}{3^3}.$$

With this we see a general recursive pattern,

$$a_n = a_{n-1} + \frac{(-1)^{n-1}}{3^n},$$

where the alternating sign is what guarantees us that each point is in the Cantor set, the endpoint of a removed interval, provided $a_1 = \frac{1}{3}$.

   To show that this sequence converges to $\frac{1}{4}$, notice what happens if we expand the recursive definition in each sequence term,

$$a_1 = \frac{1}{3},$$

$$a_2 = \frac{2}{9} = \frac{1}{3} - \frac{1}{9},$$

$$a_3 = \frac{7}{27} = \frac{2}{9} + \frac{1}{27} = \frac{1}{3} - \frac{1}{9} + \frac{1}{27},$$

and, in general,

$$a_n = a_{n-1} + \frac{(-1)^{n-1}}{3^n} = \frac{1}{3} - \frac{1}{9} + \frac{1}{27} - \cdots + \frac{(-1)^{n-1}}{3^n}.$$

The critical point here is that these sequence terms are exactly the partial sums of a geometric series which we can sum,

$$\sum_{n=1}^{\infty} \frac{(-1)^{n-1}}{3^n} = \frac{1/3}{1-(-1/3)} = \frac{1}{4}.$$

Using the fact that a convergent series and its sequence of partial sums converge to the same value shows that the sequence $(a_n) \to \frac{1}{4}$. Since each $a_n \in C$, the fact that $\frac{1}{4}$ is a limit point of this closed set implies that $\frac{1}{4} \in C$.

A similar approach will show that $\frac{1}{10}$ is also in the Cantor set. Moreover this strategy can be modified to show that $\frac{3}{4}, \frac{3}{10}, \frac{7}{10}$, and $\frac{9}{10}$ are all in the Cantor set. In fact, the above technique can be extended to a large class of rational numbers which have the form $m/(3^n + 1)$ where $m, n \in \mathbb{N}$. You are asked to ponder this in the exercises. A recent article that considers this class and others can be found in [20].

When we discussed the Cantor set earlier, we devised an addressing scheme in order to show that C is uncountable. In that discussion we mentioned that the address assigned to $1/4$ is the sequence $\overline{02}$. To see why this is valid, recall that each digit in the string assigned to a number in the Cantor set indicated what interval the point is in at the corresponding stage of the construction. This addressing actually has a very strong connection with series. In fact, if $x \in C$ has address assigned by the string $a_1 a_2 a_3 \ldots$, then

$$x = \sum_{n=1}^{\infty} \frac{a_n}{3^n};$$

such a sum clearly converges when a_n is either 0 or 2 by a comparison test argument. To see why this is the case for $1/4$, notice that

$$\frac{0}{3} + \frac{2}{9} + \frac{0}{27} + \frac{2}{81} + \cdots = \sum_{n=1}^{\infty} 2\left(\frac{1}{9}\right)^n = \frac{1}{4}.$$

Understanding which rational numbers are in the Cantor set is an interesting question and the irrational numbers in the Cantor provide an equally interesting endeavor. Clearly there are irrational numbers in C since the set it uncountable, but is it possible to identify any of these, say $\pi/4$, $\sqrt{2}/2$, or $\sqrt{3}/2$?

Exercises

Exercise 3.3.1. Determine whether each of the given series converges absolutely, converges conditionally, or diverges.

(a) $\displaystyle\sum_{n=1}^{\infty} \frac{(-1)^n}{2n+1}$ (e) $\displaystyle\sum_{n=1}^{\infty} \frac{\cos(n\pi)}{n}$

(b) $\displaystyle\sum_{n=1}^{\infty} \frac{e^n}{n!}$ (f) $\displaystyle\sum_{n=1}^{\infty} \frac{(-2)^n}{n^n}$

(c) $\displaystyle\sum_{n=1}^{\infty} (-1)^n \frac{3n-1}{2n+1}$ (g) $\displaystyle\sum_{n=1}^{\infty} \frac{n^{10} 10^n}{n!}$

(d) $\displaystyle\sum_{n=1}^{\infty} (-1)^n \frac{n}{n^2+1}$ (h) $\displaystyle\sum_{n=1}^{\infty} (-1)^{n+1} \frac{n!}{n^n}$

Exercise 3.3.2. Suppose a_n is a sequence converging to $\frac{1}{2}$. Determine whether the given series is absolutely convergent.

(a) $\displaystyle\sum_{n=1}^{\infty} \frac{a_n \cos(n\pi)}{n^2}$ (b) $\displaystyle\sum_{n=1}^{\infty} \frac{n^n}{n! a_1 a_2 a_3 \cdots a_n}$

Exercise 3.3.3. Let $\sum a_n = 1 - 1 + 1/2 - 1/2 + 1/3 - 1/3 + \cdots$. Show that the series converges conditionally.

Exercise 3.3.4. (a) Suppose $\sum a_n$ is an absolutely convergent series. Show that $\sum a_n^2$ also converges (absolutely). Is the conclusion guaranteed if only conditional convergence is assumed?

(b) Consider the converse of the above statement, i.e. if $a_n \geq 0$ and $\sum a_n$ converges, can we conclude anything about the convergence of $\sum \sqrt{a_n}$?

Exercise 3.3.5. Suppose $\sum a_n$ is an absolutely convergent series and let (b_n) be a bounded sequence. Show that $\sum a_n b_n$ converges absolutely. Does the above result remain true if we only assume conditional convergence?

Exercise 3.3.6. Let (b_n) be a sequence of nonnegative terms and let (a_n) be a sequence satisfying

$$\left| \frac{a_{n+1}}{a_n} \right| \leq \frac{b_{n+1}}{b_n}.$$

If $\sum b_n$ converges, show that $\sum a_n$ converges absolutely.

Exercise 3.3.7. Follow the steps below to complete the proof of part (c) of Theorem 3.3.6.

(a) First verify that $\lim n^{1/n} = 1$. To do this, begin by defining a new sequence $a_n = n^{1/n} - 1$. Then show that $a_n > 0$ (for $n \geq 2$) and $n = (a_n + 1)^n$. Next, find a suitable upper bound for a_n using the Binomial Expansion Theorem and singling out a single term. To complete the proof, use the Squeeze Theorem to show that $(a_n) \to 0$.

(b) Now show that the three examples stated in the text satisfy $L = 1$.

Exercise 3.3.8. Prove Theorem 3.3.7(b).

Exercise 3.3.9. Supply a proof for the Ratio Test using the Root Test.

Exercise 3.3.10. Use the technique exhibited in the text to show that $\frac{3}{4}, \frac{1}{10}, \frac{3}{10}, \frac{7}{10}$, and $\frac{9}{10}$ are in the Cantor set. Can you generalize this to a statement about rational numbers of the form $m/(3^n + 1)$ where $m, n \in \mathbb{N}$.

3.4 Sequence Spaces

In this section we focus on two particular spaces of sequences which are closely related to ℓ^∞. In actuality, these new spaces and ℓ^∞ are generalizations of normed linear spaces which arise by imposing different norms on $\mathbb{R}^n$. The two examples which are the focus of this section are extensions of Example 1.5.11.

Example 3.4.1. Define the space ℓ^1 to be the collection of all absolutely summable sequences of real numbers. Symbolically, we write

$$\ell^1 = \left\{ (a_n)_{n=1} = (a_1, a_2, a_3, \ldots) : \sum_{n=1}^{\infty} |a_n| < \infty \right\}.$$

This set becomes a vector space if we define addition and scalar multiplication in the same manner as we did in ℓ^∞; we will use this same definition of addition and scalar multiplication for all subsequent spaces of sequences. To establish a norm on ℓ^1, we define

$$\|(a_n)\|_1 = \sum_{n=1}^{\infty} |a_n|;$$

the exercises will ask you to verify that this rule is in fact a norm. We will suppress the notation and use the symbol ℓ^1 to represent the normed linear space $(\ell^1, \|\cdot\|_1)$; similar notation will be used for ℓ^2 and ℓ^p below.

Example 3.4.2. Define the space ℓ^2 to be the collection of sequences of real numbers which are square-summable,

$$\ell^2 = \left\{ (a_n)_{n=1} = (a_1, a_2, a_3, \ldots) : \sum_{n=1}^{\infty} |a_n|^2 < \infty \right\}.$$

Furthermore, we define

$$\|(a_n)\|_2 = \left(\sum_{n=1}^{\infty} |a_n|^2\right)^{1/2}.$$

It is true that the absolute value bars are unnecessary for this norm, but common practice is to include them. Exactly why this is the case will become apparent later in this section.

When encountering new spaces, it's always beneficial to focus on understanding what objects are or are not in the spaces. For example, the sequence $(1/n)$ is not in ℓ^1 though it is in ℓ^2. The sequence $(1/\sqrt{n})$ is in neither space, while the sequence $(1/n^2)$ is in both. These are simple examples and though they may not be intensely insightful, it is also always important to think about containment relationships between similar spaces. Continuing with this line of thinking, is it possible to come up with a sequence (a_n) which is in ℓ^1 but not in ℓ^2? The answer is explored in Exercise 3.4.5.

Before showing that these new spaces are complete, we must first verify that they are in fact normed linear spaces. To do this, we must show that they are both vector spaces and that the rules defined in the previous examples are norms. We leave the proof for ℓ^1 as an exercise. For ℓ^2, we will actually show that it is an inner product space.

Theorem 3.4.3. *The space ℓ^2 is an inner product space.*

Proof. To show ℓ^2 is a vector space, we will show that the set is closed under the operation of sequence addition. The other vector space axioms then follow in a manner completely analogous to the ℓ^1 case. Let (a_n) and (b_n) be sequences in ℓ^2. By definition, $(a_n) + (b_n) = (a_n + b_n)$ and thus showing that $(a_n) + (b_n) \in \ell^2$ is equivalent to showing that

$$\sum_{n=1}^{\infty} |a_n + b_n|^2 < \infty.$$

The key here is to use the inequality

$$|ab| \leq \frac{a^2 + b^2}{2}.$$

(Why is this true?) Two immediate facts follow from this observation. First, by the Comparison Test we see that the series $\sum_{n=1}^{\infty} a_n b_n$ is absolutely convergent since the dominating sequence $(a_n^2 + b_n^2)/2$ gives rise to a convergent series. Furthermore,

$$|a_n + b_n|^2 \leq a_n^2 + 2|a_n b_n| + b_n^2 \leq 2(a_n^2 + b_n^2)$$

for all $n \in \mathbb{N}$, and another application of the Comparison Test guarantees us that the sequence $(a_n + b_n) \in \ell^2$.

Next we must show that there is an inner product on ℓ^2 which gives rise to the previously defined norm. It's easy to see that the rule we want is given by

$$\langle (a_n), (b_n) \rangle = \sum_{n=1}^{\infty} a_n b_n.$$

For two sequences in ℓ^2, this function is well defined by the observation above that the series $\sum_{n=1}^{\infty} a_n b_n$ converges absolutely. Inner product axiom (a) follows immediately by commutativity of real number multiplication. Property (b) is also fairly obvious. Properties (c) and (d) then follow from the Algebraic Limit Theorem for series. The fact that the function defined in Example 3.4.2 is a norm on ℓ^2 now follows from Proposition 1.5.15.   □

Proposition 3.4.4. *The spaces ℓ^1 and ℓ^2 are complete.*

Proof. We present the proof for ℓ^2 and encourage the reader to reread the proof of Theorem 2.5.7 before proceeding. Assume $\big((a_{nk})_{k=1} \big)_{n=1}$ is a Cauchy sequence in ℓ^2. As before, we must first identify a limit candidate, then show that it is in our space and that the given sequence converges to it. Notice first that for each $k \in \mathbb{N}$

$$|a_{nk} - a_{mk}| = (|a_{nk} - a_{mk}|^2)^{1/2} \leq \left(\sum_{k=1}^{\infty} |a_{nk} - a_{mk}|^2 \right)^{1/2} = \|(a_{nk}) - (a_{mk})\|_2.$$

The fact that we are dealing with a Cauchy sequence (indexed by n) together with the above estimate shows that for each $k \in \mathbb{N}$ the sequence $(a_{nk})_{n=1}$ is a Cauchy sequence in $\mathbb{R}$. Let $(a_k)_{k=1}$ be the sequence generated by these Cauchy sequences, that is, let a_k be the limit of $(a_{nk})_{n=1}$ as $n \to \infty$.

To see that this new sequence is in ℓ^2, we again appeal to the result that Cauchy sequences are bounded. Let $M > 0$ be such a bound for our original Cauchy sequence. Then

$$\sum_{k=1}^{\infty} |a_{nk}|^2 \leq M^2$$

for each $n \in \mathbb{N}$. Now if we fix $K \in \mathbb{N}$, the inequality above implies that

$$\sum_{k=1}^{K} |a_{nk}|^2 \leq M^2$$

for each $n \in \mathbb{N}$. With this point of view, we are dealing with a finite sum; applying the Algebraic and Order Limit Theorems as $n \to \infty$ implies that

$$\sum_{k=1}^{K} |a_k|^2 \leq M^2.$$

Since this estimate holds for every $K \in \mathbb{N}$, Theorem 3.1.7 implies that the series $\sum_{k=1}^{\infty} |a_k|^2$ converges to a finite value, no larger than M^2, and thus (a_k) is a sequence in ℓ^2.

Finally, we need to show that $\|(a_{nk}) - (a_k)\|_2 \to 0$ as $n \to \infty$. Let $\varepsilon > 0$ and choose $N \in \mathbb{N}$ such that

$$\|(a_{nk}) - (a_{mk})\|_2 < \varepsilon/2$$

for $n, m \geq N$. Then

$$\sum_{k=1}^{\infty} |a_{nk} - a_{mk}|^2 < \varepsilon^2/4$$

for $n, m \geq N$ and thus

$$\sum_{k=1}^{K} |a_{nk} - a_{mk}|^2 < \varepsilon^2/4$$

for a fixed $K \in \mathbb{N}$ and $n, m \geq N$. Keeping $n \geq N$ and K fixed, but letting $m \to \infty$ we see that

$$\sum_{k=1}^{K} |a_{nk} - a_k|^2 \leq \varepsilon^2/4.$$

Notice that this estimate also relies on both the Algebraic and Order Limit Theorems. Again, since this estimate holds for an arbitrary $K \in \mathbb{N}$, we conclude that the series $\sum_{k=1}^{\infty} |a_{nk} - a_k|^2$ converges (for $n \geq N$) by Theorem 3.1.7 and the Order Limit Theorem confirms that

$$\sum_{k=1}^{\infty} |a_{nk} - a_k|^2 \leq \varepsilon^2/4$$

whenever $n \geq N$. Therefore $\|(a_{nk}) - (a_k)\|_2 \leq \varepsilon/2 < \varepsilon$ for $n \geq N$ completing the proof.   □

More generally, we can define a range of sequence spaces based on these two examples. For $p \geq 1$, define the space ℓ^p by

$$\ell^p = \left\{ (a_n)_{n=1} = (a_1, a_2, a_3, \ldots) : \sum_{n=1}^{\infty} |a_n|^p < \infty \right\}.$$

The rule

$$\|(a_n)\|_p = \left(\sum_{n=1}^{\infty} |a_n|^p \right)^{1/p}$$

then defines a norm on the space. Verifying norm properties (a) and (b) is not difficult, but the triangle inequality provides more of a challenge. Once we know that we do indeed have a norm, the verification that these spaces are complete follows in a manner completely analogous to the technique employed to show that ℓ^2 is complete. We will confirm that the triangle inequality does hold in Sect. 9.4. In a more general context, these spaces can be defined for sequences of complex numbers. In this setting, taking the absolute value (or *modulus*) of each sequence term is necessary to guarantee that the norm function maps into $[0, \infty)$.

At this point it's natural to wonder why we should care about so many spaces. One answer is simply that we should because we can, but this is somewhat unsatisfying. A second would be to say that an investigation of these spaces provides more insight into the concept of series convergence; the intuition is similar to dealing with p-series, the higher the index p, the more likely it is that a given series is in the space. This is explored in Exercise 3.4.5. A final answer, and the least apparent at the moment is that these spaces are all unique and present some interesting structural properties which depend on the index p. One such distinction is the focus of Exercise 3.4.3 where you are asked to show that among all the ℓ^p spaces, ℓ^2 is the only one which can be realized as a Hilbert space with norm satisfying Proposition 1.5.15. This structural difference is of great importance and we close this chapter with a demonstration of the geometry available in the Hilbert space setting.

Let (X, d) be a metric space. We say that a subset $A \subseteq X$ is *open* if for every $x \in A$, there is an $\varepsilon > 0$ such that $B_\varepsilon(x) \subseteq A$; the notation $B_\varepsilon(x)$ was introduced in Exercise 1.5.10. For a set $M \subseteq X$, we say M is *closed* if the complement of M is open in X. These two definitions are motivated entirely by their counterparts for subsets of $\mathbb{R}$.

Example 3.4.5. For $\mathbb{R}^2$ with the Euclidean norm, the set $\{(x, y) \in \mathbb{R}^2 : x^2 + y^2 < 1\}$ is open while the sets $\{(x, y) \in \mathbb{R}^2 : x^2 + y^2 \leq 1\}$ and $\{(x, y) \in \mathbb{R}^2 : x^2 + y^2 = 1\}$ are both closed.

Definition 3.4.6. Let X be a vector space and let $M \subseteq X$. We say that M is *convex* if $(1 - t)x + ty$ is in M for every $x, y \in M$ and every $t \in [0, 1]$.

Fig. 3.2 Examples of convex
and non-convex sets in $\mathbb{R}^2$

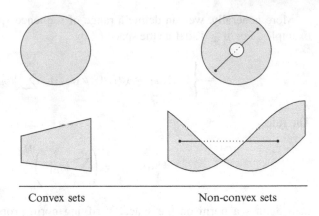

| Convex sets | Non-convex sets |

The definition above says that a set is convex if the line joining any two points
of the set lies entirely in the set. If a set M is a subspace of a vector space X,
then M is automatically convex since it is closed with respect to both addition and
scalar multiplication; the specific details are requested in the exercises. In Fig. 3.2,
the sets on the left are both convex while the sets on the right are not. To understand
the image, notice that for each region in the right-hand column we have chosen
two points and drawn the line connecting them. The fact that these lines do not lie
entirely inside the region is indicated by the dashed portion of the line.

The following result shows that convex sets have a very nice structural property
in the Hilbert space setting.

Proposition 3.4.7 (Nearest Point Property). *Every nonempty, closed, convex set*
M in a Hilbert space $(X, \langle \cdot, \cdot \rangle)$ contains a unique element of smallest norm.
Furthermore, given any $x \in X$, there is a unique $m_0 \in M$ such that

$$\|x - m_0\| = \inf\{\|x - m\| : m \in M\}.$$

Proof. For the first statement, we set $\alpha = \inf\{\|m\| : m \in M\}$ to represent the
smallest norm for an element of M; the infimum exists since we are assuming that
M is not empty together with the fact that the set used to define the infimum is
bounded below by 0. By Exercise 2.2.14, there is a sequence (x_n) in M such that
$(\|x_n\|)$ converges to α. Moreover, since M is convex, the vector $\frac{1}{2}x_n + \frac{1}{2}x_m$ is in M
for every $m, n \in \mathbb{N}$ and hence

$$\left\|\frac{x_n + x_m}{2}\right\|^2 \geq \alpha^2$$

for all $n, m \in \mathbb{N}$. Applying the parallelogram equality to the vectors $\frac{1}{2}x_n$ and $\frac{1}{2}x_m$
shows that

$$0 \le \left\| \frac{x_n - x_m}{2} \right\|^2 = \frac{1}{2} \left(\|x_n\|^2 + \|x_m\|^2 \right) - \left\| \frac{x_n + x_m}{2} \right\|^2$$

which, upon clearing denominators, becomes

$$0 \le \|x_n - x_m\|^2 = 2 \left(\|x_n\|^2 + \|x_m\|^2 \right) - \|x_n + x_m\|^2$$

$$\le 2 \left(\|x_n\|^2 + \|x_m\|^2 \right) - 4\alpha^2$$

$$= 2(\|x_n\|^2 - \alpha^2) + 2(\|x_m\|^2 - \alpha^2).$$

Since the sequence $(\|x_n\|)$ converges to α, the Algebraic Limit Theorem guarantees us that $(\|x_n\|^2)$ converges to α^2. Then, for $\varepsilon > 0$, each term in the final piece of the inequality can be bounded above by $\varepsilon^2/2$ by choosing N so that $(\|x_n\|^2 - \alpha^2) < \varepsilon^2/4$ when $n \ge N$. The previous estimate now implies that

$$\|x_n - x_m\| < \varepsilon$$

for all $n, m \ge N$ and thus the sequence (x_n) is Cauchy in X. Applying the fact that X is complete, we know that there is an $m_0 \in X$ with (x_n) converging to m_0 in the norm. The fact that M is closed then assures us that $m_0 \in M$. Finally, Exercise 2.5.4 allows us to conclude that $(\|x_n\|)$ converges to $\|m_0\|$ from which the uniqueness of limits shows that $\|m_0\| = \alpha$ as desired.

To show that there is only element of M with this property, we assume that there are two points $x, y \in M$ such that $\|x\| = \|y\| = \alpha$. The fact that M is convex again shows that $\frac{1}{2}x + \frac{1}{2}y \in M$ and

$$\left\| \frac{x + y}{2} \right\| \ge \alpha.$$

Another application of the parallelogram equality indicates that

$$0 \le \left\| \frac{x - y}{2} \right\|^2 = \frac{1}{2} \left(\|x\|^2 + \|y\|^2 \right) - \left\| \frac{x + y}{2} \right\|^2$$

$$= \alpha^2 - \left\| \frac{x + y}{2} \right\|^2$$

$$\le 0.$$

With this we see that $\|x - y\| = 0$ and conclude that $x = y$ by norm property (a).

For the second statement, fix $x \in X$ and consider the translation

$$M - x = \{m - x : m \in M\}.$$

The fact that M is nonempty, closed, and convex is sufficient to conclude that $M - x$ also has these three properties. Thus there is a unique element $k_0 \in M$ with minimal norm, i.e.

$$\|k_0\| = \inf\{\|k\| : k \in M - x\}.$$

Each element $k \in M - x$ can be written as $k = m - x$ for some $m \in M$. If we substitute this into the previous expression we find there is a unique $m_0 = k_0 + x \in M$ such that

$$\|x - m_0\| = \inf\{\|m - x\| : m \in M\}$$

completing the proof. $\square$

The Nearest Point Property does not hold if we remove the hypothesis of convex or closed, or change our space to a Banach space. The easiest case to consider is a set that is not closed. For this take the interval $(0, 1)$ in $\mathbb{R}$ with norm given by the absolute value. It should be clear that there is no element of minimal norm here. The case for non-convex sets or non-Hilbert spaces is harder. The reason for this is that every finite dimensional vector space can be realized as a Hilbert space. And, in the finite dimensional case, any closed subset of a vector space has this property though the uniqueness is lost. This forces us to remove ourselves to the infinite dimensional setting to find counterexamples for these two situations.

As a final comment, the property just demonstrated is key in proving the Projection Theorem which details the decomposition of a Hilbert space in terms of closed subspaces. The Projection Theorem is pivotal in proving the Riesz Representation Theorem (see Theorem 4.5.14.) which is a fundamental result in the study of Hilbert spaces.

Exercises

Exercise 3.4.1. Let $a, b \in \mathbb{R}$. Show that $|ab| \le (a^2 + b^2)/2$.

Exercise 3.4.2. (a) Show that ℓ^1 is a normed linear space.
(b) Show that ℓ^1 is complete.

Exercise 3.4.3. (a) Show that the sequences $(1/3^n)_{n=1}^{\infty}$ and $(1/4^n)_{n=1}^{\infty}$ satisfy the parallelogram equality in ℓ^1.
(b) On the other hand, find two sequences in ℓ^1 which do not satisfy the parallelogram equality to show that norm given for ℓ^1 cannot be described as an inner product in the sense of Proposition 1.5.15.
(c) Use this same line of reasoning to show that the norm on ℓ^p arises as an inner product as described in Proposition 1.5.15 if and only if $p = 2$.

Exercise 3.4.4. Does the rule $\|(a_n)\| = \sum_{n=1}^{\infty} |a_n|^p$ define a norm on ℓ^p? Explain.

Exercise 3.4.5. (a) Show the following containment relationships $\ell^1 \subseteq \ell^2 \subseteq c_0 \subseteq \ell^\infty$. Further, give justification that these containments are all strict.

(b) If $1 \leq p < q \leq \infty$, show that $\ell^p \subseteq \ell^q \subseteq \ell^\infty$.

(c) Show that the collection of sequences ℓ^1 is not complete with respect to the sup-norm.

Exercise 3.4.6. Verify the statements made in Example 3.4.5.

Exercise 3.4.7. Show that every subspace of a vector space is convex.

Exercise 3.4.8. Let X be a vector space.

(a) Show that the intersection of any two convex subsets of X is convex.

(b) Can the same be said for the union of two convex sets? Prove or provide a counterexample.

Exercise 3.4.9. Let X be a vector space and let M be a convex subset of X. If $x \in X$, show that the translation $M + x = \{m + x : m \in \mathcal{M}\}$ is also convex.

Chapter 4
Continuity

The goal of this chapter is to explore the limit of a function whose domain is a subset of $\mathbb{R}$ and to then use this idea to investigate the notion of continuity. In the first-year calculus sequence, these topics are usually covered at a quick pace in an effort to move on to more interesting constructs such as the derivative and the integral. However, as the derivative is defined as a functional limit, it makes sense to have a thorough understanding of limits before moving forward. Historically, the derivative and the integral came before the ideas of limits and continuity. Many of the developments of the seventeenth and eighteenth centuries were motivated by a desire to understand the behavior of series of functions, specifically Fourier series. There were, however, complications that arose, and, as the mathematics developed, the need for a more stable foundation became apparent. In order to better understand Fourier series, mathematicians were forced to consider integration of such objects, and this study eventually led, in the early nineteenth century, to the formal development of the modern concept of limits. We have already discussed one particular type of limit, that of a sequence, and the more general notion of a functional limit is an extension of those ideas.

4.1   Sequences and the Limit of a Function

In its earliest form, the concept of a functional limit (or the limit of a function) was defined by comparing the distance between points in the domain with their corresponding quantities in the range, and it is likely that you have experienced this, perhaps only to a limited degree, in your study of calculus. We will eventually come to these same ideas, but initially we will present the definition of the limit of a function in terms of sequences. One reason for this is that it allows us to make use of our thorough work with sequences. This will lead to simple proofs of many of the facts you know intuitively to be true. Another reason, though authors will forever argue about such things, is that the sequential approach seems to be more visually appealing, and more in line with how most students actually think about limits. In a

M.A. Pons, *Real Analysis for the Undergraduate: With an Invitation to Functional Analysis*,   137
DOI 10.1007/978 1 4614 9638 0_4, © Springer Science+Business Media New York 2014

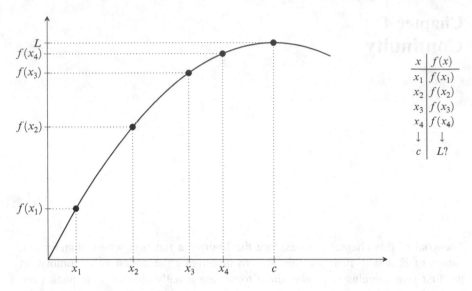

x	$f(x)$
x_1	$f(x_1)$
x_2	$f(x_2)$
x_3	$f(x_3)$
x_4	$f(x_4)$
$\downarrow$	$\downarrow$
c	$L?$

Fig. 4.1 Limits via a table of values

first year calculus course, the idea of a limit is most often introduced visually and with a table of values. The table considers domain points which are chosen ever closer to the limit point in question, call it c, and calculates the value of the function at these domain points in an effort to predict the behavior of the function. Similarly, in a visual manner instructors often imitate this process by tracing a finger along the x-axis toward the limit point, then along the graph, and finally along the y-axis toward the limit L to demonstrate the behavior expressed by the table.

It is only then that an informal definition of limit is given, properties are discussed, and finally a formal definition is supplied. However, students seem to relate best to the visual approach and the table of values. And both of these, at least it seems to be the case to this author, are essentially sequential techniques. We want to approach a point c in the domain. How can we do that? We take a sequence in the domain which converges to the point c. This is represented by the left-hand column in the table of values (Fig. 4.1) or tracing along the x-axis toward c. We then evaluate the function at each point of the sequence. This produces a sequence in the range, the right-hand column in the table of values, and we investigate its behavior. Calculus students use these techniques as a means of determining the value of the limit, but we will make this process more concrete and conclude that the limit exists if and only if the behavior matches for every approach path, i.e. along every sequence converging to c.

This discussion provides the general idea of a limit. Our definition will provide more rigor and we follow with a more thorough discussion.

Definition 4.1.1 (Sequential Definition of a Functional Limit). Let $f : A \to \mathbb{R}$ and let c be a limit point of A. We say that $\lim_{x \to c} f(x) = L$ if for every sequence (x_n) in A with $x_n \neq c$ for all $n \in \mathbb{N}$ and $\lim x_n = c$, it follows that $\lim f(x_n) = L$.

The driving motivation behind the functional limit is to gather information about what a function is doing near a point c in order to better understand possible behavior at c. Notice that the point c is not required to be in the domain of our function; this is required so that the behavior at the point does not interfere with the analysis of activity near the point. However we do require that c be a limit point of the domain. This allows us to consider domain points which are arbitrarily close to c in order to obtain best possible estimates of the activity in question. Furthermore there are two different sequences to consider when looking at such a limit. First, a sequence in the domain which converges to c. Again, these sequence terms never assume the value c so that the behavior at c causes no interference. Then we study the convergence behavior of the corresponding range sequence obtained by evaluating the function at each point in the domain sequence. If this convergence behavior agrees for all possible domain sequences, i.e. converges to a value L, then we say that the functional limit exists and is equal to L. As with sequential limits, functional limits are unique if they exist and this will be explored in the exercises.

Example 4.1.2. As a first example, consider the statement $\lim_{x \to 2} 3x - 2 = 4$. Let's begin by taking the sequence $(2 - 1/n)$, which converges to 2 and for which 2 is not a term in the sequence, and examine what happens when we evaluate the function $f(x) = 3x - 2$ along this sequence. Notice that

$$f(2 - 1/n) = 3(2 - 1/n) - 2 = (6 - 3/n) - 2 = 4 - 3/n.$$

It is clear now that this sequence converges to 4 by the Algebraic Limit Theorem for sequences. This isn't enough to say that the limit of f at 2 is equal to 4, but it does provide some evidence that the statement is true.

To verify the limit statement, our argument will look similar to those used to show sequential convergence and often require some forethought. Consider what happens if we take a sequence in $\mathbb{R}$, the domain of the function $f(x) = 3x - 2$, with $x_n \neq 2$ for all n and $\lim x_n = 2$. To show that the sequence $(f(x_n))$ converges to 4, for $\varepsilon > 0$, we need to produce an $N \in \mathbb{N}$ so that $|f(x_n) - 4| < \varepsilon$ for all $n \geq N$. In order to accomplish this, keep in mind that we know something about the behavior of (x_n) and thus our first step is to relate the quantities $|x_n - 2|$ and $|f(x_n) - 4|$. Observe that

$$|f(x_n) - 4| = |(3x_n - 2) - 4| = |3x_n - 6| = 3|x_n - 2|.$$

From this calculation it is apparent that if $\varepsilon > 0$ and we choose $N \in \mathbb{N}$ so that $|x_n - 2| < \varepsilon/3$ for all $n \geq N$, we can rest assured that $|f(x_n) - 4|$ will be less than ε.

Proof. Let (x_n) be a sequence in $\mathbb{R}$ with $x_n \neq 2$ for all n and $\lim x_n = 2$. Also let $\varepsilon > 0$. Since $(x_n) \to 2$, we can find an $N \in \mathbb{N}$ such that $|x_n - 2| < \varepsilon/3$ for all $n \geq N$. If we then consider $n \geq N$, we have

$$|f(x_n) - 4| = |(3x_n - 2) - 4| = |3x_n - 6| = 3|x_n - 2| < 3(\varepsilon/3) = \varepsilon.$$

Thus we conclude that $(f(x_n)) \to 4$ and hence $\lim_{x \to 2} 3x - 2 = 4$. □

Example 4.1.3. Let's show now that $\lim_{x \to 1} x^3 + 5 = 6$. Here we will make use of our thorough work with sequences instead of going through the definition of convergence.

Proof. Let (x_n) be a sequence in $\mathbb{R}$ with $x_n \neq 1$ for all n and $\lim x_n = 1$. Evaluating $f(x) = x^3 + 5$ along this sequence yields the range sequence $(x_n^3 + 5)$. By the Algebraic Limit Theorem for sequences and the fact that $(x_n) \to 1$, we can immediately conclude that $(x_n^3 + 5) \to 1^3 + 5 = 6$. Therefore $\lim_{x \to 1} x^3 + 5 = 6$. □

This last example highlights the effectiveness of our work with sequences. We proved many results which demonstrated that the convergence of a sequence is preserved with respect to algebraic operations and with respect to certain algebraic functions, e.g. roots, absolute value, etc. This will be of great use to us as we continue along this path of inquiry.

Just as with sequences, it will also be important to identify when functional limits do not exist and for this we have the following theorem. We have stated the theorem as having two conclusions though this is for ease of use rather than necessity which will be exhibited by the examples following the statement.

Theorem 4.1.4 (Divergence Criteria for Functional Limits). *Let $f : A \to \mathbb{R}$ and let c be a limit point of A. Then $\lim_{x \to c} f(x)$ does not exist provided one of the following occur:*

(a) *there is a sequence (x_n) in A with $x_n \neq c$ for all $n \in \mathbb{N}$ and $\lim x_n = c$ for which $\lim f(x_n)$ does not exist;*
(b) *there exist sequences (x_n) and (y_n) in A with $x_n \neq c$ and $y_n \neq c$ for all $n \in \mathbb{N}$, and $\lim x_n = \lim y_n = c$ but $\lim f(x_n) \neq \lim f(y_n)$.*

Example 4.1.5. Take $f(x) = 1/x^2$ and $g(x) = \sin(1/x)$, and consider what is happening as x approaches zero. Though these functions are not defined at zero, it is certainly a limit point of their respective domains. For f, we will use divergence criterion (a) since our intuition tells us that the values of $1/x^2$ grow without bound as x approaches 0. With this in mind, take the domain sequence $(1/n)$. Evaluating f at these points yields the sequence (n^2) which diverges to ∞. Thus we conclude that $\lim_{x \to 0} 1/x^2$ does not exist; see Fig. 4.2.

For the function $g(x) = \sin(1/x)$, we can use the oscillating behavior of $\sin x$ to our advantage (Fig. 4.3). In particular, we know that $\sin x$ repeats its output every 2π units. For example, $\sin x$ takes the value 1 when evaluated at points of the form

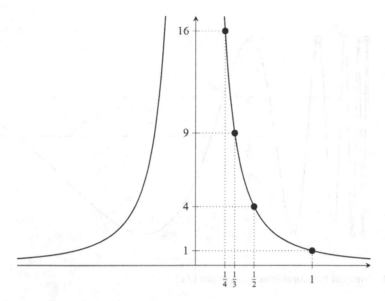

Fig. 4.2 Divergent functional limit: $f(x) = 1/x^2$

$2\pi n + \pi/2$ and the value -1 when evaluated at points of the form $2\pi n + 3\pi/2$. To make use of this information, we also need to take the inversion $1/x$ into account and we will therefore need to manipulate these domain values so as to work in our favor. We set $x_n = 1/(2\pi n + \pi/2)$ and $y_n = 1/(2\pi n + 3\pi/2)$. Evaluating g along these two sequences, both of which converge to zero,

$$g(x_n) = \sin(1/x_n) = \sin(2\pi n + \pi/2) = 1$$

while

$$g(y_n) = \sin(1/y_n) = \sin(2\pi n + 3\pi/2) = -1.$$

Thus we conclude by divergence criterion (b) that $\lim_{x \to 0} \sin(1/x)$ does not exist.

For the second example, we could also have applied divergence criterion (a). Consider evaluating g along the sequence $(x_1, y_1, x_2, y_2, \ldots)$. As a result, we obtain the sequence $(1, -1, 1, -1, \ldots)$, which diverges. Using this same reasoning, we can show that divergence criterion (b) always implies (a). However, in the first example, no matter how we choose a sequence (x_n) converging to zero, we arrive at the conclusion that $(f(x_n))$ diverges to infinity. This reveals our true reason for stating the divergence criteria in two parts—some functions do have the same *divergence behavior* no matter how you approach a limit point of the domain, but do not confuse this with a convergent situation.

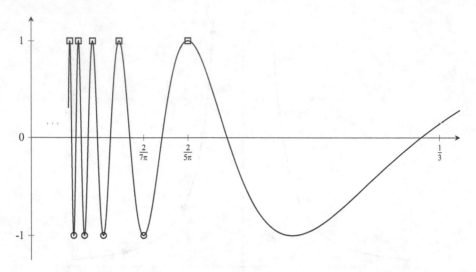

Fig. 4.3 Divergent functional limit: $g(x) = \sin(1/x)$

As a final remark on these two examples, notice again that for the function $f(x) = 1/x^2$, the sequence $f(x_n)$ diverges in a very particular manner regardless of the chosen domain sequence (as long as it converges to 0). In order to be as specific as possible about limit behavior we use the familiar language and say that f has a *vertical asymptote* or an *infinite limit* at $x = 0$. Taking this example as an archetype, we have the following definition that once again relates the functional limit to a sequential limit. See Exercises 4.1.11 and 4.1.13 for more on vertical asymptotes.

Definition 4.1.6 (Divergence Criterion: ∞). Let $f : A \to \mathbb{R}$ and let c be a limit point of A. We say that $\lim_{x \to c} f(x) = \infty$ if for every sequence (x_n) in A with $x_n \neq c$ for all n and $\lim x_n = c$, it follows that $\lim f(x_n) = \infty$.

Example 4.1.7. Define a function $f : \mathbb{R} \to \mathbb{R}$ by

$$f(x) = \begin{cases} 1, & x \in \mathbb{Q}; \\ 0, & x \notin \mathbb{Q}. \end{cases}$$

This function is commonly referred to as the *Dirichlet function* and will reappear several times throughout the text. For our discussion here, the function possesses the interesting property that it does not have a limit at any point. To see this, let $c \in \mathbb{R}$. Appealing to the density property of both the rational and irrational numbers, we can find two sequences (x_n) and (y_n) with $x_n \neq c$ and rational for all $n \in \mathbb{N}$, and $y_n \neq c$ and irrational for all $n \in \mathbb{N}$, and with both sequences converging to c. Evaluating the function along the rational sequence, we have $f(x_n) = 1$ for all n and thus $\lim f(x_n) = 1$. On the other hand, $f(y_n) = 0$ for all n and hence $\lim f(y_n) = 0$. Therefore we conclude that $\lim_{x \to c} f(x)$ does not exist.

Theorem 4.1.8 (Algebraic Limit Theorem for Functional Limits). *Let* $f : A \to \mathbb{R}$ *and* $g : A \to \mathbb{R}$ *and let* c *be a limit point of* A. *If* $\lim_{x \to c} f(x) = L_1$ *and* $\lim_{x \to c} g(x) = L_2$, *then:*

(a) $\lim_{x \to c} kf(x) = kL_1$ *for all* $k \in \mathbb{R}$;

(b) $\lim_{x \to c} [f(x) + g(x)] = L_1 + L_2$;

(c) $\lim_{x \to c} [f(x)g(x)] = L_1 L_2$;

(d) $\lim_{x \to c} \left[\dfrac{f(x)}{g(x)} \right] = \dfrac{L_1}{L_2}$ *provided* $L_2 \neq 0$.

Proof. The proof of each of these statements is a simple application of the corresponding theorem for sequences. We provide a proof of part (a) as an example and leave the remaining parts as Exercise 4.1.4. Let $f : A \to \mathbb{R}$ be a function with limit L_1 at c, a limit point of A. Also let $k \in \mathbb{R}$. Applying the definition of functional limit, if (x_n) is any sequence in A with $x_n \neq c$ for all n and $\lim x_n = c$, then the range sequence $(f(x_n))$ converges to L_1. By the Algebraic Limit Theorem for sequences, it follows immediately that the sequence $(kf(x_n)) \to kL_1$. Since the sequence (x_n) was arbitrary in A, we conclude that $\lim_{x \to c} kf(x) = kL_1$. $\square$

Now we turn our attention to the more standard definition of a limit. The basic question asks if we can force values in the range to be close to L by restricting our attention to values in the domain which are close to c. For sequences, we translate this by saying that convergence in the domain implies convergence in the range. As was the case with our study of limit points in $\mathbb{R}$, there is also a way to understand this in terms of ε-neighborhoods. The following definition captures this approach.

Definition 4.1.9 ($\varepsilon - \delta$ Definition of a Functional Limit). Let $f : A \to \mathbb{R}$ and let c be a limit point of A. We say that $\lim_{x \to c} f(x) = L$ if for every $\varepsilon > 0$, there is a $\delta > 0$ such that if $0 < |x - c| < \delta$ and $x \in A$, it follows that $|f(x) - L| < \varepsilon$.

It may not be apparent that these two characterizations of a limit are equivalent, but rest assured that they are.

Theorem 4.1.10. *The sequential and* $\varepsilon - \delta$ *characterizations of a functional limit are logically equivalent.*

Proof. For our general hypothesis, let $f : A \to \mathbb{R}$ and let c be a limit point of A. Now we must show that the $\varepsilon - \delta$ definition of a function limit implies the sequential definition and vice versa.

First suppose that f has limit L according to the $\varepsilon - \delta$ definition, that is, given any $\varepsilon > 0$, there is a $\delta > 0$ such that if $0 < |x - c| < \delta$ and $x \in A$, it follows that $|f(x) - L| < \varepsilon$. To show that the sequential characterization also holds, let (x_n) be a sequence in A with $x_n \neq c$ for all $n \in \mathbb{N}$ and $\lim x_n = c$. It is now our task to show that $(f(x_n))$ converges to L. As this is a convergence statement, let $\varepsilon > 0$. By our assumption that f has $\varepsilon - \delta$ limit L, we may find $\delta > 0$ such that if $0 < |x - c| < \delta$ and $x \in A$, it follows that $|f(x) - L| < \varepsilon$. Since we also know that $(x_n) \to c$ and $x_n \neq c$ for all n, we may choose $N \in \mathbb{N}$ such that $0 < |x_n - c| < \delta$ for all $n \geq N$. Combining our choice for δ and N, we have that $|f(x_n) - L| < \varepsilon$ for all $n \geq N$ and conclude that $(f(x_n))$ converges to L. This confirms that the sequential definition of a functional limit is also satisfied.

For the converse we will work with the contrapositive, that is, we will assume that the $\varepsilon - \delta$ definition does not hold and show that the sequential definition also fails to hold. The key here is to take care with the negation of the definitions. Using symbolic logic to represent the $\varepsilon - \delta$ definition, we write

$$(\forall \varepsilon > 0)\,(\exists \delta > 0)\,(\forall x \in A \setminus \{c\})\,(0 < |x - c| < \delta \Rightarrow |f(x) - L| < \varepsilon).$$

Negating this statement, we have

$$(\exists \varepsilon > 0)\,(\forall \delta > 0)\,(\exists x \in A \setminus \{c\})\,(0 < |x - c| < \delta \ \text{and} \ |f(x) - L| \geq \varepsilon).$$

Putting this into words, L is not the limit of f at c if there is an $\varepsilon > 0$ such that for all $\delta > 0$ there is an $x \in A$ with $x \neq c$ and $0 < |x - c| < \delta$ for which $|f(x) - L| \geq \varepsilon$. Assuming this, we will show then that we can construct a sequence (x_n) in A with $\lim x_n = c$ and $x_n \neq c$ for all n for which $(f(x_n))$ does not converge to L, which is the negation of the sequential characterization.

Let $\varepsilon > 0$ be specified by the negation of the $\varepsilon - \delta$ definition. To construct the desired sequence, consider first $\delta = 1$. Using our assumption that the $\varepsilon - \delta$ definition does not hold, choose x_1 to be a point in $A \setminus \{c\}$ such that $0 < |x_1 - c| < 1$ and $|f(x_1) - L| \geq \varepsilon$. Repeating this process, let $\delta = 1/2$ and choose x_2 to be a point in $A \setminus \{c\}$ such that $0 < |x_2 - c| < 1/2$ and $|f(x_2) - L| \geq \varepsilon$. In general, for $n \in \mathbb{N}$ and $\delta = 1/n$, we may choose x_n in $A \setminus \{c\}$ such that $0 < |x_n - c| < 1/n$ and $|f(x_n) - L| \geq \varepsilon$. The sequence (x_n) satisfies $x_n \neq c$ for all n and $(x_n) \to c$ since $(1/n) \to 0$. Moreover, for every $n \in \mathbb{N}$ we have that $|f(x_n) - L| \geq \varepsilon$. Thus $(f(x_n))$ does not converge to L. This demonstrates that the sequential definition also fails completing the proof.   □

Example 4.1.11. Here we will reprove the limit statements of Examples 4.1.2 and 4.1.3 by applying this new characterization. Consider first the statement $\lim_{x \to 2} 3x - 2 = 4$. The game here is essentially the same as before, but we have changed the look of things. Given $\varepsilon > 0$, we need to find a $\delta > 0$ so that if $x \in \mathbb{R}$ with $0 < |x - 2| < \delta$, then $|(3x - 2) - 4| < \varepsilon$. The obvious first step is to look for a relationship between the two quantities $|x - 2|$ and $|(3x - 2) - 4|$. Simple algebra shows that

$$|(3x - 2) - 4| = 3|x - 2|.$$

This is the same relationship we used before, and now, instead of choosing N we choose a positive real number δ which will force the proper outcome to occur. In particular, if $\varepsilon > 0$, we choose $\delta = \varepsilon/3$.

Proof. Let $\varepsilon > 0$ and choose $\delta = \varepsilon/3$. If we then consider $x \in \mathbb{R}$ with $0 < |x - 2| < \delta$, we see that

$$|(3x - 2) - 4| = 3|x - 2| < 3(\varepsilon/3) = \varepsilon.$$

Thus we conclude that $\lim_{x \to 2} 3x - 2 = 4$.   □

Looking back at our previous proof of this fact, the two have many similarities and this one may actually be the shorter. However, don't give up on the sequential approach just yet. Next we wish to show that $\lim_{x \to 1} x^3 + 5 = 6$. In the previous proof, we were able to use our knowledge of sequences; though the proof here will be short, the ideas behind it are more involved than our previous examination. As before, the first step is to relate the quantities $|x - 1|$ and $|(x^3 + 5) - 6|$. Factoring,

$$|(x^3 + 5) - 6| = |x^3 - 1| = |x - 1||x^2 + x + 1|.$$

The first factor here is the quantity we will restrict, but we must also control the size of the second factor and an upper bound would be ideal. The size of this term depends on the value that x assumes, and *we control* the distance between x and 1. Consider what happens if we choose $\delta_1 = 1$. Then we will assume that $0 < |x - 1| < 1$, which guarantees that x is no greater than two. The triangle inequality and properties of the absolute value then guarantee us that

$$|x^2 + x + 1| \leq |x^2| + |x| + 1 = |x|^2 + |x| + 1 < 2^2 + 2 + 1 = 7.$$

Combining this with our previous calculation, we see that if $0 < |x - 1| < 1$, then

$$|(x^3 + 5) - 6| = |x^3 - 1| < 7|x - 1|.$$

This looks more like our previous example and we can force $7|x - 1| < \varepsilon$ by choosing $\delta_2 = \varepsilon/7$. So what is δ? Do we choose $\delta = 1$ or $\delta = \varepsilon/7$? When dealing with sequences, we often had to make multiple choices for N and then we took the most restrictive of these by choosing the maximum. The idea here is the same except that the most restrictive choice for a δ value is obtained by taking a minimum.

Proof. Let $\varepsilon > 0$ and choose $\delta = \min\{\varepsilon/7, 1\}$. Then if $x \in \mathbb{R}$ with $0 < |x - 1| < \delta$, it follows that $|x| = |x - 1 + 1| \leq |x - 1| + 1 < 2$, and thus

$$|(x^3 + 5) - 6| = |x^3 - 1| = |x - 1||x^2 + x + 1|$$
$$\leq |x - 1|(|x|^2 + |x| + 1)$$
$$< 7|x - 1|$$
$$< 7(\varepsilon/7) = \varepsilon.$$

Hence we conclude that $\lim_{x \to 1} x^3 + 5 = 6$. $\qquad\square$

As a final comment, keep in mind that while both the sequential and $\varepsilon - \delta$ definitions are equivalent, it is often the case that one allows for more efficiency than the other. We will make use of the sequential definition more so than the other due to our previous work with sequences. This will enable us to prove some statements rather quickly, but be aware that all such proofs could be modified to incorporate either definition. In short, neither approach is superior to the other.

Exercises

Exercise 4.1.1. Let $f : A \to \mathbb{R}$ and let c be a limit point of A. If $\lim_{x \to c} f(x)$ exists, show that it is unique.

Exercise 4.1.2. Use the sequential definition of a functional limit to prove each of the following limit statements. You may use the Algebraic Limit Theorem for sequences as we did in Example 4.1.3, but you should also attempt to prove each statement by explicitly using the definition of sequential convergence as we did in Example 4.1.2.

(a) $\lim_{x \to c} k = k$ where $c, k \in \mathbb{R}$
(b) $\lim_{x \to 3} 2x - 4 = 2$
(c) $\lim_{x \to 0} x^4 = 0$
(d) $\lim_{x \to 2} x^4 = 16$

Exercise 4.1.3. Use the $\varepsilon - \delta$ definition of a functional limit to prove each of the following statements.

(a) $\lim_{x \to c} k = k$ where $c, k \in \mathbb{R}$
(b) $\lim_{x \to 3} 2x - 4 = 2$
(c) $\lim_{x \to 0} x^4 = 0$
(d) $\lim_{x \to 2} x^4 = 16$

Exercise 4.1.4. (a) Supply a proof for parts (b) and (c) of Theorem 4.1.8.
(b) Supply a proof of part (d) of Theorem 4.1.8 by considering part (f) of Exercise 2.2.4.

Exercise 4.1.5. Let $f : A \to \mathbb{R}$ and $g : A \to \mathbb{R}$ and let c be a limit point of A. Suppose that $\lim_{x \to c} f(x)$ exists and that $\lim_{x \to c} g(x)$ does not exist.

(a) If $k \in \mathbb{R}$ is not 0, show that $\lim_{x \to c}[k g(x)]$ does not exist.
(b) Show that $\lim_{x \to c}[f(x) + g(x)]$ does not exist.
(c) Must it be the case that $\lim_{x \to c}[f(x)g(x)]$ does not exist?

Exercise 4.1.6 (Order Limit Theorem for Functional Limits). Let f and g be functions defined on a set A and assume $f(x) \le g(x)$ for all $x \in A$. Further, let c be a limit point of A and assume that the limits for f and g exist at c. Show that

$$\lim_{x \to c} f(x) \le \lim_{x \to c} g(x).$$

Exercise 4.1.7 (Squeeze Theorem for Functional Limits). Let f, g, and h be functions defined on a set A and assume that $f(x) \le g(x) \le h(x)$ for all $x \in A$. Also, let c be a limit point of A and suppose the $\lim_{x \to c} f(x) = L = \lim_{x \to c} h(x)$. Show that $\lim_{x \to c} g(x) = L$. Explain why Exercise 4.1.6 is not applicable.

Exercise 4.1.8. Let $f : A \to \mathbb{R}$ and let c be a limit point of A.

(a) If $\lim_{x \to c} f(x) = L$, show that $\lim_{x \to c} |f(x)| = |L|$.
(b) Show that it is possible for $\lim_{x \to c} |f(x)|$ to exists even if $\lim_{x \to c} f(x)$ does not exist.

Exercise 4.1.9. Use the divergence criteria to supply a proof for each of the following.

(a) The function $f(x) = |x|/x$ does not have a limit at $x = 0$.
(b) The function $f(x) = (x - 1)/(x - 3)$ does not have a limit at $x = 3$.
(c) The function $f(x) = 1/(x^4 + x^2)$ diverges to infinity at $x = 0$.

Exercise 4.1.10. The function defined by

$$t(x) = \begin{cases} 1, & x = 0 \\ 1/n, & x = m/n \text{ in lowest terms with } n > 0; \\ 0, & x \notin \mathbb{Q}. \end{cases}$$

is typically referred to as Thomae's function. In this exercise we will explore $\lim_{x \to 1} t(x)$.

(a) Choose a sequence (x_n) which converges to 1 but does not contain 1 as a term in the sequence and find the limit of the sequence $(t(x_n))$.
(b) Construct two different sequences and repeat part (a). Make a conjecture for the value of $\lim_{x \to 1} t(x)$.
(c) To prove the claim from part (b), first consider the following. Let (x_n) be a sequence of rational numbers converging to 1 which does not contain 1 as a term, and use Exercise 1.4.5 to show that the limit of the sequence $(t(x_n))$ must be 0.
(d) Now, let (x_n) be a sequence of irrational numbers converging to 1 and show that $\lim_{n \to \infty} t(x_n) = 0$.
(e) Finally, to piece all of this together, let (x_n) be any sequence converging to 1 which does not contain 1 as a term. Use parts (c) and (d), to show that $\lim_{x \to 1} t(x) = 0$.

Exercise 4.1.11. (a) In Example 4.1.5 we showed that $\lim_{x \to 0} 1/x^2$ does not exist. Continue this line of reasoning and show that $\lim_{x \to 0} 1/x^2 = \infty$.
(b) Give a definition to describe the statement $\lim_{x \to c} f(x) = -\infty$.

Exercise 4.1.12. Let $f : A \to \mathbb{R}$ and let c be a limit point of A. The *right-hand limit* (or *limit from above*) of f at c, denoted by the symbol $\lim_{x \to c+} f(x)$ exists and is equal to L if for every sequence (x_n) in A with $x_n > c$ for all $n \in \mathbb{N}$ and $\lim x_n = c$, it follows that $\lim f(x_n) = L$.

(a) Show that the function $x/|x|$ has right-hand limit 1 at $c = 0$.
(b) Give a definition for a left-hand limit.

(c) Show that the function $x/|x|$ has left-hand limit -1 at $c = 0$.
(d) Show that $\lim_{x \to c} f(x) = L$ if and only if both one-sided limits exist and satisfy $\lim_{x \to c+} f(x) = \lim_{x \to c-} f(x)$.

Exercise 4.1.13. (a) Mimic Definition 4.1.6 to supply a definition for the statement $\lim_{x \to c+} f(x) = \infty$. Use this to show that $\lim_{x \to 0+} 1/x = \infty$.
(b) Provide definitions for the other possible infinite limits,

$$\lim_{x \to c+} f(x) = -\infty, \quad \lim_{x \to c-} f(x) = \infty, \quad \lim_{x \to c-} f(x) = -\infty.$$

(c) Show that $\lim_{x \to 0-} 1/x = -\infty$.
(d) We say that a function f has a *vertical asymptote* at c if either of the one-sided limits diverge to ∞ or $-\infty$. Show that every function of the form $f(x) = 1/x^n$, where $n \in \mathbb{N}$, has a vertical asymptote at 0.

4.2 Continuity

Definition 4.2.1. Let $f : A \to \mathbb{R}$ and let $c \in A$. We say f is *continuous at* c if for every sequence (x_n) in A with $\lim x_n = c$, it follows that $\lim f(x_n) = f(c)$.

The definition does bear some resemblance to the definition of a functional limit but the differences are substantial. Here we require c to be in A, not just a limit point, so that $f(c)$ actually exists and this value then plays the role of limit. The functional limit attempts to predict behavior of the function near c, while the notion of continuity seeks to identify points where behavior *near* the point matches what is happening *at* the point.

Another substantial difference is the fact that the terms in the domain sequence (x_n) are now allowed to assume the value c. This is permissible since we have also insisted that c be in A so that $f(c)$ exists. Supposing $x_n = c$ for some n, observe that the quantity $|f(x_n) - f(c)| = |f(c) - f(c)| = 0$. Thus this allowance causes no further complications in showing that $f(x_n)$ converges to $f(c)$. To be clear we have the following statement.

Proposition 4.2.2. Let $f : A \to \mathbb{R}$ and let $c \in A$.

(a) If c is an isolated point of A, then f is continuous at c.
(b) If c is a limit point of A, then f is continuous at c if and only if $\lim_{x \to c} f(x) = f(c)$.

Proof. For the statement in (a), consider what it means to be an isolated point. In other words, supposing c is isolated in A, what are the sequences in A which converge to c? The only choice is the constant sequence (c) since in this case c is *not* a limit point of A. It is then immediate that the range sequence $(f(c))$ is also constant converging to $f(c)$. Therefore f is continuous at any isolated point.

For (b), first assume that f is continuous at c, a limit point of A. To show that $\lim_{x \to c} f(x) = f(c)$, take an arbitrary sequence (x_n) in A with $x_n \neq c$ for all $n \in \mathbb{N}$ and $\lim x_n = c$. To show that $(f(x_n)) \to f(c)$, we simply observe that the sequence (x_n) is in A and converges to c. The definition of continuity then implies that $(f(x_n))$ converges to $f(c)$ as desired. In other words, the definition of continuity considers a larger class of domain sequences since it takes all sequences converging to c into account whereas the functional limit only considers those which do not contain c as a term.

For the reverse direction, we assume that $c \in A$ is a limit point with $\lim_{x \to c} f(x) = f(c)$. To show that f is continuous, let (x_n) be a sequence in A with $\lim x_n = c$. To show that $(f(x_n)) \to f(c)$, we consider two cases in order to deal with the fact that the functional limit only considers sequences which do not contain c as a term. Let $\varepsilon > 0$. First, if only *finitely* many of the terms of (x_n) are *not* equal to c, then choose $N \in \mathbb{N}$ so that $x_n = c$ for all $n \geq N$. It follows then that $|f(x_n) - f(c)| = |f(c) - f(c)| = 0 < \varepsilon$ for all $n \geq N$ and thus $(f(x_n)) \to f(c)$.

Next consider the case $x_n \neq c$ for infinitely many terms in the sequence (x_n). Take (x_{n_j}) to be the subsequence of (x_n) consisting of all terms not equal to c. This subsequence certainly converges to c, and hence $\lim_{j \to \infty} f(x_{n_j}) = f(c)$ since none of the terms in this subsequence are equal to c. With this, we may choose $J \in \mathbb{N}$ so that $|f(x_{n_j}) - f(c)| < \varepsilon$ for all $j \geq J$. If we now consider $n \in \mathbb{N}$ with $n \geq n_J$ (playing the role of N), then either $x_n \neq c$ or $x_n = c$. If $x_n \neq c$, then $x_n = x_{n_j}$ for some $j \in \mathbb{N}$. Notice that this implies that $n_j \geq n_J$ which in turn implies that $j \geq J$ since subsequences are defined by increasing sequences of natural numbers. Hence

$$|f(x_n) - f(c)| = |f(x_{n_j}) - f(c)| < \varepsilon.$$

On the other hand, if $x_n = c$, then

$$|f(x_n) - f(c)| = |f(c) - f(c)| = 0 < \varepsilon.$$

Combining these two statements it follows that $|f(x_n) - f(c)| < \varepsilon$ for all $n \geq n_J$ and thus $(f(x_n)) \to f(c)$ showing that f is continuous at c. $\qquad \square$

Before considering examples, we will often speak of functions which are continuous on subsets of $\mathbb{R}$. To be explicit with our meaning here, let $f : A \to \mathbb{R}$ and let $B \subseteq A$. We say f is *continuous on the set B* if f is continuous at each point of B.

Example 4.2.3. Rational functions play an important role in calculus due to the fact that they provide more interesting behavior than polynomials even though they are simply algebraic combinations of polynomials. For an example here, let's take $f(x) = (x^2 - x - 2)/(x - 2)$. The domain of this function is $A = (-\infty, 2) \cup (2, \infty)$ and if we consider any point $c \in A$, then we can show the function is continuous there. Indeed, if (x_n) is a sequence in A with $(x_n) \to c$, then we can apply the Algebraic Limit Theorem for sequences to see that

$$(f(x_n)) = \left(\frac{x_n^2 - x_n - 2}{x_n - 2} \right) \to \frac{c^2 - c - 2}{c - 2} = f(c),$$

since the denominator is never zero on A. It may seem quite natural to factor the numerator and cancel the factors of $x - 2$ which is allowable since we have acknowledged our domain. However, it should be pointed out that this function is only equal to the function $g(x) = x + 1$ on A; in other words, the graph of f looks like a line *everywhere but* 2.

Let's focus our attention now on $c = 2$. Since this point is not in the domain of f, we cannot apply the definition of continuity, though we would also not call 2 a discontinuity; we'll reserve that term for points in the domain. To remedy this problem, take h to be defined by

$$h(x) = \begin{cases} \dfrac{x^2 - x - 2}{x - 2}, & x \neq 2; \\ 3, & x = 2. \end{cases}$$

Though it is currently written in a somewhat complicated manner, this is a rather simple function. To be more specific, if x is not 2, then h is the linear function g defined above. Furthermore, at $x = 2$, h and g have the same value. Thus $h(x) = g(x)$ for every point $x \in \mathbb{R}$. In order to show h is continuous on $\mathbb{R}$, it therefore suffices to show that $g(x) = x + 1$ is continuous on $\mathbb{R}$. Let $c \in \mathbb{R}$ and let (x_n) be any sequence in $\mathbb{R}$ which converges to c. Then $g(x_n) = x_n + 1 \to c + 1 = g(c)$. This function takes the function f and extends its domain to $\mathbb{R}$ in a way that preserves the continuity of f.

To see that the value of 3 was not a random choice in the definition of the function h, consider now the new function

$$j(x) = \begin{cases} \dfrac{x^2 - x - 2}{x - 2}, & x \neq 2; \\ 1, & x = 2. \end{cases}$$

This function is defined on all of $\mathbb{R}$ and is continuous everywhere except 2. To investigate the behavior at 2, we appeal to Proposition 4.2.2 and let (x_n) be any sequence in $\mathbb{R}$ with $x_n \neq 2$ for all n and $\lim x_n = 2$. Then, we proceed as we did in the case of the function f but by using part (b) of the Algebraic Limit Theorem for sequences after factoring,

$$(j(x_n)) = \left(\frac{x_n^2 - x_n - 2}{x_n - 2} \right) = (x_n + 1) \to 3 \neq 1 = j(2);$$

in this calculation part (d) of the Algebraic Limit Theorem for sequences isn't applicable, but we can still calculate the limit due to the algebraic structures particular to this function.

In the first-year calculus sequence it is common to refer to non-domain points of certain functions as discontinuities but the use of the word in this context is somewhat loose. For example, a student feels confident in saying that $f(x) = 1/x$

has a discontinuity at $x = 0$ but would not say that $g(x) = \sqrt{x}$ has a discontinuity at $x = -1$. This seems to be due to the fact that zero is a limit point of the domain of f while -1 is not a limit point of the domain of g even though students at that level have not yet been introduced to the limit point concept. To erase this ambiguity, we will be quite specific in using the word discontinuity for *domain points* for which the limit does not exist or does not match the value of the function at the point. The following theorem will act as our definition for discontinuities.

Theorem 4.2.4. *Let $f : A \to \mathbb{R}$ and let $c \in A$. Then f has a discontinuity at c if there is a sequence (x_n) in A with $\lim x_n = c$ for which $(f(x_n))$ does not converge to $f(c)$.*

Example 4.2.5. For another example, we consider an extension of the function $g(x)$ from Example 4.1.5. Let

$$f(x) = \begin{cases} x \sin\left(\dfrac{1}{x}\right), & x \neq 0; \\ 0, & x = 0. \end{cases}$$

We can show that this function is continuous at zero whereas its predecessor did not have a limit at zero. To verify this, let (x_n) be a sequence in $\mathbb{R}$ with $x_n \neq 0$ for all n and $\lim x_n = 0$. We must now show that the sequence $(f(x_n))$ converges to $f(0)$. Let $\varepsilon > 0$ and choose $N \in \mathbb{N}$ so that $|x_n - 0| < \varepsilon$ for all $n \geq N$. Then, if $n \geq N$,

$$|f(x_n) - f(0)| = |x_n \sin(1/x_n) - 0| = |x_n \sin(1/x_n)| \leq |x_n| < \varepsilon$$

since $|\sin x|$ is bounded by 1 for all $x \in \mathbb{R}$. This guarantees us that $(f(x_n)) \to 0$ which is the desired value since $f(0) = 0$.

For another proof of this fact we can use a Squeeze Theorem argument to show that $(f(x_n))$ converges to zero if $(x_n) \to 0$. Notice that the bound $0 \leq |\sin x| \leq 1$ implies that

$$0 \leq \left| x_n \sin\left(\dfrac{1}{x_n}\right) \right| \leq |x_n|$$

for all n. Thus we see immediately that $(f(x_n)) \to 0$ as $n \to \infty$. Figure 4.4 demonstrates the squeezing behavior.

For a second example, consider a modified form of the Dirichlet function from Sect. 4.1,

$$g(x) = \begin{cases} x, & x \in \mathbb{Q}; \\ 0, & x \notin \mathbb{Q}. \end{cases}$$

Just as the Dirichlet function possessed an interesting property in terms of limits, this function has the property that it is continuous at *exactly* one point, namely zero. In the exercises you will verify that it is discontinuous at all other points in $\mathbb{R}$.

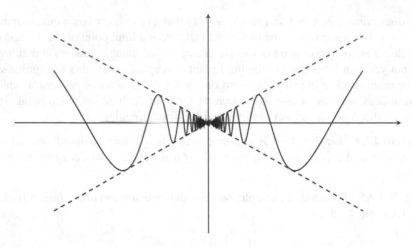

Fig. 4.4 Squeezing $x \sin(1/x)$

Let (x_n) be a sequence with $\lim x_n = 0$. To conclude that f is continuous at 0, we must show that the sequence $(g(x_n))$ converges to $g(0) = 0$. Let $\varepsilon > 0$ be given. We will need to consider two cases based on the rational and irrational numbers in the specified sequence. First, if only finitely many of the terms are rational, we may choose $N \in \mathbb{N}$ so that x_n is irrational for all $n \geq N$, in which case $|g(x_n) - g(0)| = |0 - 0| = 0 < \varepsilon$ for all $n \geq N$ by the definition of g.

For the more substantial case, suppose there are infinitely many rational terms in our sequence and let (x_{n_j}) be the subsequence consisting of all the rational terms. Since (x_n) converges to zero, we know that this subsequence also converges to zero and thus we may choose $J \in \mathbb{N}$ such that $|x_{n_j} - 0| < \varepsilon$ for all $j \geq J$. Now consider an arbitrary $n \geq n_J$ (n_J is playing the role usually played by N). If x_n is rational, then $x_n = x_{n_j}$ for some $j \in \mathbb{N}$ with $n_j \geq n_J$; as in the proof of Theorem 4.2.2, this implies that $j \geq J$ since subsequences are defined by increasing sequences of natural numbers, and thus

$$|g(x_n) - g(0)| = |g(x_{n_j}) - g(0)| = |x_{n_j} - 0| < \varepsilon.$$

On the other hand, if x_n is irrational, then $|g(x_n) - g(0)| = 0 < \varepsilon$. With these two estimates, we have that $|g(x_n) - g(0)| < \varepsilon$ for all $n \geq n_J$ and thus $(g(x_n))$ converges to $g(0)$. In either case we have that $(g(x_n)) \to g(0)$ and hence g is continuous at 0.

Thomae's function defined in Exercise 4.1.10 is an example of a function which is continuous at every irrational number and discontinuous at every rational number; the discontinuity at 1 is the result of the mentioned exercise. For more functions of this variety, we point the interested reader to the article [11].

Theorem 4.2.6 (Algebraic Limit Theorem for Continuous Functions). *Let* $f :$ $A \to \mathbb{R}$ *and* $g : A \to \mathbb{R}$ *and let* $c \in A$. *If* f *and* g *are continuous at* c, *then:*

(a) kf *is continuous at* c *for every* $k \in \mathbb{R}$;

(b) $f + g$ *is continuous at c;*

(c) fg *is continuous at c;*

(d) $\dfrac{f}{g}$ *is continuous at c if $g(c) \neq 0$.*

Proof. Using the sequential characterization of functional limit and Proposition 4.2.2, the proof follows almost immediately from Theorem 4.1.8.   □

Theorem 4.2.7. *Let $f : A \to \mathbb{R}$ and $g : B \to \mathbb{R}$ be functions with $f(A) \subseteq B$ so that $g \circ f : A \to \mathbb{R}$ is defined and let $c \in A$. If f is continuous at c and g is continuous at $f(c)$, then $g \circ f$ is continuous at c.*

Proof. Exercise 4.2.8 asks you to prove this using the sequential characterization of continuity and we will supply a different proof following Theorem 4.2.8.   □

As with functional limits, we can characterize continuity at a point in terms of ε-neighborhoods.

Theorem 4.2.8. *Let $f : A \to \mathbb{R}$ and let $c \in A$. Then f is continuous at c if and only if for every $\varepsilon > 0$, there is a $\delta > 0$ such that if $|x - c| < \delta$ and $x \in A$, it follows that $|f(x) - f(c)| < \varepsilon$.*

Proof. The proof here is nearly identical to that of Theorem 4.1.10 and we will allow the reader the chance to supply the details.   □

Example 4.2.9. While our focus in the last two sections has relied heavily on the notion of sequential limits, the theorem above is usually taken as the standard definition for continuity at a point. The two ideas are logically equivalent though again there are instances where one is simpler to use than the other. Consider the function f from Example 4.2.5. To show that this function is continuous at zero, let $\varepsilon > 0$ and set $\delta = \varepsilon$. By Theorem 4.2.8, we need to consider $x \in \mathbb{R}$ with $0 < |x - 0| < \delta$. We then see that

$$|f(x) - f(0)| = |x \sin(1/x) - 0| \leq |x| < \delta = \varepsilon$$

and therefore conclude that f is continuous at zero.

The proof is not necessarily simpler, only a bit shorter. By using the sequential definitions, we are able to exploit the time we spent investigating sequences. In fact, many of the statements we derived there were continuity statements in disguise! One benefit of exploring examples with both definitions is to see the similarities, or dissimilarities, of the techniques involved. This activity is meant to help build intuition and, when faced with a limit statement, help us in answering the question— will it be more workload efficient to work with a sequential or an $\varepsilon - \delta$ definition? Exercise 4.2.5 will ask you to use this new characterization to show that the modified Dirichlet function is continuous at zero.

For another example, let's use this new characterization to prove Theorem 4.2.7.

Proof. Considering the hypothesis of the theorem, for $\varepsilon > 0$, we must find $\delta > 0$ so that if $|x - c| < \delta$ and $x \in A$, it follows that $|g(f(x)) - g(f(c))| < \varepsilon$. One complication here is that we will need to think of the points $f(x)$ and $f(c)$ not only just as points in the range of f but also as points in the domain of g. Since g is continuous at $f(c)$, for $\varepsilon > 0$ we know there is a $\delta_1 > 0$ so that if $|y - f(c)| < \delta_1$ and $y \in B$, it follows that $|g(y) - g(f(c))| < \varepsilon$. Moreover, since f is continuous at c, we also know that there is a $\delta_2 > 0$ so that if $|x - c| < \delta_2$ and $x \in A$, it must be the case that $|f(x) - f(c)| < \delta_1$. Combining these two statements, if $|x - c| < \delta_2$ and $x \in A$, then $f(x) = y \in B$ by hypothesis and $|y - f(c)| = |f(x) - f(c)| < \delta_1$. This immediately implies that $|g(f(x)) - g(f(c))| = |g(y) - g(f(c))| < \varepsilon$. Therefore we conclude that $g \circ f$ is continuous at c. $\qquad\square$

Example 4.2.10. Let $f : \mathbb{R} \to \mathbb{R}$. We say f is an *additive function* if $f(x + y) = f(x) + f(y)$ for every $x, y \in \mathbb{R}$. An example of such a function is $f(x) = 3x$,

$$f(x + y) = 3(x + y) = 3x + 3y = f(x) + f(y).$$

Functions with this property have the interesting quality that continuity at any single point implies continuity at all other points. To see this, suppose we have an additive function that is continuous at 0. In order to show that this function is continuous at every other real number, let $c \in \mathbb{R}$ and let (x_n) be a sequence in $\mathbb{R}$ which converges to c. We must now show that the sequence $(f(x_n))$ converges to $f(c)$. To use the continuity at 0, we construct a new sequence $y_n = x_n - c$. Notice that this sequence converges to 0 by construction and thus, since f is continuous at 0, it must be the case that $(f(y_n)) \to f(0)$. Part (a) of Exercise 4.2.9 asks you to show that $f(0) = 0$ for any additive function, and thus we have that $(f(y_n)) \to 0$. By definition of y_n, this is equivalent to saying that $(f(x_n - c)) \to 0$. Now, turning our attention back to the sequence $(f(x_n))$ and using the additive property of f, we have

$$f(x_n) = f(x_n - c + c) = f(x_n - c) + f(c)$$

which converges to $f(c)$.

Continuity and Set Properties

Suppose we have a set $A \subseteq \mathbb{R}$ with a certain property, e.g. an interval, an open, closed, or compact set, etc. Given a function f, it is natural then to ask if the set $f(A) = \{f(x) : x \in A\}$, the set of all range values corresponding to domain values in A, has the same property. For a general function f, it seems highly unlikely that anything substantial can be concluded due to the fact that functions come in so many varieties. It is therefore not unreasonable to assume that f also has some desirable property. For the purpose of this discussion, we will assume that f is continuous on A.

For what appears to be the simplest of situations, assume that A is an interval. Is it true then that $f(A)$ is also an interval? The answer is yes; however, the proof of this fact is delayed until Corollary 4.3.8. The result should be familiar as it is logically equivalent to the Intermediate Value Theorem. What happens in the case when A is open? Again, assuming f is continuous, must it be the case that $f(A)$ is also open? Unfortunately this is not true. Take, for example, the function $f(x) = \sin x$ on the interval $(0, \infty)$. Computing, or considering the graph, we find that $f(A) = [-1, 1]$, a set that is not open.

Next consider the case when A is closed. Taking into account the relationship between open and closed sets and the previous example, your intuition should indicate that $f(A)$ may not necessarily be closed. This is correct, but if we try to modify our previous example to $A = [0, \infty]$ we conclude that $f(A) = [-1, 1]$, a set that is closed. Thinking a bit more creatively, consider the function $g(x) = 1/x$ on the closed interval $A = [1, \infty)$. Here we find that $f(A) = (0, 1]$ and this set is not closed. Exercise 4.2.11 asks you to show that the situation is similar for a bounded set.

The surprising fact then is that we do obtain an affirmative answer in the case of compact sets, that is, if we require the set to be *both* closed and bounded. There is no similar result for sets which are bounded and open.

Theorem 4.2.11. *Let K be a compact subset of $\mathbb{R}$ and suppose $f : K \to \mathbb{R}$ is continuous. Then $f(K)$ is compact.*

We will work with the sequential characterization of compactness since this will pair nicely with our sequential characterization of continuity. The key then will be to relate sequences in the range set $f(K)$ with sequences in the domain K and continuity will provide the link between their respective limits.

Proof. Let $K \subseteq \mathbb{R}$ be compact and let $f : K \to \mathbb{R}$ be continuous. To show $f(K)$ is a compact set, we take an arbitrary sequence (y_n) in $f(K)$. Now we must produce a subsequence (y_{n_j}) which converges to a point in $f(K)$. By definition, we can connect the sequence (y_n) to a sequence in K, call it (x_n), via the relationship $f(x_n) = y_n$. Using the fact that K is compact, we know there exists a subsequence (x_{n_j}) of (x_n) which has a limit $x \in K$. Moreover, the continuity of f at $x \in K$ implies that $(f(x_{n_j})) \to f(x)$. From the fact that $y_{n_j} = f(x_{n_j})$, we deduce that the subsequence (y_{n_j}) also converges to $f(x)$ which is in $f(K)$. Thus $f(K)$ is a compact set. $\qquad\square$

A simple consequence of this theorem is a result typically referred to as the Extreme Value Theorem; it is often used in the application of optimization problems in the first-year calculus sequence but is also a powerful theoretical tool.

Theorem 4.2.12 (Extreme Value Theorem). *Let K be a compact subset of $\mathbb{R}$ and suppose $f : K \to \mathbb{R}$ is continuous. Then there exist $c, d \in K$ such that $f(c) \le f(x) \le f(d)$ for all $x \in K$. In other words, f attains a maximum and a minimum on K.*

Proof. Let K be a compact subset of $\mathbb{R}$ and suppose $f : K \to \mathbb{R}$ is continuous. We will first produce a point $d \in K$ such that $f(d) \geq f(x)$ for every $x \in K$. To do this, we will show that there is a point $d \in K$ such that $f(d)$ is the supremum of $f(K)$. The previous theorem implies that $f(K)$ is compact, and hence bounded. Thus the supremum exists and we set $\beta = \sup f(K)$. To produce the desired domain point d, we define a sequence (y_n) in $f(K)$ as follows. For each $n \in \mathbb{N}$, choose a point $y_n \in f(K)$ such that $\beta - 1/n < y_n \leq \beta$ which is permissible by Lemma 1.2.10. By the Squeeze Theorem, it follows that $(y_n) \to \beta$. Since $y_n \in f(K)$, there is also a sequence (x_n) in K defined by the relationship $f(x_n) = y_n$. The fact that K is compact now implies that (x_n) has a subsequence (x_{n_j}) which converges to a point $d \in K$. The continuity of f further implies that $(y_{n_j}) = (f(x_{n_j})) \to f(d)$. However, the sequence (y_{n_j}) must also converge to β and hence $f(d) = \beta$. We leave the proof that f also attains a minimum as Exercise 4.2.12.   □

As a matter of terminology, we are using the terms maximum and minimum here since the supremum and infimum are actually elements of the set in question, which we know is not always the case. For compact sets, we showed in Exercise 2.4.8 this is always true and the exercises ask you to use this fact to produce a much shorter proof of the Extreme Value Theorem. Our point with the proof above was to provide another example of how to use continuity in relating sequential limits in the domain and range of a function.

Exercises

Exercise 4.2.1. (a) Show that if f is continuous on $\mathbb{R}$, then $|f|$ is also continuous.
(b) Is the converse true? Why?

Exercise 4.2.2. (a) Use the definition of continuity to show that $f(x) = x$ and $g(x) = k$, where $k \in \mathbb{R}$, are continuous on $\mathbb{R}$.
(b) Show that for each $n \in \mathbb{N}$ the function $f(x) = x^n$ is continuous on $\mathbb{R}$.
(c) Show that any polynomial is continuous on $\mathbb{R}$.
(d) Show that any rational function is continuous on its domain, i.e. the set of points where the denominator is not zero.

Exercise 4.2.3. (a) Use your knowledge of sequences to show that the function $f(x) = \sqrt{x}$ is continuous on $[0, \infty)$.
(b) Show that $g(x) = \sqrt[3]{x}$ is continuous at $c = 0$.
(c) Show that $g(x) = \sqrt[3]{x}$ is continuous at all nonzero real numbers. You may find the factorization identity $a^3 - b^3 = (a - b)(a^2 + ab + b^2)$ useful whether using the sequential or $\varepsilon - \delta$ characterization of continuity.

Exercise 4.2.4. (a) Let $f : A \to \mathbb{R}$ and let $c \in A$. Show that f is continuous at c if and only if $\lim_{x \to c}[f(x) - f(c)] = 0$.
(b) Let $f : A \to \mathbb{R}$ and let $c \in A$. Show that f is continuous at c if and only if $\lim_{h \to 0} f(c + h) = f(c)$.

Exercise 4.2.5. (a) Show that the modified Dirichlet function from Example 4.2.5 is discontinuous at every nonzero real number.

(b) Show that the modified Dirichlet function is continuous at zero using $\varepsilon - \delta$ characterization of continuity.

Exercise 4.2.6. (a) Suppose $f : \mathbb{R} \to \mathbb{R}$ is continuous on $\mathbb{R}$ and let $K = \{x : f(x) = 0\}$. Show that K is a closed set.

(b) Does the previous result still hold if we set $K = \{x : f(x) = k\}$ where $k \in \mathbb{R}$ with $k \neq 0$?

(c) Suppose $f : \mathbb{R} \to \mathbb{R}$ is continuous and satisfies $f(r) = 0$ for every $r \in \mathbb{Q}$. Show that $f(x) = 0$ on all of $\mathbb{R}$.

Exercise 4.2.7. Supply a proof of Theorem 4.2.6.

Exercise 4.2.8. Use the sequential characterization of continuity to prove Theorem 4.2.7.

Exercise 4.2.9. Suppose $f : \mathbb{R} \to \mathbb{R}$ is an additive function which is continuous at 0. This exercise will show that any such function has the form $f(x) = kx$ for some $k \in \mathbb{R}$.

(a) Show that $f(0) = 0$ and that $f(-x) = -f(x)$ for each $x \in \mathbb{R}$.

(b) In Example 4.2.10 we used the sequential characterization of continuity to show that any additive function which is continuous at 0 is necessarily continuous on all of $\mathbb{R}$. Now use the $\varepsilon - \delta$ characterization to prove this same fact.

(c) Use induction to show that $f(nx) = nf(x)$ for every $n \subset \mathbb{N}$ and every $x \in \mathbb{R}$.

(d) Let $f(1) = k$. Use part (c) to show that $f(n) = kn$ for every $n \in \mathbb{N}$. Use this information to show that $f(1/n) = k(1/n)$ for every $n \subset \mathbb{N}$.

(e) Show that $f(z) = kz$ for every $z \in \mathbb{Z}$.

(f) Show that $f(r) = kr$ for every $r \in \mathbb{Q}$.

(g) Finally, show $f(x) = kx$ for every $x \in \mathbb{R}$.

Exercise 4.2.10. Suppose $f : \mathbb{R} \to \mathbb{R}$ is continuous on $[0, \infty)$ and suppose also that f is an odd function, i.e. $f(-x) = -f(x)$ for all $x \in \mathbb{R}$.

(a) Show that $f(0) = 0$.

(b) Show that f is also continuous on $(-\infty, 0)$.

Exercise 4.2.11. (a) Give an example of a continuous function f and a bounded set A such that $f(A)$ is not bounded.

(b) Give an example of a continuous function f and an open, bounded set A such that $f(A)$ is not bounded.

(c) Give an example of a continuous function f and an open, bounded set A such that $f(A)$ is not open.

(d) Give an example of a continuous function f and an open, bounded set A such that $f(A)$ is neither open nor bounded.

Exercise 4.2.12. (a) Complete the proof of the Extreme Value Theorem by showing that f also attains a minimum.

(b) Supply a shorter proof for the Extreme Value Theorem by using Exercise 2.4.8.

Exercise 4.2.13. The goal of this exercise is to classify discontinuities. Considering the definition and Exercise 4.1.12, there are three distinct possibilities. Let $f : A \to \mathbb{R}$ and let $c \in A$.

(a) We say f has a *removable discontinuity* at c if $\lim_{x \to c} f(x) = L$ and $L \neq f(c)$.

(b) We say f has a *jump discontinuity* at c if both the left- and right-hand limits exist but $\lim_{x \to c^-} f(x) \neq \lim_{x \to c^+} f(x)$.

(c) We say f has an *essential discontinuity* at c if either of the one-sided limits does not exist.

Another common classification says f has a discontinuity of the *first kind* at c if either (1) or (2) occur, and of the *second kind* if it has an essential discontinuity.

(a) Show that the Dirichlet function has an essential discontinuity at each point of $\mathbb{R}$.

(b) Show that the function

$$f(x) = \begin{cases} \dfrac{x^2 - 1}{x + 1}, & x \neq -1; \\ 0, & x = -1 \end{cases}$$

has a removable discontinuity at $x = -1$.

(c) Provide an example of a function with a removable discontinuity at each natural number.

(d) Provide an example of a function with a jump discontinuity at each natural number.

(e) Provide an example of a function with an essential discontinuity at each natural number.

Exercise 4.2.14. Let $I \subseteq [a, b]$ be an open interval and $f : [a, b] \to \mathbb{R}$. We define the *oscillation* of f on I by

$$\mathrm{osc}(f, I) = \sup\{f(x) : x \in I\} - \inf\{f(x) : x \in I\}.$$

(a) Compute $\mathrm{osc}(x, [-1, 3])$, $\mathrm{osc}(x^2, [-1, 3])$, and $\mathrm{osc}(\sin x, [0, 2\pi])$.

(b) Show that $\mathrm{osc}(f, I) \geq 0$.

(c) Show that $\mathrm{osc}(f, I) = \sup\{|f(x) - f(y)| : x, y \in I\}$.

Exercise 4.2.15. Let $f : [a, b] \to \mathbb{R}$. For $c \in (a, b)$ we define the oscillation of f at c by

$$\mathrm{osc}(f, c) = \inf\{\mathrm{osc}(f, I) : I \text{ is an open interval containing } c\}.$$

(a) Compute $\mathrm{osc}(x, 1)$, $\mathrm{osc}(x^2, 2)$, $\mathrm{osc}(g, 0)$ where $g(x) = \sin(1/x)$ for $x \neq 0$ and $g(0) = 0$, and $\mathrm{osc}(f, 0)$ where f is the Dirichlet function.

(b) Show that $\mathrm{osc}(f, c) \geq 0$ for every $c \in (a, b)$.

(c) Show that f is continuous at c if and only if $\mathrm{osc}(f, c) = 0$.

4.3   The Intermediate Value Theorem

The Intermediate Value Theorem (Fig. 4.5) is one of the most important theorems in the study of continuous functions defined on intervals. If you recall, sign charts are often used in calculus to identify where a derivative is positive or negative in order to understand where a function is increasing and decreasing. The construction of the sign chart itself is simply an application of this theorem. We will explore an example after stating and proving the theorem in order to make this connection more explicit.

Theorem 4.3.1 (Intermediate Value Theorem). *Let $f : [a,b] \to \mathbb{R}$ be continuous. If L is a real number satisfying $f(a) < L < f(b)$ or $f(a) > L > f(b)$, then there exists a real number $c \in (a,b)$ with $f(c) = L$.*

Proof. We will prove the theorem first for the special case when $L = 0$ and so we assume that $f(a) < 0 < f(b)$. This form of the theorem states that if a continuous function attains at least one positive and one negative value, then the function must be zero at some point in between the points where the function has opposing signs. Our proof technique here will resemble that used to prove the Bolzano–Weierstrass Theorem. To begin, set $a_1 = a$ and $b_1 = b$ and let M denote the length of $[a_1, b_1]$. Now consider the midpoint of this interval, $c_1 = (b_1 + a_1)/2$. If $f(c_1) < 0$, set $a_2 = c_1$ and $b_2 = b_1$. On the other hand, if $f(c_1) \geq 0$, set $a_2 = a_1$ and $b_2 = c_1$. Notice that in either case, $f(a_2) < 0$ and $f(b_2) \geq 0$ and the new interval $[a_2, b_2]$ is contained in $[a_1, b_1]$ with length $M/2$.

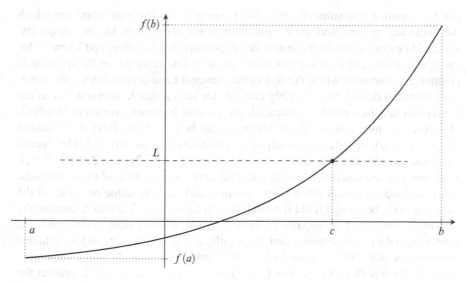

Fig. 4.5 Intermediate Value Theorem

Continuing this process, we construct two sequences (a_n) and (b_n) with $f(a_n) < 0$ for all n, $f(b_n) \geq 0$ for all n, and $[a_1, b_1] \supseteq [a_2, b_2] \supseteq [a_3, b_3] \supseteq \ldots$. Moreover, the length of $[a_n, b_n]$ is $M/2^{n-1}$. Applying Exercise 2.1.7, the intersection of these closed, bounded intervals contains a unique point c. And, since $(b_n - a_n) \to 0$, it must be the case that both sequences converge to c; the Monotone Convergence Theorem guarantees us that both sequences do in fact converge. By continuity, we have $(f(a_n)) \to f(c)$ and $(f(b_n)) \to f(c)$. The Order Limit Theorem now indicates that $f(c) \leq 0$ since $f(a_n) < 0$ for all n, and $f(c) \geq 0$ since $f(b_n) \geq 0$ for all n. This immediately implies that $f(c) = 0$.

For the more general statement, consider the new function $g(x) = f(x) - L$. The inequality $f(a) < L < f(b)$ can be rewritten as $f(a) - L < 0 < f(b) - L$, which implies that $g(a) < 0 < g(b)$. Furthermore, g is continuous by construction, and thus the previous case asserts the existence of a point $c \in (a, b)$ with $g(c) = 0$. From the definition of g we conclude that $f(c) = L$. $\square$

In the exercises you will consider several problems asking you to apply the Intermediate Value Theorem; the technique used in the general case will be a helpful thing to consider there.

Example 4.3.2. Solving equations and inequalities is a very important mathematical skill. Solutions to equations are usually found via algebraic manipulation while solving inequalities is often a more complicated task. To illustrate, suppose we are asked to find the solutions of

$$\frac{x+2}{x-3} = 0 \qquad \text{and} \qquad \frac{x+2}{x-3} > 0.$$

For the equation, the solution set is $\{-2\}$, since this is the only point for which the numerator is zero, and the denominator is not zero. To tackle the inequality, we could determine where the numerator is positive and negative, and likewise for the denominator, and then use this information to determine where the quotient is positive, i.e. determine where the sign of the numerator and denominator is the same. This approach doesn't seem terribly difficult for such a simple quotient, but as the expression becomes more complicated, the process becomes ever more involved. However, the Intermediate Value Theorem can help us here. First, the function $f(x) = (x+2)/(x-3)$ is zero only at –2 and discontinuous only at 3. This breaks the domain of the function into three components, $(-\infty, -2) \cup (-2, 3) \cup (3, \infty)$. We know that the function is continuous and never zero on each of these intervals. Thus the function cannot attain both a positive and negative value on either of the three intervals, because if it did it would have to have a zero. Then to determine the sign of the function f on any one of these intervals, we only need to determine the output sign of a single domain point. Explicitly, if $x = -3$, then $f(-3) = 1/6$, and we conclude that $f(x) > 0$ on $(-\infty, -2)$. Testing one value from the remaining intervals, we find that $f(x) < 0$ on $(-2, 3)$ and $f(x) > 0$ on $(3, \infty)$. To answer the initial question then, we conclude that $(x+2)/(x-3) > 0$ on $(-\infty, -2) \cup (3, \infty)$.

As mentioned earlier, this strategy is used in calculus to construct sign charts for the derivative of a function and then this information is used to determine the increasing and decreasing behavior of the function. In this context, the strategy is exactly the same as in our example above. First identify where the derivative is zero or discontinuous. This partitions the domain into smaller subintervals, from which one test value is used to determine the sign on that subinterval. We state the basic idea as a corollary.

Corollary 4.3.3. *If $f : (a, b) \to \mathbb{R}$ is continuous and $f(x) \neq 0$ for all $x \in (a, b)$, then $f(x) > 0$ for every $x \in (a, b)$ or $f(x) < 0$ for every $x \in (a, b)$.*

Connected Sets

In Chaps. 1 and 2 we discussed several set structures in $\mathbb{R}$: open, closed, and compact sets. Here we discuss one more type of set which turns out to have a rather simple nature, though this is not totally apparent at first. This simplicity is due to the structure of $\mathbb{R}$ itself, and in higher dimensions or more abstract settings such sets become much more interesting.

Definition 4.3.4. Let $E \subseteq \mathbb{R}$. We say that E is *disconnected* if there exist open sets $U, V \subseteq \mathbb{R}$ such that

(a) $U \cap V = \emptyset$;
(b) $U \cap E \neq \emptyset$ and $V \cap E \neq \emptyset$;
(c) $(U \cap E) \cup (V \cap E) = E$.

A set is called *connected* if it is not disconnected.

Common usage of these terms will aid us as we try to think about which sets are connected or disconnected. The finite set $\{1, 2\}$ should be disconnected since the elements are separated by a noticeable distance while the interval $(1, 2)$ should be connected since we can move freely between the points in the set without leaving the set. Before exploring examples more thoroughly, we point out a common strategy. To determine whether a set is connected or disconnected, the best approach is to start with a sort of visual to provide inspiration for your choice. Then, to show a set is disconnected, we must demonstrate the existence of sets U and V which satisfy the definition. To show a set is connected, the best strategy is often a proof by contradiction due to the fact that the definition is given as a negation.

Example 4.3.5. As a first example, take $E = \{1, 2\}$. Thinking visually for a moment, we need to find open sets U and V which are disjoint with each set containing exactly one element of E. There are many sets which will work and we will take $U = (0, 3/2)$ and $V(3/2, 3)$. Notice that $U \cap V = \emptyset, U \cap E = \{1\} \neq \emptyset$, $V \cap E = \{2\} \neq \emptyset$, and $(U \cap E) \cup (V \cap E) = E$.

Next, let $E = (0, 1) \cup (1, 2)$. There is no distance between these two sets, but the fact that 1 is not included in either interval provides a break in the set and thus we will aim to show this is also disconnected. The choice for U and V is a simple matter here since the two sets in this union are themselves open. We let $U = (0, 1)$ and $V = (1, 2)$, and leave it to the reader to verify that these sets satisfy the definition of disconnected.

For a final example, let's show that $\mathbb{Q}$ is also disconnected. The structure of $\mathbb{Q}$ is somewhat more complicated than the previous two examples, but we can use the fact that the irrational numbers provide gaps to construct the sets U and V. Again, there are many options which will work, one such pair being $U = (-\infty, e)$ and $V = (e, \infty)$. We could replace e with any other irrational number and the solution would still be valid. Observe that these two sets satisfy the definition of disconnected and that in this case we were required to take unbounded open sets so as to cover the rational numbers.

Example 4.3.6. Showing that a set is connected in $\mathbb{R}$ is a bit more involved than the previous examples. As mentioned earlier, it seems likely that any interval should be connected. This is in fact true as we will now show. Let $E \subseteq \mathbb{R}$ be an interval. To show this set is connected, we argue by contradiction and therefore assume that it is disconnected. With this assumption, there exist open sets $U, V \subseteq \mathbb{R}$ with the three properties from Definition 4.3.4. By property (b), choose $a \in U \cap E$ and $b \in V \cap E$ and, without loss of generality, assume $a < b$. To obtain a contradiction, we will now produce a point c such that $c \in E$ and $c \notin E$. Take $c = \sup(U \cap [a, b])$, which exists since $U \cap [a, b]$ is nonempty and bounded above. To show that $c \in E$, it is easy to see that $a \leq c \leq b$ since $a \in U \cap E$, and $b \notin U$ with $a < b$. If $c = a$ or $c = b$, then $c \in E$. Or, if $a < c < b$, then $c \in E$ since E is an interval. In any case, $c \in E$.

To show that c is also not in E, we first show that $c \notin U \cap E$ and $c \notin V \cap E$. For the first point, assume $c \in U \cap E$. This implies that $c \neq b$ and $c \in [a, b)$. From the fact that U is open, we can find an $\varepsilon > 0$ such that the set $B_\varepsilon(c) \subseteq U$. Keeping in mind that $c \in [a, b)$ and using the fact that the open interval $B_\varepsilon(c)$ extends to the right of c, we can produce a point $d \in B_\varepsilon(c) \cap [a, b) \subseteq U \cap [a, b]$ with $d > c$. But this contradicts the definition of c and hence $c \notin U \cap E$.

Next, assume that $c \in V \cap E$. Using the fact that V is open, we again produce an ε-neighborhood about c, this time with $B_\varepsilon(c) \subseteq V$. However, the fact that c is defined as a supremum implies that there exist $d \in U \cap [a, b]$ with $c - \varepsilon < d \leq c$. This means that $d \in U$ and $d \in B_\varepsilon(c) \subseteq V$, but this cannot happen since the sets U and V are disjoint. Thus $c \notin V \cap E$. With these two statements, it follows that $c \notin (U \cap E) \cup (V \cap E) = E$. At this point we have our contradiction and therefore conclude that E is connected.

As it turns out, intervals are the *only* connected subsets of $\mathbb{R}$. The following theorem states this formally and the proof supplies evidence for the remaining logical implication.

Theorem 4.3.7. *Let $E \subseteq \mathbb{R}$. Then E is connected if and only if E is an interval.*

Proof. The previous example handles one direction. To confirm the remaining statement, we will argue by contrapositive and assume that E is not an interval. Now we must show that E is disconnected. Using the fact that E is not an interval, we can find $a, b \in E$ and $c \in \mathbb{R}$ with $a < c < b$ and $c \notin E$. Set $U = (-\infty, c)$ and $V = (c, \infty)$. It is clear that $U \cap V = \emptyset$. Also, $a \in U \cap E$ and $b \in V \cap E$, and therefore neither set is empty. Furthermore, each element $x \in E$ satisfies exactly one of $x < c$ or $x > c$ guaranteeing us that $E \subseteq (U \cap E) \cup (V \cap E)$. The reverse inclusion is obvious since each set $U \cap E$ and $V \cap E$ is a subset of E. Our conclusion is then that E must be disconnected. □

The final result of this section returns to the discussion of continuity and set properties from Sect. 4.2. In light of the previous theorem, the corollary confirms the claim made early that continuous functions always map intervals to intervals.

Corollary 4.3.8. *Let $E \subseteq \mathbb{R}$ be connected and $f : E \to \mathbb{R}$ be continuous. Then $f(E)$ is connected.*

Proof. By the preceding theorem, we know that E is interval. To show that $f(E)$ is also an interval, let $x, y \in f(E)$ with $x < y$ and $z \in \mathbb{R}$ with $x < z < y$. These hypotheses imply the existence of $a, b \in E$ with $f(a) = x$ and $f(b) = y$ and thus we obtain the inequality $f(a) < z < f(b)$. Furthermore, since E is an interval, we know that $[a, b] \subseteq E$ and hence the Intermediate Value Theorem guarantees us that there is a point $c \in (a, b) \subseteq E$ with $f(c) = z$. Therefore $z \in f(E)$ and $f(E)$ is also an interval. □

Exercises

Exercise 4.3.1. (a) Is there a function which is continuous on all of $\mathbb{R}$ and has $(-\infty, 2) \cup (2, \infty)$ as its range? Explain.
(b) Is there a function which is continuous on all of $\mathbb{R}$ and has $\mathbb{Q}$ as its range? Explain.

Exercise 4.3.2. (a) Show that the function $f(x) = 3x^3 + \sin x - 1$ has a root in $(-1, 1)$.
(b) Show that the equation $x^4 + x = 3$ has a solution in $(1, 2)$.
(c) Show that the equation $\sqrt[3]{x} = 1 - x$ has a real solution.
(d) Show that there is real number b with $\cos b = b$.

Exercise 4.3.3. (a) Suppose $f : [0, 1] \to [0, 1]$ is continuous. Show that there is a point $c \in [0, 1]$ with $f(c) = c$.
(b) Suppose f is continuous on $[0, 2]$ with $f(0) = f(2)$. Show that there is a point $c \in [0, 1]$ such that $f(c) = f(c + 1)$.

Exercise 4.3.4. Let $a > 0$ and let n be a natural number. Show that there exists $b \in \mathbb{R}$ such that $b^n = a$. **Note:** Do not simply take b to be an nth root. The point here is to assert the existence of such objects.

Exercise 4.3.5. If $I \subseteq \mathbb{R}$ is an interval and $f : I \to \mathbb{R}$, then we say f is *strictly increasing* on I if $f(x) < f(y)$ whenever $x < y$ in I and *strictly decreasing* if $f(x) > f(y)$ whenever $x < y$ in I. Now suppose $f : I \to \mathbb{R}$ is continuous.

(a) Show that f is injective (1-to-1) if and only if f is either strictly increasing or decreasing on I.
(b) If f is injective, show that the function $g : f(I) \to I$ defined by $g = f^{-1}$ is continuous on $f(I)$. Keep in mind that $f(I)$ is also an interval.

Exercise 4.3.6. Use the definition of disconnected to show that the natural numbers, the integers, and the irrational numbers are disconnected.

4.4 Continuity on a Set and Uniform Continuity

Recall that if $f : A \to \mathbb{R}$ and $B \subseteq A$, then we say f is continuous on B if f is continuous at each point of B. Continuity itself is a *local property*, i.e. its focus is on behavior around a single domain point, and the notion of continuity on a set is an attempt to expand this to a *global property*. However, a reduction to the local behavior doesn't accomplish much. Our goal is to examine a truly global continuity property. We begin with a simple case considering functions $f : \mathbb{R} \to \mathbb{R}$. It may be helpful to review the discussion preceding Proposition 1.1.13. Throughout the section we work with open intervals (a, b) and assume $a < b$.

Theorem 4.4.1. *Let $f : \mathbb{R} \to \mathbb{R}$. Then f is continuous on $\mathbb{R}$ if and only if $f^{-1}((a, b))$ is an open set for every $a, b \in \mathbb{R}$.*

The set $f^{-1}((a, b))$ may seem mysterious at first glance. By definition of the inverse function acting on sets, we see that $f^{-1}((a, b)) = \{x \in \mathbb{R} : a < f(x) < b\}$. This second form is more tangible and we will use it when calculating explicitly. Before proving the theorem, let's take a moment to think about why the statement seems reasonable. To do this, it's best to consider the $\varepsilon - \delta$ definition of continuity. It may seem as though that definition begins in the domain and then moves to the range, but this is not true. The definition of continuity begins by assuming $\varepsilon > 0$, thereby producing an ε-neighborhood about the point $f(c)$ in the range of f. The task is then to find a corresponding δ-neighborhood around c in the domain which maps into the ε-neighborhood. The theorem above is playing this same strategy, but in a more explicit fashion; begin with *any* open interval in the codomain and show that it pulls back to an open set in the domain.

For more evidence, take $f(x) = x^2$. It is certainly not the case that $f(V)$ is open whenever $V \subseteq \mathbb{R}$ is open. Take $V = (-1, 1)$, for example. Then $f((-1, 1)) = [0, 1)$. However, $f^{-1}((-1, 1)) = \{x \in \mathbb{R} : -1 < f(x) < 1\} = \{x \in \mathbb{R} : -1 < x^2 < 1\} = (-1, 1)$. More specifically, if (a, b) is any open interval, then

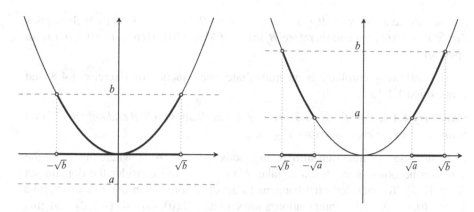

Fig. 4.6 $f(x) = x^2$ continuous on $\mathbb{R}$

$$f^{-1}((a,b)) = \begin{cases} \emptyset, & a < b \le 0; \\ (-\sqrt{b}, \sqrt{b}), & a < 0 < b; \\ (-\sqrt{b}, -\sqrt{a}) \cup (\sqrt{a}, \sqrt{b}), & 0 \le a < b. \end{cases}$$

In any case, the result is an open set (see Fig. 4.6). Observe that the sets on the last row are not open intervals. So, while continuous functions always map intervals to intervals, it is not the case that the inverse image of an interval is an interval.

Two quick facts to keep in mind as we begin the proof. First, by definition of f^{-1} as a map on sets, we have $a \in f^{-1}(V)$ if and only if $f(a) \in V$. Second, we will use ε-neighborhood notation as the statement of the proof requires us to work with sets. Hence we must be able to switch freely between the notations $|x - c| < \varepsilon$ and $x \in B_\varepsilon(c)$.

Proof (Theorem 4.4.1). To begin we assume that $f^{-1}((a,b))$ is an open set for every $a, b \in \mathbb{R}$. To show that f is continuous on $\mathbb{R}$, choose an arbitrary $c \in \mathbb{R}$ and let $\varepsilon > 0$. We must now choose $\delta > 0$ such that $|x - c| < \delta$ implies $|f(x) - f(c)| < \varepsilon$. This is equivalent to the ε-neighborhood implication: $x \in B_\delta(c)$ implies $f(x) \in B_\varepsilon(f(c))$. Now, we know that $B_\varepsilon(f(c))$ is an open interval, and thus $f^{-1}(B_\varepsilon(f(c)))$ is also open by our assumption. Moreover, $c \in f^{-1}(B_\varepsilon(f(c)))$ since $f(c) \in B_\varepsilon(f(c))$. Thus there is a $\delta > 0$ such that $B_\delta(c) \subseteq f^{-1}(B_\varepsilon(f(c)))$ by the definition of open set. If we now choose $x \in B_\delta(c) \subseteq f^{-1}(B_\varepsilon(f(c)))$ it follows immediately that $f(x) \in B_\varepsilon(f(c))$ proving the desired implication.

For the converse, assume that f is continuous on $\mathbb{R}$ and let (a, b) be an open interval. To show that $f^{-1}((a,b))$ is open, for each $c \in f^{-1}((a,b))$, we must produce a $\delta > 0$ such that $B_\delta(c) \subseteq f^{-1}((a,b))$. Let $c \in f^{-1}((a,b))$. Then it must be the case that $f(c) \in (a, b)$, an open interval. Thus there is an $\varepsilon > 0$ such that $B_\varepsilon(f(c)) \subseteq (a, b)$. The continuity of f assures us that there is a $\delta > 0$ such that $x \in B_\delta(c)$ implies $f(x) \in B_\varepsilon(f(c))$. To show that this δ-neighborhood is the one we seek, let $x \in B_\delta(c)$. Then $f(x) \in B_\varepsilon(f(c)) \subseteq (a,b)$ which implies that $x \in f^{-1}((a,b))$ and therefore $B_\delta(c) \subseteq f^{-1}((a,b))$. Hence $f^{-1}((a,b))$ is an open set. $\square$

The following corollary is an immediate consequence of Exercise 1.4.8 and Proposition 1.1.13

Corollary 4.4.2. *Let* $f : \mathbb{R} \to \mathbb{R}$. *Then* f *is continuous on* $\mathbb{R}$ *if and only if* $f^{-1}(V)$ *is an open set for every open set* $V \subseteq \mathbb{R}$.

Restricting our attention now to functions $f : A \to \mathbb{R}$ where $A \subseteq \mathbb{R}$, the situation becomes more delicate. Take $f(x) = x^2$ and consider the domain set $A = [1,3]$. To consider what happens to an open interval in the codomain, take $V = (0,4)$. A naive computation shows that $f^{-1}((0,4)) = (-2,2)$ and this observation immediately produces several problems. First, this set is not even contained in our domain. This is because we naturally think of $f(x) = x^2$ as a function on $\mathbb{R}$. For situations such as this one, we need to take into account that we are *only* considering our function as defined on A. With this, we have $f^{-1}((0,4)) = \{x \in [1,3] : 0 < f(x) < 4\} = [1,2)$, which is contained in A, but this set is not open. The problem here is due to the fact that we have restricted our domain to a set that is itself not open. There are several technical remedies and we will use the notion of relative open sets.

Definition 4.4.3. Let $A \subseteq \mathbb{R}$. We say that $B \subseteq A$ is *open relative to* A if there is an open set $O \subseteq \mathbb{R}$ such that $B = A \cap O$.

With this definition in mind, we can write $[1,2) = [1,3] \cap (0,2)$ and this is enough to conclude that $f^{-1}((0,4))$ is open relative to the domain set A; notice $(0,2)$ is not the only open set for which this equality holds. This is the framework needed to consider continuity in this context and we have the following theorem.

Before proving this statement, review again Theorem 4.2.8. Notice in the implication there we have the phrase "if $|x - c| < \delta$ and $x \in A$." It is this second condition "$x \in A$" which differentiates this theorem and its proof from the previous statement.

Theorem 4.4.4. *Let* $f : A \to \mathbb{R}$. *Then* f *is continuous on* A *if and only if* $f^{-1}((a,b))$ *is open relative to* A *for every* $a,b \in \mathbb{R}$. *If* A *is open, then* f *is continuous on* A *if and only if* $f^{-1}((a,b))$ *is open for every* $a,b \in \mathbb{R}$.

Proof. We first suppose that f is continuous on A and let (a,b) be an open interval. To show that $f^{-1}((a,b))$ is open relative to A we must construct an open set O and show that $f^{-1}((a,b)) = A \cap O$. To construct this open set, let c be an arbitrary point in $f^{-1}((a,b))$. Then $f(c)$ is in the open interval (a,b) and hence there is an $\varepsilon_c > 0$ such that $B_{\varepsilon_c}(f(c)) \subseteq (a,b)$. Since f is continuous, there is also a $\delta_c > 0$ such that if $x \in A$ and $x \in B_{\delta_c}(c)$, then $f(x) \in B_{\varepsilon_c}(f(c)) \subseteq (a,b)$. We now define O as

$$O = \bigcup B_{\delta_c}(c)$$

where the union is over all points $c \in f^{-1}((a,b))$. Clearly this set is open as it is the union of open sets (Theorem 1.4.9 can be extended to uncountable unions).

To show that $f^{-1}((a,b)) = A \cap O$, we will show both subset inclusions. First, suppose $x \in f^{-1}((a,b))$. In this case we have $x \in A$ since $f^{-1}((a,b)) = \{x \in A : a < f(x) < b\}$. Also, $x \in O$ by the construction above. Thus $f^{-1}((a,b)) \subseteq A \cap O$. Next assume that $x \in A \cap O$. Then $x \in A$ and $x \in O$, or $x \in A$ and $x \in B_{\delta_c}(c)$ for some $c \in f^{-1}((a,b))$; we could take $c = x$ in this case. By choice of δ_c, we then see that $f(x) \in B_{\varepsilon_c}(f(c)) \subseteq (a,b)$, which implies that $x \in f^{-1}((a,b))$. Thus $A \cap O \subseteq f^{-1}((a,b))$.

Conversely assume that $f^{-1}((a,b))$ is open relative to A for any open interval $(a,b) \subseteq \mathbb{R}$. To show f is continuous on A, let $c \in A$ and $\varepsilon > 0$. The task is now to find a suitable δ. We know that the set $B_\varepsilon(f(c))$ is an open interval and hence $f^{-1}(B_\varepsilon(f(c)))$ is open relative to A, i.e. there is an open set $O \subseteq \mathbb{R}$ such that $f^{-1}(B_\varepsilon(f(c))) = A \cap O$. The fact that $c \in f^{-1}(B_\varepsilon(f(c)))$ and the set equality just given implies $c \in O$. The fact that O is open supplies a $\delta > 0$ such that $B_\delta(c) \subseteq O$. To see that this δ is the one we seek, let $x \in B_\delta(c)$ with $x \in A$. Then $x \in A \cap B_\delta(c) \subseteq A \cap O = f^{-1}(B_\varepsilon(f(c)))$ which means $f(x) \in B_\varepsilon(f(c))$. In other words, if $x \in A$ with $|x - c| < \delta$, then $|f(x) - f(c)| < \varepsilon$. □

Example 4.4.5. For the function g defined on $\mathbb{R}$ by

$$g(x) = \begin{cases} 1/x^2, & x \neq 0; \\ 0, & x = 0, \end{cases}$$

we know there is a discontinuity at 0 and therefore g is not continuous on $\mathbb{R}$. Computing $g^{-1}((a,b))$ to confirm this, we have

$$g^{-1}((a,b)) = \begin{cases} \emptyset, & a < b \leq 0; \\ (-\infty, -1/\sqrt{b}) \cup \{0\} \cup (1\sqrt{b}, \infty), & a < 0 < b; \\ (-1/\sqrt{a}, -1/\sqrt{b}) \cup (1/\sqrt{b}, 1/\sqrt{a}), & 0 \leq a < b. \end{cases}$$

The fact that the second row here contains non-open sets verifies that g is not continuous. What if we restrict our attention to the domain $A = (-\infty, 0) \cup (0, \infty)$? For this domain, we have

$$g^{-1}((a,b)) = \begin{cases} \emptyset, & a < b \leq 0; \\ (-\infty, -1/\sqrt{b}) \cup (1\sqrt{b}, \infty), & a < 0 < b; \\ (-1/\sqrt{a}, -1/\sqrt{b}) \cup (1/\sqrt{b}, 1/\sqrt{a}), & 0 \leq a < b. \end{cases}$$

It should be clear that the sets present here are all open relative to A, as illustrated in Fig. 4.7, and thus g is continuous on A.

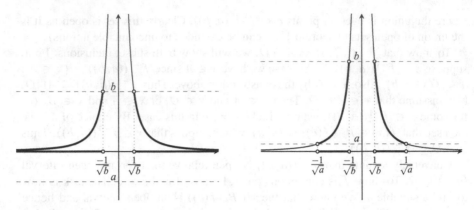

Fig. 4.7 $g(x) = 1/x^2$ continuous on $(-\infty, 0) \cup (0, \infty)$

Uniform Continuity

The topic we discuss now is a strengthened version of continuity and is defined for subsets of $\mathbb{R}$. This is an instance where a sequential approach is less intuitive at the outset, though we will employ a sequential characterization to determine when the property does not hold.

Definition 4.4.6. Let $f : A \to \mathbb{R}$. We say f is *uniformly continuous* on A if for every $\varepsilon > 0$, there is a $\delta > 0$ such that if $x, y \in A$ with $|x - y| < \delta$, it follows that $|f(x) - f(y)| < \varepsilon$.

Take a moment to compare this to the definition of continuity at a point. Notice here that we do not pre-specify a domain point c as is the case for continuity. The key here is that the choice of δ depends only on ε whereas with continuity, the choice for δ is allowed to depend on both ε and the predetermined domain point c. To say this more succinctly, uniform continuity requires that, for $\varepsilon > 0$ given, the same choice of $\delta > 0$ supplies the desired implication for all domain points. This is a subtle change and is caused by simply switching the order of quantifiers. And as we will see, many of our most familiar functions do not have this new property.

Example 4.4.7. To understand these dependencies, let $f(x) = x^2$ and consider $c \in \mathbb{R}$. To show f is continuous at c, for $\varepsilon > 0$, we must find $\delta > 0$ so that $|x^2 - c^2| < \varepsilon$ whenever $|x - c| < \delta$. First notice that

$$|x^2 - c^2| = |x + c||x - c|.$$

Thus if $\delta_1 = 1$, then $|x - c| < \delta_1$ implies that

$$|x| = |x - c + c| \leq |x - c| + |c| < 1 + |c|$$

and hence we can bound $|x + c|$ by

$$|x + c| \le |x| + |c| \le (1 + |c|) + |c| = 2|c| + 1.$$

If we then choose $\delta_2 = \varepsilon / (2|c|+1)$, and set $\delta = \min\{\delta_1, \delta_2\}$, we see that $|x-c| < \delta$ implies

$$|x^2 - c^2| = |x + c||x - c| < (2|c| + 1)|x - c| < (2|c| + 1)\delta < \varepsilon.$$

This is not a formal proof, but our goal is to understand the choice of δ. In explicit form, we have

$$\delta = \min\left\{1, \frac{\varepsilon}{2|c| + 1}\right\}.$$

This choice depends on both c and ε and observe that for larger and larger values of c, we would necessarily require smaller and smaller values of δ. This is due to the second term in the definition of δ. Though this is not enough to conclude that this particular function is not uniformly continuous on $\mathbb{R}$, because there may be a better choice for δ that we haven't considered, it does supply evidence that, when considered as a function on $\mathbb{R}$, f is not uniformly continuous. The following result gives us a sequential criterion for determining when a function is not uniformly continuous and we will return to this example shortly.

Theorem 4.4.8 (Criterion for Nonuniform Continuity). *Let $f : A \to \mathbb{R}$. Then f is not uniformly continuous on A if there exist sequences (x_n) and (y_n) in A and $M > 0$ such that $(|x_n - y_n|) \to 0$ as $n \to \infty$ while*

$$|f(x_n) - f(y_n)| \ge M$$

for all n.

Example 4.4.9. Returning now to $f(x) = x^2$, we will *show* that this function is not uniformly continuous on $\mathbb{R}$. From the discussion above, we know that the problem lies in the fact that for a fixed $\varepsilon > 0$, larger values of c require smaller values of δ. We can exploit this observation and consider what happens when we choose two domain sequences which grow without bound, but get closer and closer to each other. Take $x_n = n$ and $y_n = n + 1/n$. Then $|x_n - y_n| = 1/n$ which converges to 0, but

$$|f(x_n) - f(y_n)| = |x_n^2 - y_n^2| = |n^2 - (n^2 + 2 + 1/n^2)| = 2 + 1/n^2 \ge 2$$

for all n. Thus, if we take $M = 2$, we see that this function is not uniformly continuous on $\mathbb{R}$.

Now, consider what happens if we turn our attention to a smaller domain, $A = [0, 3]$, for instance. If we take points $x, y \in A$, then we have $|x + y| \leq 6$. Thus if $\varepsilon > 0$ and $\delta = \varepsilon/6$, we can show that $|x^2 - y^2| < \varepsilon$. We have the following formal proof.

Proof. Let $A = [0, 3]$ and define $f : A \to \mathbb{R}$ by $f(x) = x^2$. To show f is uniformly continuous on A, let $\varepsilon > 0$ and choose $\delta = \varepsilon/6$. If $|x - y| < \delta$, we see that

$$|f(x) - f(y)| = |x^2 - y^2| = |x + y||x - y| \leq (|x| + |y|)|x - y| \leq 6|x - y| < 6\delta = \varepsilon$$

and thus f is uniformly continuous on A. $\square$

For this function, a similar result would hold if we considered f on *any* bounded set but this is a rare circumstance. However, there is a general relationship between uniform continuity and continuous functions defined on compact subsets of $\mathbb{R}$.

Theorem 4.4.10. *Let $K \subseteq \mathbb{R}$ be a compact set. If $f : K \to \mathbb{R}$ is continuous on K, then f is uniformly continuous on K.*

Proof. Let K be a compact subset of $\mathbb{R}$ and suppose $f : K \to \mathbb{R}$ is continuous. We have several formulations of compact sets and each definition leads to a slightly different proof. We will use the sequential characterization and a proof by contradiction. Exercise 4.4.9 will ask you to use the open cover characterization of compactness to prove the statement. Continuing on, we will use a proof by contradiction because we have a simple characterization describing when a set is not uniformly continuous in terms of sequences. Towards a contradiction, we now assume that f is not uniformly continuous on K. By Theorem 4.4.8 there exist an $M > 0$ and sequences (x_n) and (y_n) in K such that $(|x_n - y_n|) \to 0$ as $n \to \infty$ and

$$|f(x_n) - f(y_n)| \geq M$$

for all n. To reach a contradiction, we will show that there are sequence terms which violate this last inequality.

Using the compactness of K, we know there is a subsequence (x_{n_j}) of (x_n) which converges to a point $c \in K$. This provides us with two pieces of information. First, since f is continuous on K, we know that $(f(x_{n_j})) \to f(c)$ as $j \to \infty$. Second, if we consider the corresponding subsequence (y_{n_j}), then it must also converge to c. To see this, notice that

$$|y_{n_j} - c| \leq |y_{n_j} - x_{n_j}| + |x_{n_j} - c|;$$

given $\varepsilon > 0$, we can make each term on the right less than $\varepsilon/2$ by the hypothesis on (x_n) and (y_n) and the fact that $(x_n) \to c$. We leave the particular details as an exercise. The continuity of f now implies that $(f(y_{n_j}))$ also converges to $f(c)$.

However, these two statements imply that it must be the case that the sequence $(f(x_{n_j}) - f(y_{n_j})) \to 0$. Thus there exists $J \in N$ such that

$$|f(x_{n_j}) - f(y_{n_j})| < M$$

for $j \geq J$ contradicting our hypothesis above. Thus we conclude that f must be uniformly continuous on K.   □

Example 4.4.11. To see that the hypothesis that K be compact is crucial, consider the function $g(x) = 1/x$ on the interval $(0, 1)$. This function is certainly continuous here, but its graph becomes increasingly steep as we approach 0. This steepness indicates that small intervals in the domain are stretched out in the range, forcing range values further and further apart as we move closer to the origin. With this intuition, take two sequences converging to zero, say $(x_n) = (1/n)$ and $(y_n) = (1/(n + 1))$. Then $(|x_n - y_n|) \to 0$ but

$$|f(x_n) - f(y_n)| = |n - (n + 1)| = 1$$

for all n. Thus this function is not uniformly continuous on the given domain.

Exercises

Exercise 4.4.1. (a) Let $n \subset N$. Use Theorem 4.4.1 to show that $f(x) = x^n$ is
    continuous on $\mathbb{R}$.
(b) Show that $f(x) = 1/x$ is continuous on $(-\infty, 0) \cup (0, \infty)$.
(c) Show that

$$f(x) = \begin{cases} 1/x, & x \neq 0; \\ 0, & x = 0, \end{cases}$$

is not continuous on $\mathbb{R}$.

Exercise 4.4.2. (a) Show that f is continuous on $\mathbb{R}$ if and only if $f^{-1}((a, \infty))$ is
    an open set for every $a \in \mathbb{R}$.
(b) Supply a proof for Corollary 4.4.2.
(c) Show that $f : \mathbb{R} \to \mathbb{R}$ is continuous on $\mathbb{R}$ if and only if $f^{-1}(C)$ is a closed set
    for every closed set $C \subseteq \mathbb{R}$.

Exercise 4.4.3. Let $f(x) = mx + b$ and show f is uniformly continuous on $\mathbb{R}$.

Exercise 4.4.4. Show that $f(x) = \sin(1/x)$ is not uniformly continuous on $(0, 1)$.

Exercise 4.4.5. Let $f(x) = x^3$.

(a) Argue that f is not uniformly continuous on $\mathbb{R}$.
(b) Show that f is uniformly continuous on any bounded subset of $\mathbb{R}$.

Exercise 4.4.6. Show that $g(x) = 1/x^2$ is uniformly continuous on $[1, \infty)$ but not on $(0, 1]$.

Exercise 4.4.7. (a) Suppose $f : A \to \mathbb{R}$ is uniformly continuous and let (x_n) be a Cauchy sequence in A. Show that the sequence $(f(x_n))$ is also Cauchy.

(b) Show that the statement above is not necessarily true if we remove the hypothesis of uniform continuity.

Exercise 4.4.8. A function $f : A \to \mathbb{R}$ is called *Lipschitz* if there exists a bound $M > 0$ such that

$$\left| \frac{f(x) - f(y)}{x - y} \right| \leq M$$

for all $x, y \in A$. Show that if $f : A \to \mathbb{R}$ is Lipschitz, then it is uniformly continuous on A.

Exercise 4.4.9. Reprove Theorem 4.4.10 using the open cover characterization of compactness. You can use a direct approach here.

4.5   Bounded Linear Operators

To extend the ideas of this chapter to normed linear spaces, we must first investigate general linear operators, which are typically referred to as linear transformations in the finite dimensional setting of most linear algebra texts. While we don't explicitly use the notion of basis, it will be useful to recall that the dimension of a vector space is taken to be the number of elements in a basis; of course, this only makes sense after showing that, in the finite dimensional setting, any two bases for a vector space contain the same number of elements.

Definition 4.5.1. Let X and Y be real vector spaces. A function $T : X \to Y$ is called a *linear operator* if

(a) $T(x + y) = T(x) + T(y)$ for all $x, y \in X$;
(b) $T(cx) = cT(x)$ for all $x \in X$ and $c \in \mathbb{R}$.

When using linear operators, it is standard practice to forgo the use of parentheses unless absolutely necessary. For example, the second condition above is typically written as $T(cx) = cTx$.

Example 4.5.2. If X is a vector space, we define the *identity operator* $I : X \to X$ by $Ix = x$ for every $x \in X$. The fact that this operator maps a vector space to itself immediately implies that it satisfies the two hypotheses above. Though a simple example it is very important in the general study of linear operators.

Example 4.5.3. The study of linear transformations in linear algebra centers around understanding functions of the following form: let $X = \mathbb{R}^n$, $Y = \mathbb{R}^m$, and let

A be any $m \times n$ matrix. The map $T : X \to Y$ given by $Tx = Ax$ is a linear transformation. In this setting a more general statement can be made which indicates why understanding the matrix transformations is so important. Let X and Y be finite dimensional vector spaces and let $T : X \to Y$ be a linear transformation. If B_1 is a basis for X and B_2 is a basis for Y, then there is a matrix A_T, depending on B_1 and B_2, such that $(Tx)_{B_2} = A_T(x)_{B_1}$ for all $x \in X$, where $(x)_{B_1}$ represents the coordinates of x with respect to the basis B_1 and likewise for the notation concerning B_2. This relationship ties many of the properties of the general transformation to those of the matrix providing a concrete setting for exploration.

Example 4.5.4. Moving into larger spaces, let $X, Y = \ell^2$. The operator $T : \ell^2 \to \ell^2$ defined by

$$T(a_1, a_2, a_3, \ldots) = (0, a_1, a_2, a_3, \ldots)$$

is called the *forward shift operator*. To confirm that T is linear, let $(a_n), (b_n) \in \ell^2$. Then

$$T((a_n) + (b_n)) = T(a_n + b_n) = (0, a_1 + b_1, a_2 + b_2, a_3 + b_3, \ldots)$$
$$= (0, a_1, a_2, a_3, \ldots) + (0, b_1, b_2, b_3, \ldots)$$
$$= T(a_n) + T(b_n),$$

and, for $c \in \mathbb{R}$,

$$T(c(a_n)) = T(ca_n) = (0, ca_1, ca_2, ca_3, \ldots) = c(0, a_1, a_2, a_3, \ldots) = cT(a_n).$$

Similarly we can define the *backward shift operator* $S : \ell^2 \to \ell^2$ by

$$S(a_1, a_2, a_3, \ldots) = (a_2, a_3, a_4, \ldots).$$

For T and S it is fairly easy to see that the operator maps into ℓ^2. For an example where this isn't quite as obvious, fix a sequence $\lambda = (\lambda_1, \lambda_2, \lambda_3, \ldots)$ in ℓ_∞ and define $M_\lambda : \ell^2 \to \ell^2$ by

$$M_\lambda(a_1, a_2, a_3, \ldots) = (\lambda_1 a_1, \lambda_2 a_2, \lambda_3 a_3, \ldots).$$

To verify that M_λ maps into ℓ^2, we must show that $M_\lambda(a_n) \in \ell^2$ for each sequence $(a_n) \in \ell^2$. To show this, keep in mind that $|\lambda_n| \leq \|\lambda\|_\infty$ for all n. Then, if $(a_n) \in \ell^2$, the Algebraic Limit Theorem for series shows that

$$\sum_{n=1}^{\infty} \|\lambda\|_\infty^2 |a_n|^2 = \|\lambda\|_\infty^2 \sum_{n=1}^{\infty} |a_n|^2 < \infty$$

and the Comparison Test shows then that the series $\sum_{n=1}^{\infty} |\lambda_n a_n|^2$ must also converge. Thus $M_\lambda(a_n) \in \ell^2$. You will be asked to verify that M_λ is linear in the exercises.

The most basic properties of linear operators are given in the following Proposition. In the statement $\mathbf{0}_X$ and $\mathbf{0}_Y$ represent the additive identity in X and Y, respectively.

Proposition 4.5.5. *Let X and Y be vector spaces and let $T : X \to Y$ be a linear operator. The following hold:*

(a) $T\mathbf{0}_X = \mathbf{0}_Y$;
(b) $T(-x) = -Tx$ *for every* $x \in X$;
(c) $T(cx + dy) = cTx + dTy$ *for every* $x, y \in X$ *and* $c, d \in \mathbb{R}$.

Example 4.5.6. Let $X = C([0, 1])$ denote the collection of all functions which are continuous on the closed interval $[0, 1]$. We define addition and scalar multiplication in the standard pointwise manner,

$$(f + g)(x) = f(x) + g(x) \quad \text{and} \quad (kf)(x) = kf(x).$$

Theorem 4.2.6 shows that the closure axioms of a vector space hold for $C([0, 1])$; the other vector space axioms follow by definition of the addition and scalar multiplication. Now we define $\| \cdot \|_\infty : C([0, 1]) \to [0, \infty)$ by

$$\|f\|_\infty = \sup \{|f(x)| : x \in [0, 1]\}.$$

If f is continuous on $[0, 1]$, then the supremum is guaranteed to exist by the Extreme Value Theorem and norm property (a) follows immediately. Properties (b) and (c) then follow quickly from properties of supremum and the absolute value. We will discuss the completeness of this space in Sect. 6.5.

The space $C([0, 1])$ can be defined more generally on any closed, bounded interval $[a, b]$, but we will typically work on $[0, 1]$. As a matter of caution, notice that we are using the same symbol for this norm as we did with the sup-norm in Sect. 2.5, but this should cause no confusion. Naively, it should always be clear from context whether we are talking about a sequence of real numbers or about a function defined on an interval. From a more rigorous point of view, remember that a sequence is defined as a function $f : \mathbb{N} \to \mathbb{R}$. Thus these two norms are actually the same; the difference in appearance is due to the fact, that on the one hand, we are considering functions defined on an interval while, on the other, we are dealing with a function defined on a countable set.

There are a multitude of operators to consider on these types of spaces though many of these must be delayed until we have discussed differentiation and integration. Here we will focus on two types of operators that arise from the operations of function multiplication and composition. First, let g be any function which is continuous on $[0, 1]$. We then define the *multiplication operator* $M_g : C([0, 1]) \to C([0, 1])$ by

$$M_g f = gf.$$

This operator is certainly linear due to the associative, distributive, and commutative properties of real number multiplication; foregoing the formal function notation, we summarize this algebraically by

$$M_g(f + h) = g(f + h) = gf + gh = M_g f + M_g h$$

and

$$M_g(cf) = g(cf) = c(gf) = cM_g f.$$

For the second class of operators, let φ be any function which is continuous on $[0, 1]$ and whose range is a subset of $[0, 1]$. We then define the *composition operator* C_φ by

$$C_\varphi f = f \circ \varphi.$$

Notice that we have to be sensitive to the range of φ so that the composition given above is valid. For the domain $[0, 1]$, examples of valid φ's include $\varphi(x) = x^2$ or $\varphi(x) = \sin x$. When working with these types of operators, multiplication or composition, keep in mind that once a g or φ is chosen, it does not change. In short, the chosen function *defines* the operator.

As opposed to a basic course in linear algebra, we will seek to understand the analytic properties of linear operators. We are allowed to do this as long as we work within the construct of a normed linear or inner product space where we have a vector space structure and a metric. Previously we discussed how to use this structure to extend the ideas of sequence convergence and Cauchy sequences in $\mathbb{R}$. Here we will explore the extension of continuous functions. In $\mathbb{R}$ we discussed three distinct means of defining continuous functions: the sequential characterization, the $\varepsilon - \delta$ characterization, and the open set characterization of Sect. 4.4. It turns out that these three concepts are also equivalent in metric spaces (see Chap. 1 of [25]), and thus we have the freedom to choose which characterization we will use. The $\varepsilon - \delta$ characterization aids in simplifying the proof of Theorem 4.5.9.

Definition 4.5.7. Let X and Y be a normed linear spaces and let $T : X \to Y$ be a linear operator. We say T is *continuous at* $x \in X$ if for every $\varepsilon > 0$, there is a $\delta > 0$ such that if $y \in X$ with $\|x - y\|_X < \delta$, it follows that $\|Tx - Ty\|_Y < \varepsilon$.

The statement above is identical to our $\varepsilon - \delta$ characterization of continuity at a point $c \in \mathbb{R}$ except for the fact that our domain and codomain spaces may be different, meaning that we work with $\| \cdot \|_X$ when considering domain points and $\| \cdot \|_Y$ for range points. Demonstrating continuity of linear operators is often more difficult than for the functions we have encountered on $\mathbb{R}$ due to various complications that arise in the infinite dimensional setting. However, the linearity provides us with a way around this which certainly does not hold for general functions mapping $\mathbb{R}$ into $\mathbb{R}$. To explore this new approach, we first need a definition.

Definition 4.5.8. Let X and Y be normed linear spaces and let $T : X \rightarrow Y$ be a linear operator. We say that T is *bounded* if there is an $M > 0$ so that $\|Tx\|_Y \leq M\|x\|_X$ for all $x \in X$.

Theorem 4.5.9. *Let X and Y be normed linear spaces and let $T : X \rightarrow Y$ be a linear operator. The following are equivalent.*

(a) T *is a bounded operator;*
(b) T *is continuous at $\mathbf{0}_X$ in X;*
(c) T *is continuous at x for each $x \in X$.*

Proof. $(a) \Rightarrow (b)$. Suppose T is a bounded operator with $M > 0$ satisfying $\|Tx\|_Y \leq M\|x\|_X$. To show T is continuous at $\mathbf{0}_X$, let $\varepsilon > 0$ and choose $\delta = \varepsilon/M$. Then, if $x \in X$ with $\|x - \mathbf{0}_X\|_X < \delta$ we have

$$\|Tx - T\mathbf{0}_X\|_Y = \|Tx - \mathbf{0}_Y\|_Y = \|Tx\|_Y \leq M\|x\|_X = M\|x - \mathbf{0}_X\|_X < M(\varepsilon/M) = \varepsilon.$$

Therefore T is continuous at $\mathbf{0}_X$.

$(b) \Rightarrow (c)$. This argument is identical to that given in Example 4.2.10 where we showed that any additive function on $\mathbb{R}$ which is continuous at 0 is in fact continuous on all of $\mathbb{R}$.

$(c) \Rightarrow (a)$. Suppose that T is continuous at each point of X. In particular, T is continuous at $\mathbf{0}_X$ and thus, for $\varepsilon = 1$, there is a $\delta > 0$ such that if $\|x\|_X = \|x - \mathbf{0}_X\|_X < \delta$, it follows that $\|Tx\|_Y = \|Tx - \mathbf{0}_Y\|_Y = \|Tx - T\mathbf{0}_X\|_Y < 1$. If $x \in X$, not the zero vector, and $\alpha = \delta/2$, then $x/\|x\|_X$ has norm 1 and thus

$$\left\| \alpha \left(\frac{x}{\|x\|_X} \right) \right\|_X = \alpha < \delta.$$

If we now use the linear properties of the operator with the homogeneity of the norm, observe that

$$\|Tx\|_Y = \left\| \frac{\|x\|_X}{\alpha} T \left(\alpha \frac{x}{\|x\|_X} \right) \right\|_Y = \frac{\|x\|_X}{\alpha} \left\| T \left(\alpha \frac{x}{\|x\|_X} \right) \right\|_Y < \frac{\|x\|_X}{\alpha} \cdot 1 = \left(\frac{1}{\alpha} \right) \|x\|_X.$$

Therefore, if we set $M = 1/\alpha$ we see that $\|Tx\|_Y \leq M\|x\|_X$ for all nonzero $x \in X$. To complete the proof, we observe that the inequality holds in the trivial sense for the zero vector in X. $\square$

Example 4.5.10. All of the operators discussed earlier in this section are bounded. The easiest example of this is the forward shift on ℓ^2 since

$$\|T(a_n)\|_2 = \|(0, a_1, a_2, \dots)\|_2 = \left(\sum_{n=1}^{\infty} |a_n|^2 \right)^{1/2} = \|(a_n)\|_2;$$

this says that T is a *norm-preserving operator* or an *isometry*. From this calculation we see that if we set $M = 1$, then we have $\|T(a_n)\|_2 \leq M\|(a_n)\|_2$ for all $(a_n) \in \ell^2$.

For finite dimensional spaces, it can be shown that any linear operator is bounded and therefore continuous though we will not explore that here except in the special case of matrix transformations on $\mathbb{R}^2$; see Exercise 4.5.3. You will also be asked to show that the other sequence space operators discussed earlier are bounded.

Example 4.5.11. For an example of a linear operator that is not continuous, we will produce a linear operator that is not bounded; this approach is simpler than showing the definition of continuity is not satisfied. Take $X, Y = p^\infty$ with the sup-norm (see Example 2.5.12) and define an operator $T : p^\infty \to p^\infty$ by

$$T(a_1, a_2, a_3, \ldots, a_n, \ldots) = (1a_1, 2a_2, 3a_3, \ldots, na_n, \ldots).$$

Observe that the operator certainly maps into p^∞ since elements of the space have only finitely many nonzero terms. Exercise 4.5.8 asks you to show that this operator is linear, and that it does *not* map ℓ^∞ into itself.

To show that T is not bounded we suppose that it is, i.e. suppose there exists $M > 0$ such that $\|T(a_n)\|_\infty \leq M\|(a_n)\|_\infty$ for all $(a_n) \in p^\infty$. Now choose $N \in \mathbb{N}$ such that $N > M$. If we then consider the vector e_N which has a 1 in the Nth position and zeros everywhere else, then $\|e_N\|_\infty = 1$ and we have

$$\|Te_N\|_\infty = N = N\|e_N\|_\infty > M\|e_N\|_\infty$$

which is a contradiction. Thus T cannot be bounded.

Example 4.5.12. As a final example, we consider the class of linear operators whose codomain is $\mathbb{R}$ with norm given by the absolute value. These operators receive special attention and, for the sake of clarity, we have a definition: if X is a normed linear space, then any linear operator $L : X \to \mathbb{R}$ is called a *linear functional*. As a result of Theorem 4.5.9, a linear functional is continuous if and only if there exists $M > 0$ such that $|Lx| \leq M\|x\|_X$. The language *bounded linear functional* is typically used.

For an example, take $X = C([0, 1])$ with the sup-norm and define $K_0 : C([0, 1]) \to \mathbb{R}$ by

$$K_0 f = f(0),$$

the functional which evaluates a function at the origin. Of course, we could do this for any point $x \in [0, 1]$ giving a general functional K_x. To see that this is linear,

$$K_0(f + g) = (f + g)(0) = f(0) + g(0) = K_0 f + K_0 g$$

and

$$K_0(cf) = (cf)(0) = cf(0) = cK_0 f.$$

This is called a *point evaluation functional*.

Example 4.5.13. If X is a Hilbert space and $x_0 \in X$ we can define a linear functional $L : X \to \mathbb{R}$ by

$$Lx = \langle x, x_0 \rangle.$$

The fact that this functional is linear follows from properties of the inner product; that it is also bounded is a consequence of the Cauchy–Schwarz Inequality which shows that $M = \|x_0\|_X$ will satisfy the definition. We close this chapter with one of the most fundamental theorems of Hilbert space theory which states that these are the only bounded linear functionals on a Hilbert space.

Theorem 4.5.14 (Riesz Representation Theorem). *Let X be a Hilbert space and let $L : X \to \mathbb{R}$ be a bounded linear functional. Then there is a unique vector $x_0 \in X$ such that $Lx = \langle x, x_0 \rangle$ for every $x \in X$.*

We will only prove the theorem in the special case of $X = \mathbb{R}^2$ with the Euclidean inner product though we will discuss the extension to general Hilbert spaces after the proof. In our particular case *any* linear functional is continuous, and so our hypothesis that L is bounded is redundant. Before beginning recall that for vector spaces X and Y, the *kernel* and *range* of an operator $T : X \to Y$ are defined by

$$\ker(T) = \{x \in X : T(x) = \mathbf{0}_Y\}$$

and

$$\mathrm{range}(T) = \{y \in Y : y = Tx \text{ for some } x \in X\}.$$

These sets are subspaces of X and Y, respectively, and therefore have dimensions associated with them. In the finite dimensional setting, there is a familiar result, sometimes referred to as the dimension theorem, relating these to the dimension of the domain space. If $T : X \to Y$ and X has dimension n, then

$$n = \dim \ker(T) + \dim \mathrm{range}(T). \tag{4.1}$$

Proof (Theorem 4.5.14). Take $X = \mathbb{R}^2$ with the Euclidean inner product and suppose $L : \mathbb{R}^2 \to \mathbb{R}$ is a linear functional (necessarily continuous by the previous comments). If $Lx = 0$ for every $x \in \mathbb{R}^2$, then $x_0 = \mathbf{0}$ satisfies the conclusion of the theorem. Thus we may assume that L is not the zero functional. We will build the desired vector x_0 in several stages to prove the existence. By Exercise 4.5.12 we know that L maps onto $\mathbb{R}$ and hence $\dim \mathrm{range}(T) = 1$. From Eq. (4.1) it follows that $\dim \ker(L) = 1$ which in turn implies that there is a vector $(x_1, x_2) \in \mathbb{R}^2$ with

$$\ker(L) = \{c(x_1, x_2) : c \in \mathbb{R}\};$$

we are really saying that $\{(x_1, x_2)\}$ is a basis for the kernel of L. Now take the vector $z_0 = (x_2, -x_1)$. This vector is not in $\ker(L)$ and thus $Lz_0 \neq 0$. Also this vector is orthogonal to every vector in $\ker(L)$, i.e. $\langle y, z_0 \rangle = 0$ for every $y \in \ker(L)$, where again our inner product is the Euclidean inner product on $\mathbb{R}^2$.

Next, define $y_0 = z_0/(Lz_0)$. The homogeneity of the inner product shows that y_0 is also orthogonal to every vector in $\ker(L)$. Moreover, since L respects the scalar multiplication,

$$Ly_0 = L\left(\frac{z_0}{(Lz_0)}\right) = \left(\frac{1}{(Lz_0)}\right)Lz_0 = 1.$$

At this point, take any $x \in \mathbb{R}^2$ and form the new vector $(Lx)y_0 - x$ (keep in mind that Lx is a scalar). Evaluating L, we see that

$$L\left[(Lx)y_0 - x\right] = L\left[(Lx)y_0\right] - Lx = LxLy_0 - Lx = 0.$$

This calculation shows that $(Lx)y_0 - x \in \ker(L)$ for every $x \in X$ and the fact that y_0 is orthogonal to each vector in the kernel shows that

$$\langle (Lx)y_0 - x, y_0 \rangle = 0$$

for every $x \in X$. Our goal is now to solve this equation for Lx. Utilizing the properties of the inner product we have

$$0 = \langle (Lx)y_0 - x, y_0 \rangle = \langle (Lx)y_0, y_0 \rangle - \langle x, y_0 \rangle$$
$$= (Lx)\langle y_0, y_0 \rangle - \langle x, y_0 \rangle$$
$$= (Lx)\|y_0\|^2 - \langle x, y_0 \rangle$$

for every $x \in X$ and solving,

$$Lx = (1/\|y_0\|^2)\langle x, y_0 \rangle = \langle x, y_0/\|y_0\|^2 \rangle.$$

Finally, if we take $x_0 = y_0/\|y_0\|^2$, we have $Lx = \langle x, x_0 \rangle$ for every $x \in X$. By substituting the original expression for y_0 into the equation for x_0 we can also formulate x_0 in terms of z_0,

$$x_0 = \frac{Lz_0}{\|z_0\|^2}z_0.$$

We must now show that x_0 is unique. At first this may seem conflicting since the construction of x_0 depends on our initial choice of the point z_0; however, it is the case that x_0 is unique. We will discuss this in more detail shortly. To prove uniqueness we assume that there is another vector $w_0 \in X$ such that $Lx = \langle x, w_0 \rangle$ for all $x \in X$. However, this provides the equation

$$\langle x, x_0 \rangle = \langle x, w_0 \rangle$$

which we can manipulate to see that

$$0 = \langle x, x_0 \rangle - \langle x, w_0 \rangle = \langle x, x_0 - w_0 \rangle$$

for every $x \in X$. Exercise 1.5.1 shows that $x_0 - w_0 = \mathbf{0}$ which implies that $x_0 = w_0$.

$\square$

We aim now to extend our proof to a general Hilbert space X. The first step in the proof consists of making a choice for the vector z_0 with the properties that z_0 is not in $\ker(L)$ and is orthogonal to every vector in $\ker(L)$. We were able to make this choice due to the simple structure in $\mathbb{R}^2$ and the fact that vector spaces of dimension one have very simple bases. In particular, we were able to write down an explicit choice for z_0 depending on the basis vector for $\ker(L)$. If we were to extend this proof to $\mathbb{R}^3$, the basis for $\ker(L)$ would have two vectors and we would be able to use the cross product to produce a vector orthogonal to both, and hence to any vector in the kernel. However, once we move to $\mathbb{R}^4$ applying our technique becomes cloudier and harder still if we attempt to proceed to an infinite dimensional space like ℓ^2 or $C([0, 1])$. But this is where the continuity of the linear functional comes into play; if you pay close attention, we never used this hypothesis in our proof. Thinking in terms of a set theoretic inverse we can write $L^{-1}(\{0\}) = \ker(L)$ and the continuity of L will guarantee us that this set is closed. However, such an approach requires that we know more about the topological approach to continuity, i.e. in terms of inverse images of open or closed sets. The idea is similar to Exercise 4.4.2 in the context of functions $f : \mathbb{R} \to \mathbb{R}$.

Once we know that $\ker(L)$ is closed the Projection Theorem guarantees the existence of a vector $z_0 \in X$ such that z_0 is not in $\ker(L)$ and is orthogonal to every vector in $\ker(L)$. The Projection Theorem is fundamental to understanding the geometry and decomposition structures available in a Hilbert space and, while not beyond our grasp, is beyond the scope of our current course of investigation. The interested reader can visit Sect. 1.3 in [18]. For the general case of the Riesz Representation Theorem, the proof is identical to ours after the choice of z_0 has been justified.

In the special case proved above, there is a far simpler proof (which applies to any finite dimensional inner product space) and we point the reader to Theorem 7.3 in [2]. However, this proof does not generalize readily to the infinite dimensional setting. A common theme when moving from special cases to the more general is finding the "right" technique. Thus we have included the lengthier argument so as to provide some idea as to how the result can be generalized.

Example 4.5.15. To better understand the Riesz Representation Theorem, let's consider a specific linear functional on $\mathbb{R}^2$, again with the Euclidean inner product. Define $L : \mathbb{R}^2 \to \mathbb{R}$ by

$$L(x_1, x_2) = x_1 + x_2.$$

Taking the definition of the given inner product into account, it should be obvious that $x_0 = (1, 1)$ satisfies the conclusion of the theorem,

$$L(x_1, x_2) = x_1 + x_2 = \langle (x_1, x_2), (1, 1) \rangle$$

for every $(x_1, x_2) \in \mathbb{R}^2$. To see how the proof applies, observe that

$$\ker(L) = \{(x_1, x_2) \in \mathbb{R}^2 : x_1 = -x_2\} = \{(-x_2, x_2) : x_2 \in \mathbb{R}\}.$$

One basis for this subspace is $\{(-1, 1)\}$ and hence the proof indicates that we should take $z_0 = (1, 1)$. Notice that $Lz_0 = 2$ and $\|z_0\|^2 = 2$ from which it follows that

$$x_0 = \frac{Lz_0}{\|z_0\|^2} z_0 = (1, 1)$$

as we suspected.

Consider now another choice for z_0. Notice that $\{(-2, 2)\}$ is also a basis for $\ker(L)$ and the proof indicates that we should now choose $z_0 = (2, 2)$. Computing again, we see that $Lz_0 = 4$ and $\|z_0\|^2 = 8$ so that

$$x_0 = \frac{Lz_0}{\|z_0\|^2} z_0 - \frac{1}{2}(2, 2) = (1, 1).$$

This is the uniqueness in action. No matter what basis we take for $\ker(L)$, the choice for z_0 will have the form (x_1, x_1) and the result for x_0 will be $(1, 1)$. Of course this is also true in the general case (as shown by the uniqueness), though, from the computational perspective it is less obvious since the general proof does not provide an explicit choice for z_0.

Exercises

Exercise 4.5.1. Let $(X, \| \cdot \|)$ be a normed linear space. A function $f : X \to \mathbb{R}$ is *uniformly continuous* on X if for every $\varepsilon > 0$ there is a $\delta > 0$ such that $|f(x) - f(y)| < \varepsilon$ whenever $x, y \in X$ with $\|x - y\| < \delta$. Show that the function $f : X \to [0, \infty)$ given by $f(x) = \|x\|$ is uniformly continuous on X.

Exercise 4.5.2. Prove Proposition 4.5.5.

Exercise 4.5.3. Let A be a 2×2 matrix and let T be the linear operator $T : \mathbb{R}^2 \to \mathbb{R}^2$ defined by $Tx = Ax$. Use Theorem 4.5.9 to show that T is continuous with respect to each of the given norms.

(a) Consider $\mathbb{R}^2$ with the norm $\|x\| = |x_1| + |x_2|$.

(b) Consider $\mathbb{R}^2$ with the Euclidean norm $\|x\| = \sqrt{(x_1)^2 + (x_2)^2}$. You may find the inequality $(x + y)^2 \leq 4(x^2 + y^2)$ helpful.

Exercise 4.5.4. (a) Show that the operators S and M_λ from Example 4.5.4 are linear.
(b) Show that these operators are bounded.

Exercise 4.5.5. Show in detail that $C([0, 1])$ is a vector space and that the rule $\|\cdot\|_\infty$ defines a norm.

Exercise 4.5.6. Let $g : [0, 1] \to \mathbb{R}$ be continuous on $[0, 1]$. Show that the multiplication operator M_g is bounded on $C([0, 1])$.

Exercise 4.5.7. Let $\varphi : [0, 1] \to [0, 1]$ be continuous.

(a) Show that $C_\varphi : C([0, 1]) \to C([0, 1])$ is a linear operator. Make sure to mention why we know the operator maps into the given codomain.
(b) Show that C_φ is bounded.

Exercise 4.5.8. (a) Show that the operator T from Example 4.5.11 is linear.
(b) Show that when the operator T is defined on ℓ^∞ it does not map into ℓ^∞.

Exercise 4.5.9. Show that the point evaluation functional K_x from Example 4.5.12 is bounded.

Exercise 4.5.10. Let X be a Hilbert space and let $x_0 \in X$. Show that the functional $L : X \to \mathbb{R}$ defined by

$$Lx = \langle x, x_0 \rangle$$

is a bounded linear functional.

Exercise 4.5.11. Let X and Y be vector spaces and $T : X \to Y$ a linear operator. Show that $\ker(T)$ and $\mathrm{range}(T)$ are subspaces of X and Y, respectively.

Exercise 4.5.12. The *zero functional* on a normed linear space X satisfies $\mathbf{0}(x) = 0$ for every $x \in X$. If $T : X \to \mathbb{R}$ is a linear functional which is not the zero functional, show that T maps onto $\mathbb{R}$.

Chapter 5
The Derivative

Though many of the topics we have covered to this point were explained in an intuitive manner in your previous calculus courses, it is likely that the concept of differentiation was covered in a more rigorous fashion. This is due to the fact that most of the interesting properties and consequences of the derivative are easy to prove though this does not detract from their importance. The reason for this can be attributed to the fact that the derivative is defined as a limit and, even with only a limited understanding of limits, many results about the derivative are direct consequences of the properties of functional limits and continuity. This begs the question of why we should revisit this topic if you have already seen it presented in detail. The first answer to this question is simply that ignoring the theory of differentiation would leave our story with a large plot hole. The derivative has been a key player in the study of analysis for several hundred years and therefore deserves a place of prominence. The second answer is somewhat more satisfying. Yes, we will explore ideas you are familiar with, but our perspective will be of a theoretic nature rather than the computational and application-driven perspectives of calculus. And, we will delve deeper and consider functions with much more bizarre behavior than those you have previously attempted to differentiate.

5.1 The Definition of the Derivative

As it originated, the derivative was concerned with investigating the instantaneous rate of change of a function. Geometrically, this is interpreted as identifying the slope of the line tangent to the graph of a function at domain point c. This is done by considering a point in the domain x and finding the slope of the secant line through $(x, f(x))$ and $(c, f(c))$. The limit is then taken as $x \to c$. This is illustrated in Fig. 5.1 as the points x_1, x_2, x_3 are moving closer to c. Though we will not employ the notion of secant and tangent lines explicitly, visuals of this sort will help support and confirm our intuition. Also, in order to guarantee that this limit process is producing only the most accurate estimates for the derivative,

M.A. Pons, *Real Analysis for the Undergraduate: With an Invitation to Functional Analysis*, 183
DOI 10.1007/978-1-4614-9638-0_5, © Springer Science+Business Media New York 2014

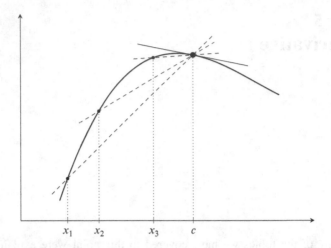

Fig. 5.1 Sequence of secant lines converging to tangent line

for a given domain point c, we will require that our function be defined in an open neighborhood around c. This is most easily accomplished by restricting our attention to functions which are defined on an interval. Throughout (a, b) will denote a non-degenerate open interval.

Definition 5.1.1. Let $f : (a, b) \to \mathbb{R}$ and let $c \in (a, b)$. We say that f is *differentiable* at c if the limit

$$\lim_{x \to c} \frac{f(x) - f(c)}{x - c}$$

exists. In this case we denote the limit above by the symbol $f'(c)$ and call it the *derivative of f at c*. If f is differentiable at each point of (a, b), we say that f is *differentiable* on (a, b) and the function $f' : (a, b) \to \mathbb{R}$ is called the *derivative* of f.

Keep in mind that we have several formulations for a functional limit. In terms of sequences, $f'(c)$ is interpreted as follows: a function f is differentiable at c if for every sequence (x_n) in (a, b) with $x_n \neq c$ for all n and $\lim x_n = c$, it follows that

$$\lim \frac{f(x_n) - f(c)}{x_n - c}$$

exists and converges to the common value $f'(c)$. Notice here that this only makes sense with the assumption provided by the definition of a limit that $x_n \neq c$ for all n. Before investigating examples, recall that the derivative has a second form which is typically used in the computational setting. As we explore the examples we will discuss why the second form is used more frequently for such tasks.

Proposition 5.1.2. *A function $f : (a, b) \to \mathbb{R}$ is differentiable at a point $c \in (a, b)$ if and only if the limit*

$$\lim_{h \to 0} \frac{f(c + h) - f(c)}{h}$$

exists. Moreover, this quantity is equal to $f'(c)$.

Example 5.1.3. Taking f to be the function $f(x) = 3x + 2$, we know that the graph of f is a line with slope 3. From a tangent line point of view, it makes sense that the only line which is "tangent" to another line is the line itself, and thus it seems reasonable to conjecture that $f'(c) = 3$ for all $c \in \mathbb{R}$. To verify this observe that

$$\lim_{x \to c} \frac{(3x + 2) - (3c + 2)}{x - c} = \lim_{x \to c} \frac{3(x - c)}{x - c} = \lim_{x \to c} 3 = 3;$$

cancelation of the factor $x - c$ is allowable since the limit requires that $x \neq c$. This basic example highlights the process of calculating derivatives from the definitions. The key is to algebraically manipulate the numerator in such a way that generates a factor of $x - c$, after which cancelation is possible. We then compute the remaining limit. This process becomes more complicated as the function grows in complexity, and, in order to improve efficiency, we will review the algebraic properties of the operation of differentiation in the following section. The fact that the derivative shortcuts exist should by no means take away from the importance of the definition as it is our foundation. In the exercises you will be asked to compute several more derivatives from the definition to encourage an appreciation for this particular operation.

Example 5.1.4. Let $f(x) = x^2$ and let $c \in \mathbb{R}$. To calculate the derivative, we factor the numerator and cancel,

$$f'(c) = \lim_{x \to c} \frac{x^2 - c^2}{x - c} = \lim_{x \to c} \frac{(x + c)(x - c)}{x - c} = \lim_{x \to c} x + c = 2c.$$

The last equality is due to the continuity of the polynomial $p(x) = x + c$.

Let's also compute the derivative using the second form from Proposition 5.1.2. The result is the same, but the algebraic strategy involved is not. We have

$$f'(c) = \lim_{h \to 0} \frac{(c + h)^2 - c^2}{x - c} = \lim_{h \to 0} \frac{c^2 + 2ch + h^2 - c^2}{x - c} = \lim_{h \to 0} \frac{h(2c + h)}{h}$$

$$= \lim_{h \to 0} 2c + h = 2c$$

where the final equality here is due to the continuity of the polynomial $p(h) = 2c + h$. This computation yields the same result as the previous calculation, but the process is fundamentally different from the algebraic perspective. Here we expanded the expression $(c + h)^2$ and then canceled what we could. It may not seem like a huge difference, factoring versus expanding, but the fact that expanding polynomial expressions, though possibly tedious as the polynomials become more complicated, is always doable makes the second form more appealing for the beginner. The situation is more complicated for other types of functions, but the second form is generally the preferred.

Consider the function $g(x) = \sin x$. Though we won't prove that the derivative of this function is $g'(x) = \cos x$ here, we can get a feel for why the second form of the derivative is again a more usable form when first encountered. With the first form, we would have the expression $\sin x - \sin c$ and the task would be to relate this to $x - c$. The second form centers around the numerator $\sin(c + h) - \sin c$. In this case however, the sum rule

$$\sin(\alpha + \beta) = \sin(\alpha)\cos(\beta) + \sin(\beta)\cos(\alpha)$$

can be employed. This still leaves quite a bit of work, but the ball is rolling.

Example 5.1.5. For a more interesting example, take $f(x) = |x|$. The behavior of this function depends on the sign of the input and we will have to take this into account as we seek to find its derivative. First consider $c \in (0, \infty)$ and let (x_n) be any sequence in $\mathbb{R}$ with $x_n \neq c$ for all n and $\lim x_n = c$. The fact that $c > 0$ and the convergence of (x_n) guarantees that there is an $N \in \mathbb{N}$ such that $x_n > 0$ for all $n \geq N$. Thus $f(x_n) = x_n$ if $n \geq N$. From this, we deduce that

$$\frac{f(x_n) - f(c)}{x_n - c} = \frac{x_n - c}{x_n - c} = 1$$

for all $n \geq N$, i.e. the sequence above is eventually a constant sequence. With this it is clear that

$$f'(c) = \lim \frac{f(x_n) - f(c)}{x_n - c} = 1.$$

A similar argument shows that $f'(c) = -1$ for all $c \in (-\infty, 0)$.

The behavior at 0 is more intriguing since the behavior of the function changes *at* 0. When $c = 0$, the expression used to compute the derivative takes the form

$$\lim_{x \to 0} \frac{f(x) - f(0)}{x - 0} = \lim_{x \to 0} \frac{|x|}{x};$$

in part (a) of Exercise 4.1.9 you showed that this limit does not exist. Thus f does not have a derivative at 0 and we conclude that $f' : (-\infty, 0) \cup (0, \infty) \to \mathbb{R}$ is given by

$$f'(x) = \begin{cases} -1, & x < 0; \\ 1, & x > 0. \end{cases}$$

Geometrically this is a result of the fact that the graph of f has a sharp corner at 0. The derivative is examining the smoothness of a given curve and abrupt changes in the behavior of the function often lead to non-smooth transitions in the graph. This reasoning gives us an intuitive means of identifying where functions are not differentiable but not all problems with respect to the existence of a derivative are caused by corners and cusps. There are much more problematic issues that will arise as we consider more extreme functions.

As a second point, when considering functions defined on an open interval, continuity is also measuring the smoothness of a curve, but in a less restrictive manner than the derivative. To understand this the function $f(x) = |x|$ is the typical example; it is continuous at 0, however it is not differentiable there. Thus continuity does not imply differentiability, but the reverse implication does hold; the proof is very insightful as to why this is true. In the proof we will make use of the sequential definition of functional limit to provide a specific context.

Theorem 5.1.6. *Let $f : (a, b) \to \mathbb{R}$ and let $c \in (a, b)$. If f is differentiable at c, then f is continuous at c.*

Proof. To show that f is continuous at c, let (x_n) be any sequence in (a, b) with $x_n \neq c$ for all $n \in \mathbb{N}$ and $\lim x_n = c$. By Exercise 4.2.4(a), it suffices to show that $(f(x_n) - f(c)) \to 0$. Notice this expression is exactly the numerator of the quotient used to define the derivative. Also, the assumption that $x_n \neq c$ for all n implies that $x_n - c \neq 0$ for all n. Hence we may introduce this factor in the denominator, and also in the numerator. Employing the fact that f is differentiable at c and the Algebraic Limit Theorem, we have

$$\lim[f(x_n) - f(c)] = \lim \left[\left(\frac{f(x_n) - f(c)}{x_n - c} \right) (x_n - c) \right]$$

$$= \lim \left[\frac{f(x_n) - f(c)}{x_n - c} \right] \lim(x_n - c)$$

$$= f'(c) \cdot 0 = 0.$$

Therefore we conclude that f is continuous at c.   □

The next natural question to ask is whether the new function f' is necessarily continuous. To be more precise here, let $f : (a, b) \to \mathbb{R}$ be differentiable on (a, b), i.e. $f'(x)$ exists for every $x \in (a, b)$. Then, is f' continuous on (a, b)? Unfortunately the answer is no. To provide an explicit example of a function with a noncontinuous derivative, it is tempting to return to our investigation of $f(x) = |x|$ from above, but this example doesn't hold any meaning in this context since the derivative does *not* exist at 0. In other words, the domain of f' is not the same as that of f. In fact, many of the functions you've been exposed to thus far have the property that their derivatives are continuous where they exist. This is largely due to the fact that the most common elementary functions have second derivatives which force their first derivative to be continuous. We will discuss this in more detail and

define higher derivatives in the next section. We close this section with a variant of
the absolute value function which is differentiable at 0.

Example 5.1.7. Define a function $g : \mathbb{R} \to \mathbb{R}$ by

$$g(x) = \begin{cases} -x^2, & x < 0; \\ x^2, & x \geq 0. \end{cases}$$

To compute the derivative here, we will use our previous calculation from
Example 5.1.4. If $c > 0$, then we know that $g'(c) = 2c$. Also, if $c < 0$,
then computing via the definition, we find that $g'(c) = -2c$. To calculate the
derivative for $c = 0$, we will employ the notion of left- and right-hand limits from
Exercise 4.1.12. We have

$$\lim_{x \to 0+} \frac{g(x) - g(0)}{x - 0} = \lim_{x \to 0+} \frac{x^2}{x} = 0$$

and

$$\lim_{x \to 0-} \frac{g(x) - g(0)}{x - 0} = \lim_{x \to 0-} \frac{-x^2}{x} = 0.$$

Thus we conclude that $g'(0) = 0$. These three calculations show that

$$g'(x) = \begin{cases} -2x, & x < 0; \\ 0, & x = 0; \\ 2x, & x > 0; \end{cases} = 2|x|.$$

This function is continuous at 0; however, it does not have a derivative there.

Exercise 5.1.7 asks you to come up with examples of functions which are not
differentiable on point sets of a certain size. At first it may seem that there is
some upper bound to the number of non-differentiable points that can exist for a
continuous function. However this is due to our intuitive notion that relates non-
differentiable points to corners on a graph. The truth of the matter is that there exist
continuous functions which are not differentiable at any domain point. Section 6.4
provides an example of such a function and we will also consider whether such
an occurrence is an oddity or if they occur more frequently among the continuous
functions than not.

Exercises

Exercise 5.1.1. Extend Example 5.1.3 by showing that the derivative of the function $f(x) = mx + b$, where $m, b \in \mathbb{R}$, is given by $f'(x) = m$.

Exercise 5.1.2. Use the definition of derivative to calculate f' for $f(x) = ax^2 + bx + c$ where $a, b, c \in \mathbb{R}$.

Exercise 5.1.3. Let $n \in \mathbb{N}$ and set $f(x) = x^n$. Show that $f'(x) = nx^{n-1}$. **Note:** You may use either form of the definition of the derivative but keep in mind the difference between these. If using the first form, you will need to be able to factor the expression $x^n - c^n$ in order to generate a factor of $x - c$. If using the second form, you will need to expand $(c + h)^n$ in which case you should consider the binomial expansion theorem.

Exercise 5.1.4. Use the definition of the derivative to calculate f' for each of the following.

(a) $f(x) = k$ where $k \in \mathbb{R}$
(b) $f(x) = \sqrt{x + 3}$
(c) $f(x) = 1/x$
(d) $f(x) = 4x - 2/x^2$

Exercise 5.1.5. Apply the sequential characterization of a functional limit to prove Proposition 5.1.2.

Exercise 5.1.6. Construct a function which is differentiable at *exactly* one point. **Hint:** Consider a modification of the Dirichlet function.

Exercise 5.1.7. (a) Construct a function which is continuous on $\mathbb{R}$ and is non-differentiable at exactly two points.
(b) Construct a function which is continuous on $\mathbb{R}$ and is non-differentiable at countably many points.

Exercise 5.1.8. Show that Thomae's function (Exercise 4.1.10) is not differentiable on the set of irrational numbers. To do this consider the sequential characterization of functional limit for a derivative and consider sequences consisting only of irrational numbers and sequences consisting only of rational numbers. The article [5] generalizes this result.

Exercise 5.1.9. Suppose $f : \mathbb{R} \to \mathbb{R}$ with $f(x + y) = f(x)f(y)$. Also suppose that f is not the zero function.

(a) Show that $f(0) = 1$.
(b) If f is differentiable on $\mathbb{R}$, show that $f'(x) = f'(0)f(x)$ for all $x \in \mathbb{R}$.

Exercise 5.1.10. Suppose $f : (0, \infty) \to \mathbb{R}$ with $f(\frac{x}{y}) = f(x) - f(y)$.

(a) Show that $f(1) = 0$.
(b) Show that f is continuous on $(0, \infty)$ if and only if f is continuous at $c = 1$.
(c) If f is differentiable on (a, b), show that $f'(x) = \frac{f'(1)}{x}$ for all $x \in (a, b)$.

Exercise 5.1.11. Suppose $f : (a, b) \to \mathbb{R}$ and let $x \in \mathbb{R}$. Show that f is differentiable at x if and only if there exists a function $g : (a, b) \to \mathbb{R}$ such that $f(y) = f(x) + (y - x)g(y)$ for all $y \in (a, b)$. In this case, show that $f'(x) = g(x)$.

5.2   Properties of the Derivative

To improve efficiency in calculating derivatives, we now investigate the basic algebraic properties of differentiation. While these follow from their Algebraic Limit Theorem cousins, several of the results are more complicated due to the more complex definition of the derivative. We are of course referring to the product, quotient, and chain rules. In the first-year calculus sequence students are often annoyed by the fact that the derivative of a product is not simply the product of the derivatives, not to mention the confusion surrounding the quotient and chain rules. From a more advanced perspective, we will take heart in the fact that, while the forms are not as simple as first anticipated, there are concise algebraic relationships intertwining these concepts.

Theorem 5.2.1 (Algebraic Limit Theorem for Derivatives). *Let $f : (a, b) \to \mathbb{R}$ and $g : (a, b) \to \mathbb{R}$, and suppose both are differentiable at $c \in (a, b)$. Then*

(a) kf is differentiable at c with $(kf)'(c) = kf'(c)$ for all $k \in \mathbb{R}$;
(b) $f + g$ is differentiable at c with $(f + g)'(c) = f'(c) + g'(c)$;
(c) (Product Rule) fg is differentiable at c with $(fg)'(c) = f'(c)g(c) + f(c)g'(c)$;
(d) (Quotient Rule) f/g is differentiable at c with $\left(\dfrac{f}{g}\right)'(c) = \dfrac{f'(c)g(c) - f(c)g'(c)}{[g(c)]^2}$
    provided $g(c) \neq 0$.

Proof. The proof of parts (a) and (b) are simple consequence of the Algebraic Limit Theorem. For (b), we have

$$(f + g)'(c) = \lim_{x \to c} \frac{(f + g)(x) - (f + g)(c)}{x - c}$$

$$= \lim_{x \to c} \frac{f(x) + g(x) - f(c) - g(c)}{x - c}$$

$$= \lim_{x \to c} \left[\frac{f(x) - f(c)}{x - c} + \frac{g(x) - g(c)}{x - c} \right]$$

$$= \lim_{x \to c} \left[\frac{f(x) - f(c)}{x - c} \right] + \lim_{x \to c} \left[\frac{g(x) - g(c)}{x - c} \right]$$

$$= f'(c) + g'(c).$$

The proof of (a) is considered in Exercise 5.2.2.

For (c), the result isn't as clean as one would hope; this seems to be due to the fact that the quantity $f(x)g(x) - f(c)g(c)$ doesn't factor as $[f(x) - f(c)][g(x) - g(c)]$, but even this wouldn't give us $(fg)'(c) = f'(c)g'(c)$. To obtain the product rule above, we will use a technique similar to that used to show that the product of convergent sequences converges by adding and subtracting an appropriate term. Observe that

$$(fg)'(c) = \lim_{x \to c} \frac{(fg)(x) - (fg)(c)}{x - c}$$

$$= \lim_{x \to c} \frac{f(x)g(x) - f(c)g(c)}{x - c}$$

$$= \lim_{x \to c} \frac{f(x)g(x) - f(c)g(x) + f(c)g(x) - f(c)g(c)}{x - c}$$

$$= \lim_{x \to c} \left[\frac{f(x)g(x) - f(c)g(x)}{x - c} + \frac{f(c)g(x) - f(c)g(c)}{x - c} \right]$$

$$= \lim_{x \to c} \left[g(x) \left(\frac{f(x) - f(c)}{x - c} \right) + f(c) \left(\frac{g(x) - g(c)}{x - c} \right) \right].$$

Using the fact that f and g are differentiable at c and Theorem 5.1.6 applied to g, the Algebraic Limit Theorem shows that the last expression above is equal to

$$\left[\lim_{x \to c} g(x) \right] \left[\lim_{x \to c} \frac{f(x) - f(c)}{x - c} \right] + f(c) \left[\lim_{x \to c} \frac{g(x) - g(c)}{x - c} \right] = g(c)f'(c) + f(c)g'(c)$$

which is equivalent to the desired conclusion.

For (d), the proof is similar and we provide some algebraic insights leaving the details as an exercise. Observe that

$$\frac{\frac{f(x)}{g(x)} - \frac{f(c)}{g(c)}}{x - c} = \frac{f(x)g(c) - f(c)g(x)}{g(x)g(c)(x - c)}$$

$$= \frac{f(x)g(c) - f(c)g(c) + f(c)g(c) - f(c)g(x)}{g(x)g(c)(x - c)}$$

$$= \frac{1}{g(x)g(c)} \left[g(c) \left(\frac{f(x) - f(c)}{x - c} \right) - f(c) \left(\frac{g(x) - g(c)}{x - c} \right) \right]. \quad \square$$

Theorem 5.2.2 (Chain Rule). *Let $A, B \subseteq \mathbb{R}$ be open intervals. Also let $f : A \to \mathbb{R}$ and $g : B \to \mathbb{R}$ with $f(A) \subseteq B$. If f is differentiable at $c \in A$ and g is differentiable at $f(c) \in B$, then $g \circ f$ is differentiable at c and*

$$(g \circ f)'(c) = g'(f(c))f'(c).$$

The proof of the chain rule stands in contrast to the previous algebraic properties of the derivative in that it requires a more precise argument. And, while it may take some time to absorb the argument that is to follow, the only tools employed are the definition of the derivative, Theorems 4.2.7 and 5.1.6, and algebraic manipulation. To understand why the proof is more complicated even though the tools are essentially the same as those used in the previous theorem, first consider the following naive approach for calculating the derivative of a composition. We will attempt to multiply and divide by an appropriate term and apply the Algebraic Limit Theorem,

$$(g \circ f)'(c) = \lim_{x \to c} \frac{g(f(x)) - g(f(c))}{x - c}$$

$$= \lim_{x \to c} \left[\left(\frac{g(f(x)) - g(f(c))}{f(x) - f(c)} \right) \left(\frac{f(x) - f(c)}{x - c} \right) \right]$$

$$= g'(f(c)) f'(c).$$

However, in order to divide here, we must guarantee that $f(x) - f(c) \neq 0$. This is not supplied by our hypothesis, and adding this restriction (for instance, we could assume that f is one-to-one) detracts from the full flavor of the theorem. This simple complication means that we cannot apply our multiply and divide technique. With a bit more savvy, however, we can work our way around this. As a final comment, do not feel downtrodden if the proof seems overly difficult. The first proof of the chain rule was given by Cauchy with the flaw mentioned above!

Proof. We begin with a new function G defined on B by

$$G(y) = \begin{cases} \dfrac{g(y) - g(f(c))}{y - f(c)} - g'(f(c)), & y \neq f(c); \\ 0, & y = f(c). \end{cases}$$

For $y \neq f(c)$, G calculates the difference between the quotient used to define the derivative of g and the value of g' at $f(c)$, and avoids the zero denominator by only considering points $y \in B$ with $y \neq f(c)$. First observe that G is continuous at $f(c)$ by the Algebraic Limit Theorem,

$$\lim_{y \to f(c)} G(y) = \lim_{y \to f(c)} \left[\frac{g(y) - g(f(c))}{y - f(c)} - g'(f(c)) \right]$$

$$= g'(f(c)) - g'(f(c)) = 0 = G(f(c))$$

since g is differentiable at $y = f(c)$. Moreover, since f is differentiable at c, it is therefore also continuous at c, and Theorem 4.2.7 guarantees us that $G \circ f$ is continuous at c with

$$\lim_{x \to c} G(f(x)) = G(f(c)) = 0.$$

If $y \in B$ with $y \neq f(c)$, we can rearrange the first line in the definition of G showing that

$$g(y) - g(f(c)) = [G(y) + g'(f(c))] (y - f(c)). \tag{5.1}$$

Furthermore, this relationship also holds when $y = f(c)$ since both sides of the equation are zero in this case. It is this observation that alleviates the problematic issue discussed before the proof. By considering this new function, we are able to move a potential zero to the numerator where it is no longer a worry. Now, consider $x \in A$ with $x \neq c$. Then $f(x) \in B$ and, substituting from Eq. (5.1), we have

$$\frac{g(f(x)) - g(f(c))}{x - c} = \frac{[G(f(x)) + g'(f(c))] (f(x) - f(c))}{x - c}$$

$$= [G(f(x)) + g'(f(c))] \left(\frac{f(x) - f(c)}{x - c} \right).$$

At this point, we take a limit as $x \to c$ and apply the Algebraic Limit Theorem with our previous observations about G to conclude that

$$(g \circ f)'(c) = \lim_{x \to c} \frac{g(f(x)) - g(f(c))}{x - c}$$

$$= \lim_{x \to c} \left[[G(f(x)) + g'(f(c))] \left(\frac{f(x) - f(c)}{x - c} \right) \right]$$

$$= [0 + g'(f(c))] f'(c)$$

$$= g'(f(c)) f'(c). \qquad \square$$

The exercises will ask you to apply the algebraic properties above to some elementary functions. To provide for a more interesting assortment of examples, we will take it on faith that the derivative of the function $f(x) = \sin x$ is $f'(x) = \cos x$.

We now turn our attention to a discussion of higher derivatives. Given a function f, its derivative, provided it exists, is a new function to which we could again apply the process of differentiation. When considered as an iterative process, the need for more definitions becomes unnecessary. On the other hand, the benefit of having a definition is that it provides a cohesive set of agreed upon terms which aid in clarifying our conversation. Thus we have the following.

Definition 5.2.3. Let $f : (a, b) \to \mathbb{R}$. If f' exists on (a, b), then we say that f is *twice-differentiable at* $c \in (a, b)$ if $(f')'(c)$ exists. If $(f')'(c)$ exists for all $c \in (a, b)$, then we say f is *twice-differentiable on* (a, b) and the function $f'' = (f')'$ is called the *second derivative of* f.

Continuing in a recursive fashion, if $n \in \mathbb{N}$ and $f^{(n-1)}$ exists on (a, b), then we say that f is *n-times differentiable at* $c \in (a, b)$ if $(f^{(n-1)})'(c)$ exists. If $(f^{(n-1)})'(c)$ exists for all $c \in (a, b)$, then we say f is *n-times differentiable on* (a, b) and the function $f^{(n)} = (f^{(n-1)})'$ is called the nth *derivative of* f.

Don't let the notation wear you down. If the first derivative exists in an open neighborhood of a point c, then we can attempt to compute the second derivative, and so on. The use of tick-marks for derivatives become excessive after the third, typically, and then we switch to the standard parenthetical notation. The parentheses are used to distinguish a derivative from a power of the function.

Example 5.2.4. For the sake of computation, let $f(x) = 4x^3 + 6x^2 - x + 1$. This function is defined on $\mathbb{R}$ and we apply Theorem 5.2.1 with Exercise 5.1.3 to find the following derivatives,

$$f(x) = 4x^3 + 6x^2 - x + 1$$
$$f'(x) = 12x^2 + 12x - 1$$
$$f''(x) = 24x + 12$$
$$f'''(x) = 24$$
$$f^{(4)}(x) = 0.$$

As another example, take the function g from Example 5.1.7. We found that $g'(x) = 2|x|$. We also showed that the absolute value function is not differentiable at 0. The function g is thus an example of a function which has a continuous first derivative, but which is not twice-differentiable at 0.

The fact that this process is defined in an iterative fashion means that any result we have that provides a relationship between a function and its derivative applies to this process. For instance, if we are asked to find a function which does not have a second derivative at a certain point, it suffices to find a function whose first derivative is not continuous at that point, for if the second derivative did exist, then the first derivative would have to be continuous there. With this manner of thinking, we can describe a hierarchy for considering higher derivatives. Let $f : (a, b) \to \mathbb{R}$. Beginning in the top left corner, if you answer affirmatively, proceed according to the arrow to the next station. If your answer is negative, you cannot move forward.

Is f continuous on (a, b)? $\to$ Does f' exist on (a, b)?

Is f' continuous on (a, b)? $\to$ Does f'' exist on (a, b)?

Is f'' continuous on (a, b)? $\to$ Does f''' exist on (a, b)?

$$\vdots$$

For polynomials, the process continues indefinitely because we eventually come to the zero function. For many other functions, we lose domain points as we take higher and higher derivatives, but this process tries to preserve the domain of the original function. For a final example, consider the function $h : \mathbb{R} \to \mathbb{R}$ defined by

$$h(x) = \begin{cases} x^2 \sin(1/x), & x \neq 0; \\ 0, & x = 0. \end{cases}$$

This is reminiscent of the function f from Example 4.2.5 which we showed to be continuous at 0. You are asked to verify that this function is also continuous at 0 in the exercises. We will take the continuity of the sine function for granted and attempt to compute h'. If $c \neq 0$, then there is a neighborhood about c where the function is defined by the top row. Using our derivative properties, Exercise 5.1.4(b), and our assumption that we know the derivative of $\sin x$, we find that $h'(c) = 2c \sin(1/c) - \cos(1/c)$. If $c = 0$, we use the definition of derivative to calculate

$$h'(0) = \lim_{x \to 0} \frac{h(x) - h(0)}{x - 0} = \lim_{x \to 0} \frac{x^2 \sin(1/x)}{x} = \lim_{x \to 0} x \sin(1/x) = 0$$

where the last equality is due to Exercise 4.2.5. Thus we have

$$h'(x) = \begin{cases} 2x \sin(1/x) - \cos(1/x), & x \neq 0; \\ 0, & x = 0. \end{cases}$$

In order to consider the second derivative, we first consider the continuity of h'. Notice the top row in the definition of h' contains the term $\cos(1/x)$. As was the case with $\sin(1/x)$, this function does not have a limit as $x \to 0$. However, the first term $2x \sin(1/x)$ has limit 0 as $x \to 0$. With these two bits of information and Exercise 4.1.5(b), we conclude that

$$\lim_{x \to 0} 2x \sin(1/x) - \cos(1/x)$$

does not exist. Therefore f' is not continuous at 0, and hence not twice-differentiable there.

Of all of our examples, this one is the most complicated. It was fairly easy to modify the absolute value function to come up with a function that had a continuous first derivative and which was not twice-differentiable at 0 (see Example 5.1.7), but we had to work harder to come up with a function whose first derivative exists but has a discontinuity. The reason for this is that while derivatives need not be continuous, the types of discontinuities that can occur are of the most bizarre type. Although we have not classified discontinuities of functions formally in our discussion (see Exercise 4.2.13), there are three basic types. Graphically, the possibilities are a hole in the graph, a jump in the graph, or something more severe, e.g. an asymptote or something akin to the zany behavior exhibited by $\sin(1/x)$ around 0. We will see in the next section that only this third category is permissible for a derivative.

Exercises

Exercise 5.2.1. Find the derivative for each of the following functions.

(a) $f(x) = (4x^3 + x^2 + 6x - 3)(3x^6 - x^4)$

(b) $g(x) = \dfrac{x^2 - 2}{x^3 + 4x}$

(c) $h(x) = x^2 \sin^2(2x)$

Exercise 5.2.2. Let $f : (a, b) \to \mathbb{R}$ and $g : (a, b) \to \mathbb{R}$, and suppose both are differentiable at $c \in (a, b)$.

(a) Show that kf is differentiable at c with $(kf)'(c) = kf'(c)$ for all $k \in \mathbb{R}$.
(b) Show that $f - g$ is differentiable at c with $(f - g)'(c) = f'(c) - g'(c)$.
(c) Supply a formal proof of Theorem 5.2.1(d).

Exercise 5.2.3. Let $n \in \mathbb{N}$ and set $f(x) = x^n$. Reprove that $f'(x) = nx^{n-1}$ by using induction and the product rule.

Exercise 5.2.4. Suppose $p : \mathbb{R} \to \mathbb{R}$ is a polynomial of degree n, where $n \in \mathbb{N}$. Show that $p^{(n+1)}$ is the zero function.

Exercise 5.2.5. For $n \in \mathbb{N}$, define $f_n : \mathbb{R} \to \mathbb{R}$ by

$$f_n(x) = \begin{cases} -x^n, & x < 0; \\ x^n, & x \geq 0. \end{cases}$$

When $n = 1$, $f_1(x) = |x|$ and we know that this function is differentiable everywhere except zero. When $n = 2$, the function has a continuous first derivative, but it is not twice-differentiable at 0 as discussed in Examples 5.1.7 and 5.2.4.

(a) Show that f_3 has a continuous second derivative, but is not thrice-differentiable at 0.
(b) Show that f_n has a continuous $(n - 1)$-derivative, but that it is not n-times differentiable at 0.

Exercise 5.2.6. (a) Show that the function h from Example 5.2.4 is continuous at 0.
(b) Show that $\lim_{x \to 0} \cos(1/x)$ does not exist.

Exercise 5.2.7. For $p \geq 0$, define $g : \mathbb{R} \to \mathbb{R}$ by

$$g_p(x) = \begin{cases} x^p \sin(1/x), & x \neq 0; \\ 0, & x = 0. \end{cases}$$

When $p = 0$ we know that this function is continuous everywhere except 0. When $p = 1$, we know that the function is continuous on $\mathbb{R}$.

(a) Show that g_p is not differentiable at 0 when $p = 1$.
(b) Find a value of p so that g_p is differentiable on $\mathbb{R}$ and g'_p is continuous but not differentiable at zero. Please prove that your choice of p works (you may take for granted the differentiability of the sine function).

5.3   Value Theorems for the Derivative

The theorems in this section are called value theorems because they all assert the existence of a domain point corresponding to a pre-specified output value; here, the function under consideration is a derivative. We begin with a familiar lemma.

Lemma 5.3.1 (Interior Extremum Theorem). *Suppose* $f : [a,b] \to \mathbb{R}$ *is continuous on* $[a,b]$ *and differentiable on* (a,b). *If* $c \in (a,b)$ *satisfies* $f(x) \le f(c)$ *for all* $x \in (a,b)$ *or* $f(x) \ge f(c)$ *for all* $x \in (a,b)$, *then* $f'(c) = 0$.

Proof. Suppose $f : [a,b] \to \mathbb{R}$ is continuous on $[a,b]$ and differentiable on (a,b). Further suppose $c \in (a,b)$ satisfies $f(x) \le f(c)$ for all $x \in (a,b)$; this is equivalent to saying that f has a maximum at c. Now, since f is differentiable at c, if (x_n) is any sequence in (a,b) with $x_n \ne c$ for all n and $\lim x_n = c$, it must follow that

$$\lim \frac{f(x_n) - f(c)}{x_n - c} = f'(c).$$

We will consider two particular types of sequences converging to c, one which shows that $f'(c) \ge 0$ and one which shows that $f'(c) \le 0$. The conclusion that $f'(c) = 0$ will then follow.

First take a sequence (x_n) in (a,b) with $x_n > c$ for all n and $\lim x_n = c$. This is allowable since c is in the *open* interval (a,b). The hypotheses on x_n guarantee us that $x_n - c > 0$ for all n while the hypothesis that c is a maximum for f guarantees us that $f(x_n) - f(c) \le 0$ for all n. This implies that the quotient

$$\frac{f(x_n) - f(c)}{x_n - c} \le 0$$

for all n. The differentiability of f at c then implies that this sequence converges to $f'(c)$, and the Order Limit Theorem assures us that $f'(c) \le 0$.

Now, take a sequence (y_n) in (a,b) with $y_n < c$ for all n and $\lim y_n = c$. With this arrangement, we have that $y_n - c < 0$ for all n, and $f(y_n) - f(c) \le 0$ for all n. Thus we have

$$\frac{f(y_n) - f(c)}{y_n - c} \ge 0$$

for all n. This quotient also converges to $f'(c)$, and applying the Order Limit Theorem again shows that $f'(c) \ge 0$. Thus we conclude that $f'(c) = 0$.

The result for the case that f has a minimum at $c \in (a,b)$ follows similarly. $\square$

Our first main result is related to the Intermediate Value Theorem which we discussed in Sect. 4.3. There we showed that any continuous function has the intermediate value property. We will now prove that any derivative must satisfy the intermediate value property even though a derivative may be discontinuous. The power of this theorem is that it classifies the possible discontinuities that can afflict a derivative. Considering the classification of discontinuities in Exercise 4.2.13, it should be easy to see that any function which satisfies the intermediate value property cannot have a removable or jump discontinuity. The conclusion is then that any discontinuity for a derivative is of the essential category, meaning that the problem runs deep in nature of the function. The function h from Example 5.2.4 exhibits this phenomenon.

Theorem 5.3.2 (Darboux's Theorem). *Let $f : [a,b] \to \mathbb{R}$ be continuous on $[a,b]$ and differentiable on (a,b). If L is real number satisfying $f'(a) < L < f'(b)$ or $f'(a) > L > f'(b)$, then there exists a point $c \in (a,b)$ with $f'(c) = L$.*

Proof. As with the proof of the Intermediate Value Theorem, we will reduce the task to the situation when $L = 0$. We also restrict ourselves to the case $f'(a) < L < f'(b)$. Define a function $g : [a,b] \to \mathbb{R}$ by $g(x) = f(x) - Lx$. Notice that $g'(x) = f'(x) - L$ and that $g'(a) < 0 < g'(b)$ by our hypothesis on L. We will now use Lemma 5.3.1 to show that there is a point $c \in (a,b)$ such that $g'(c) = 0$ which will imply that $f'(c) = L$.

By definition, g is continuous on $[a,b]$ and thus the Extreme Value Theorem indicates that g attains a minimum value on $[a,b]$. Our goal is to show that the minimum cannot occur at an endpoint, i.e. the minimum is an interior point, in which case we are free to apply the aforementioned lemma. To show that the minimum cannot occur at a, take (x_n) to be a sequence in (a,b) which converges to a. Then $x_n > a$ for all n and $g'(a) < 0$. We also know that

$$g'(a) = \lim \frac{g(x_n) - g(a)}{x_n - a}.$$

In order for this limit to be negative, it must be the case that $g(x_N) - g(a) < 0$ for some $N \in \mathbb{N}$ since the denominator is positive for every n (we could make the stronger conclusion that $g(x_n) - g(a) < 0$ for infinitely many $n \in \mathbb{N}$, but we only need one for our purposes). Thus there is a point $x \in (a,b)$ with $g(x) - g(a) < 0$, or $g(x) < g(a)$. Hence a is not the minimum for g. A similar argument with some sign adjustment works to show that the minimum cannot occur at b.

From these two cases we conclude that there is a point $c \in (a,b)$ with $f(c) \le f(x)$ for every $x \in [a,b]$. By Lemma 5.3.1 we conclude that $g'(c) = 0$, or $f'(c) = L$ as desired. $\square$

The beauty of Darboux's Theorem has been discussed, but we now consider how subtle the result truly is. Take the function f to be defined by

$$f(x) = \begin{cases} 1, & x \ge 0; \\ -1, & x < 0. \end{cases}$$

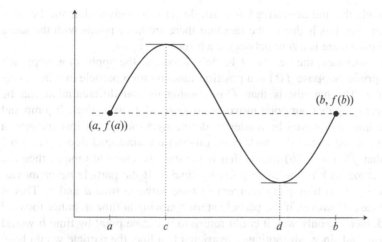

Fig. 5.2 Rolle's Theorem

Is there a function that has this as its derivative? The answer is no, because the function does not attain the values between 1 and −1. However, at first glance you should be reminded of the derivative of the absolute value function. The problem is the pesky fact that f has 0 in its domain whereas the absolute value function is not differentiable at 0. See the recent article [21] for a proof of Darboux's Theorem which has a geometric flavor.

We now turn our attention to a progression of theorems which form the core of differential calculus. The first two of these should be familiar from the calculus sequence.

Theorem 5.3.3 (Rolle's Theorem). *Let* $f : [a, b] \to \mathbb{R}$. *If* f *is continuous on* $[a, b]$ *and differentiable on* (a, b) *with* $f(a) = f(b)$, *then there is a point* $c \in (a, b)$ *with* $f'(c) = 0$.

Proof. Let f satisfy the hypothesis above. As a first case, suppose f is constant on (a, b). By Exercise 5.1.4(a), we have that $f'(x) = 0$ for all $x \in (a, b)$. In this case any $c \in (a, b)$ will satisfy the desired conclusion.

Now suppose that f is not constant. By the Extreme Value Theorem, we know that f achieves a maximum and a minimum on this interval. Observe that it cannot be the case that both the maximum and minimum happen at a, for this would force the function to be constant with $f(x) = f(a)$ for every $x \in (a, b)$. Also, the fact that $f(a) = f(b)$ means that it is not the case that the maximum occurs at a and the minimum at b, or vice versa, as this would also result in a constant function. Thus, either the maximum or the minimum (or both) occurs at an interior point $c \in (a, b)$, and, in either situation, Lemma 5.3.1 states that $f'(c) = 0$.   □

In Fig. 5.2 we have exhibited a function which has two points where the derivative is zero. The theorem only guarantees the existence of one such point, but it is possible that the derivative has multiple zeros in a given domain. For this particular function, this is due to the fact that there are three points with the same output, and hence there is a zero between each successive pair.

To better understand the beauty of Rolle's Theorem, the application approach seems appropriate. Suppose $f(t)$ is a position function for a particle moving along a straight line. The hypothesis that f is continuous and differentiable can be interpreted as saying that our point moves with a smooth motion, doesn't jump and stays on the line (this would be harder to define rigourously, but just imagine a marble rolling along a straight crack in an otherwise undamaged floor). The final hypothesis that $f(a) = f(b)$ means that if we start the clock at time a, then the particle has come back to the same point by time b. If the particle never moves, then its velocity f' is 0 at every moment in time between time a and b. This is the constant case. However, if the particle started moving at time a, either forward or backward, then the only way it could return to the same point by time b would require a turn. And since our motion is restricted to a line, the particle would have to stop to make this turn. At this moment, its velocity would necessarily have to be 0. With these types of intuitive explanations, we are unintentionally adding an extra hypothesis that the second derivative exists, or that f' is continuous at the very least. This is attributed to the fact that most of the real-world processes that calculus was invented to explain abide by such rules. The fact that Rolle's Theorem holds even without assuming that the derivative is continuous is a remarkable feat.

Example 5.3.4. When we explored the Intermediate Value Theorem, there were several exercises that use the result to assert the exist of zeros of functions and solutions to equations. In certain special cases, Rolle's Theorem can be used to show that a function has a unique root. Take $f(x) = x^3 + x^2 + x - 5$ as a function on $[0, 2]$. To find the roots of this equation would require us to know how to factor a cubic. However, we can show that the function has a unique root in this interval without specifically identifying it. First, notice that $f(0) = -5$ and $f(2) = 9$, and thus the Intermediate Value Theorem indicates that there is a point $c \in (0, 2)$ with $f(c) = 0$. This says that there is one zero.

To show that the zero is unique, we suppose that there is a second point $d \in (0, 2)$ with $c \neq d$ and $f(d) = 0$. If we consider f on the smaller interval (c, d) (or (d, c)), then $f(c) = 0 = f(d)$ and this polynomial is continuous and differentiable. Thus by Rolle's Theorem, there is a point $b \in (c, d) \subseteq (0, 2)$ such that $f'(b) = 0$. However upon inspection of the derivative of f, we see that $f'(x) = 3x^2 + 2x + 1 > 0$ on the interval $(0, 2)$. This is a contradiction since $b \in (0, 2)$ and hence it must be the case that $c = d$, i.e. there is only one root in the interval $(0, 2)$.

As with the using the Intermediate Value Theorem, we are given no information as to the location of the zero in $(0, 2)$. To the theoretical mathematician, existence is typically just as useful as having a concrete value. However, we can repeat this process by considering a smaller domain, say $(1, 2)$, and show that the zero lies in this subinterval. We can then continue this process as a means of approximating the zero. This is a useful technique for locating zeros of polynomials as there are not general forms for the zeros past the fourth degree.

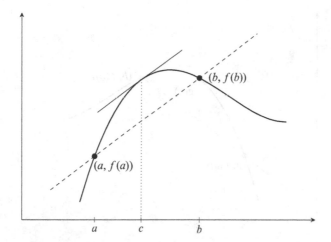

Fig. 5.3 Mean Value Theorem

Our next result is the Mean Value Theorem which can be seen geometrically as a rotated version of Rolle's Theorem (see Fig. 5.3). This observation won't be of much use in the one-dimensional setting, but it gives us a fairly intuitive way to think about the theorem in terms of graphs.

Theorem 5.3.5 (Mean Value Theorem). *Let* $f : [a, b] \to \mathbb{R}$. *If* f *is continuous on* $[a, b]$ *and differentiable on* (a, b), *then there is a point* $c \in (a, b)$ *with*

$$f'(c) = \frac{f(b) - f(a)}{b - a}.$$

To prove the statement we will begin by considering the line which passes through the graph of f at the points $(a, f(a))$ and $(b, f(b))$ and use this to construct a new function which satisfies the conditions of Rolle's Theorem. The original proof of the Mean Value Theorem is due to Cauchy who did assume continuity of the derivative. In the next two sections we will see that the Mean Value Theorem leads to very short proofs of some of the most common results concerning derivatives. It also plays a powerful role in proving the Fundamental Theorem of Calculus. For these reasons it is considered one of the, if not the, most important results in the differential calculus. For an interesting article questioning this prominent role, see [32].

Proof. To begin, consider the linear function $h : (a, b) \to \mathbb{R}$ given by

$$h(x) = \left[\left(\frac{f(b) - f(a)}{b - a} \right) (x - a) + f(a) \right]$$

which passes through the points $(a, f(a))$ and $(b, f(b))$. Now take $g : (a, b) \to \mathbb{R}$ to be the function $g = f - h$. As defined, g considers the difference between f

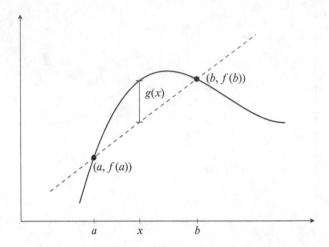

Fig. 5.4 Mean Value Theorem, part 2

and h, and, since both functions are continuous, so is g. For a visual as to what the function g is measuring, see Fig. 5.4.

By Exercises 5.1.1 and 5.2.2(b) we know that g must also be differentiable on (a, b). Moreover, since both f and h pass through the points $(a, f(a))$ and $(b, f(b))$, it must be the case that $g(a) = g(b) = 0$. We can of course show this algebraically,

$$g(a)=f(a)-h(a)=f(a)-\left[\left(\frac{f(b)-f(a)}{b-a}\right)(a-a)+f(a)\right]=f(a)-f(a)=0$$

and

$$g(b)=f(b)-h(b)=f(b)-\left[\left(\frac{f(b)-f(a)}{b-a}\right)(b-a)+f(a)\right]=f(b)-f(b)=0.$$

Thus we conclude that g satisfies the hypothesis of Rolle's Theorem and hence there is a point $c \in (a, b)$ with $g'(c) = 0$. Differentiating g, we see that

$$0 = g'(c) = (f - h)'(c) = f'(c) - h'(c) = f'(c) - \left(\frac{f(b) - f(a)}{b - a}\right).$$

Rearranging this last equation shows that c satisfies

$$f'(c) = \frac{f(b) - f(a)}{b - a},$$

completing the proof.   □

The final result of this section is a stronger version of the Mean Value Theorem.

Theorem 5.3.6 (Generalized Mean Value Theorem). *Let* $f, g : [a, b] \to \mathbb{R}$. *If* f *and* g *are continuous on* $[a, b]$ *and differentiable on* (a, b), *then there is a point* $c \in (a, b)$ *with*

$$f'(c)[g(b) - g(a)] = g'(c)[f(b) - f(a)].$$

The proof here also uses a reduction to Rolle's Theorem. Exercise 5.3.9 provides the motivation.

Exercises

Exercise 5.3.1. Complete the proof of Darboux's Theorem by showing the minimum for the function g cannot happen at the right endpoint b.

Exercise 5.3.2. Use Rolle's Theorem to show that each of the following functions has exactly one real root.

(a) $f(x) = 4x^5 + 3x^3 + 3x - 2$ on the open interval $(0, 1)$
(b) $f(x) = 2x - 1 - \sin x$ on $\mathbb{R}$

Exercise 5.3.3. Provide an application-based explanation for the Mean Value Theorem. For an example, see the discussion immediately following Rolle's Theorem.

Exercise 5.3.4. Use the Mean Value Theorem to show that $|\sin x - \sin y| \leq |x - y|$ for each $x, y \in \mathbb{R}$.

Exercise 5.3.5. Recall that a function f is *Lipschitz on a set A* if there exists a constant $M > 0$ such that

$$\left| \frac{f(x) - f(y)}{x - y} \right| \leq M$$

for all $x, y \in A$. Recall also that a function which is Lipschitz on a set A is uniformly continuous on A (see Exercise 4.4.8).

(a) Show that the converse to the previous statement is not true. In other words, construct a function which is uniformly continuous on a set A but which is not Lipschitz on A. Before beginning, consider the negation of the definition of Lipschitz.
(b) With stronger assumptions, we can prove a sufficient condition for a function to be Lipschitz. Suppose that f is differentiable on set A and further suppose that f' is bounded on A. Show that f is Lipschitz on A.
(c) Give an example of a Lipschitz function which is not differentiable at some point of its domain.

Exercise 5.3.6. A function $f : A \to \mathbb{R}$ is said to be *Hölder continuous of order* α if there exists $\alpha, M > 0$ such that

$$|f(x) - f(y)| \le M|x - y|^{\alpha}$$

for all $x, y \in A$. Notice that for $\alpha = 1$ this is simply the condition that f is a Lipschitz function.

(a) If f is Hölder continuous with order $\alpha > 0$, show that f is uniformly continuous on $[a, b]$.
(b) If f is Hölder continuous with order $\alpha > 1$, show that f is differentiable on $[a, b]$.
(c) If f is Hölder continuous with order $\alpha > 1$ and A is an open interval, show that f is constant on $[a, b]$.

Exercise 5.3.7. Let f be a function which is continuous on $[0, 3]$, differentiable on $(0, 3)$ and which satisfies $f(0) = 1$, $f(1) = 2$, and $f(3) = 2$.

(a) Show that the equation $f(x) = x$ has at least one solution in $[0, 3]$.
(b) Show that there exists a point $c \in [0, 3]$ with $f'(c) = 0$.
(c) Let $L \in (0, 1)$. Show that there is a point $d \in [0, 3]$ where $f'(d) = L$.

Exercise 5.3.8. Let $f : (a, b) \to \mathbb{R}$ and $g : (a, b) \to \mathbb{R}$ be differentiable functions. We define the *Wronskian determinant* of f and g by

$$W(f, g)(x) = \begin{vmatrix} f(x) & g(x) \\ f'(x) & g'(x) \end{vmatrix} = f(x)g'(x) - f'(x)g(x).$$

Prove that if the Wronskian determinant is never zero on (a, b), then between any two roots of f there must be a root of g. As a first observation, notice that f and g are never zero at the same point. Now consider Rolle's Theorem and the function f/g.

Exercise 5.3.9. Supply a proof of the Generalized Mean Value Theorem by considering the function $h(x) = f(x)[g(b) - g(a)] - g(x)[f(b) - f(a)]$.

Exercise 5.3.10. Use the Generalized Mean Value Theorem to provide a proof of the Mean Value Theorem.

Exercise 5.3.11. Let $f : (a, b) \to \mathbb{R}$ be differentiable and let $c \in (a, b)$. Show that there is a sequence (x_n) in (a, b) with $x_n \ne c$ for all n and $\lim x_n = c$ such that $(f'(x_n)) \to f'(c)$.

5.4 Consequences of the Value Theorems

The Mean Value Theorem leads to many of the results frequently used in the first-year calculus sequence to better understand the behavior of graphs and in applied problems such as optimization exercises. We present only a sampling here. As we work through this section we will make connections to ideas we have previously encountered and provide a few forward glimpses as well.

Theorem 5.4.1 (Zero Derivative and Constant Difference Theorems). *Let f and g be functions which are differentiable on an open interval (a, b).*

(a) *The function f is a constant function on (a, b) if and only if $f'(x) = 0$ for all $x \in (a, b)$.*
(b) *If $f'(x) = g'(x)$ for all $x \in (a, b)$, then there is a real number C such that $f(x) = g(x) + C$.*

Proof. We will prove part (a), of which the forward direction is obvious. For the converse, suppose $f'(x) = 0$ for all $x \in (a, b)$ and let x_0, y_0 be arbitrary points in (a, b). Our hypothesis that f is differentiable on (a, b) guarantees us that f is continuous on $[x_0, y_0]$ and differentiable on (x_0, y_0) (the interval depends on the order relationship between x_0 and y_0 but either will work). Thus the Mean Value Theorem implies that there is a c in (x_0, y_0) with

$$\frac{f(y_0) - f(x_0)}{y_0 - x_0} = f'(c).$$

But we also know that $f'(c) = 0$ and hence it must be the case that the numerator of the quotient above is 0, i.e. $f(x_0) = f(y_0)$. The fact that this relationship holds no matter what points we choose in (a, b) assures us that f is constant. □

The second part of the theorem indicates that if two functions have the same derivative along an interval, then, while the functions may not be equal, they only differ by a fixed constant. This result has a beautiful interpretation in terms of antiderivatives.

Definition 5.4.2. Let $f : (a, b) \to \mathbb{R}$. We call $F : (a, b) \to \mathbb{R}$ an *antiderivative* of f if $F'(x) = f(x)$ for every $x \in (a, b)$.

With this terminology, the previous theorem states that any two antiderivatives for a function differ by a constant. This may not seem like a tremendous result, but suppose you are asked to find *all* antiderivatives for a given function, assuming it has antiderivatives. The theorem then says that it suffices to find a single antiderivative, and then all the others are obtained by adding a constant. In short, it reduces the question to finding a single antiderivative. If the given function f is continuous, then the Fundamental Theorem of Calculus will show that f has an antiderivative by demonstrating existence. Thus, for any continuous function, we will be able to identify all antiderivatives!

Our next result relates the sign of a derivative to the monotonic behavior of the function.

Definition 5.4.3. Let $f : [a,b] \to \mathbb{R}$. We say f is *increasing* on $[a,b]$ if $f(x) \le f(y)$ whenever $x \le y$ in $[a,b]$. Similarly, f is *decreasing* on $[a,b]$ if $f(x) \ge f(y)$ whenever $x \le y$ in $[a,b]$.

Theorem 5.4.4 (Increasing Function Theorem). *Let $f : (a,b) \to \mathbb{R}$ be differentiable on (a,b). If $f'(x) \ge 0$ for every $x \in (a,b)$, then f is increasing on (a,b).*

Proof. The proof here is quite similar to the proof of Theorem 5.4.1 and the technique is worth remembering. To show f is increasing, let $x_0, y_0 \in (a,b)$ with $x_0 \le y_0$. This implies that $y_0 - x_0 \ge 0$. Our hypothesis that f is differentiable on (a,b) guarantees us that f is continuous on $[x_0, y_0]$ and differentiable on (x_0, y_0), and thus the Mean Value Theorem implies that there exists $c \in (x_0, y_0)$ with

$$f'(c) = \frac{f(y_0) - f(x_0)}{y_0 - x_0}.$$

We also know that $f'(c) \ge 0$. This fact together with our assumption that the denominator of the above quotient is positive implies that the numerator must satisfy $f(y_0) - f(x_0) \ge 0$. This implies that $f(x_0) \le f(y_0)$ completing the proof. $\square$

Understanding when a function is increasing and decreasing provides information about where the function has maxima and minima. Given a function, it can be difficult to identify monotonic behavior without a graph, and this is why the theorem above is so important. We saw in Example 4.3.2 how the Intermediate Value Theorem can aid in determining where a function is positive and negative. Combining that technique with this theorem gives us an efficient means of identifying where a function is increasing and decreasing, and hence where it has maxima and minima.

The following theorem relates the above principle to two functions. We remark that in the hypothesis we work on the interval $[0, \infty)$ but the result holds on any interval of the form $[a, b]$.

Theorem 5.4.5 (RaceTrack Principle). *Let f and g be functions which are differentiable on $(0, \infty)$ and continuous on $[0, \infty)$. If $f'(x) \le g'(x)$ for every $x \in (0, \infty)$ and $f(0) = g(0)$, then $f(x) \le g(x)$ for every $x \in [0, \infty)$.*

Proof. To begin, define a new function $h = g - f$ with which we can reinterpret our hypothesis as $h'(x) = g'(x) - f'(x) \ge 0$ for all $x \in (0, \infty)$. Now we fix $x_0 \in (0, \infty)$ and apply the Mean Value Theorem to h on the interval $(0, x_0)$ showing that there is a point $c \in (0, x_0)$ with

$$h'(c) = \frac{h(x_0) - h(0)}{x_0 - 0}.$$

Here we know that the quotient and the denominator are nonnegative and thus it must be the case that $h(x_0) - h(0) \geq 0$. Substituting now for f and g, we have

$$0 \leq h(x_0) - h(0) = g(x_0) - f(x_0) - (g(0) - f(0)) = g(x_0) - f(x_0)$$

which implies $f(x_0) \leq g(x_0)$. □

One of the reasons for including these three results is (though they are important in and of themselves!) to again impress upon the reader the need to experience proof techniques again and again. While the results are all related, they get at different pieces of the puzzle that is differential calculus, but their proofs are nearly identical, which is really what ties them together.

We move now to what is likely the most popular result from differential calculus. L'Hospital's Rule provides a door to examining limits which prove arduous with purely algebraic methods. It is therefore hailed as a celebrity among students. We present one case.

Theorem 5.4.6 (L'Hospital's Rule 0/0). *Let $c \in (a, b)$ and let f and g be functions which are differentiable on (a, b) with $g'(x) \neq 0$ for all $x \in (a, b)$ except possibly at c. Suppose also that $\lim_{x \to c} f(x) = \lim_{x \to c} g(x) = 0$. If*

$$\lim_{x \to c} \frac{f'(x)}{g'(x)} = L, \quad then \quad \lim_{x \to c} \frac{f(x)}{g(x)} = L.$$

Proof. Let f, g, and c be as in the statement of the theorem. To verify the conclusion we will use the sequential characterization of functional limit. To this end, let (x_n) be a sequence in (a, b) with $x_n \neq c$ for all n and $\lim x_n = c$. By the Generalized Mean Value Theorem, for each $n \in \mathbb{N}$ there is a real number y_n between x_n and c, (in either (x_n, c) or (c, x_n) depending on the order relationship between x_n and c) with

$$f'(y_n)[g(x_n) - g(c)] = g'(y_n)[f(x_n) - f(c)]$$

or

$$f'(y_n)[g(c) - g(x_n)] = g'(y_n)[f(c) - f(x_n)],$$

again depending on the relationship between x_n and c. Here we observe that $g'(y_n) \neq 0$ by hypothesis. Also, $g(x_n) - g(c) = g(x_n)$ since g is continuous at c with limit 0. Rolle's Theorem guarantees us that it must also be the case that $g(x_n) \neq 0$, otherwise there would be a point d between x_n and c with $g'(d) = 0$. These two observations allow us to write the previous equations as

$$\frac{f'(y_n)}{g'(y_n)} = \frac{f(x_n) - f(c)}{g(x_n) - g(c)} = \frac{f(x_n)}{g(x_n)}$$

where we have used the fact that $f(c) = 0$ by continuity. Notice now that $y_n \neq c$ for all n and $\lim y_n = c$ by the Squeeze Theorem. Thus applying the sequential characterization of functional limit,

$$L = \lim_{x \to c} \frac{f'(x)}{g'(x)} = \lim \frac{f'(y_n)}{g'(y_n)} = \lim \frac{f(x_n)}{g(x_n)}.$$

From this equation, we deduce that

$$\lim_{x \to c} \frac{f(x)}{g(x)} = L$$

since (x_n) was an arbitrary sequence in (a, b) with $x_n \neq c$ for all n and $\lim x_n = c$.
$\square$

The final result of this section concerns the derivative of the inverse of a function. An application of this sort is only valid for a particular class of functions but is worth our attention as the example following the proof demonstrates. Suppose for the moment that $f : (a, b) \to \mathbb{R}$ is invertible and differentiable on an open interval I. Then the function $f^{-1} : f(I) \to I$ exists. To calculate $(f^{-1})'$ we will investigate the difference quotient first to get a feel for what should happen. Consider two points $y_1, y_2 \in f(I)$. Then there exist points $x_1, x_2 \in I$ with $f(x_1) = y_1$ and likewise for x_2 and y_2. Also,

$$\frac{f^{-1}(y_1) - f^{-1}(y_2)}{y_1 - y_2} = \frac{x_1 - x_2}{f(x_1) - f(x_2)} = \frac{1}{\frac{f(x_1) - f(x_2)}{x_1 - x_2}}.$$

This computation shows that derivative of the inverse if related to the reciprocal of the derivative and hence we will need to be wary of zero denominators. The statement of the theorem takes this into account in a subtle way without assuming the invertibility of f.

Theorem 5.4.7 (Inverse Function Theorem). *Let $I \subseteq \mathbb{R}$ be an open interval and suppose $f : (a, b) \to \mathbb{R}$ is differentiable on I. If $f'(x) \neq 0$ on I, then the function $f^{-1} : f(I) \to I$ is well defined and satisfies*

$$(f^{-1})'(x) = \frac{1}{f'(f^{-1}(x))}$$

for all $x \in f(I)$.

Proof. Suppose f is as above. By Darboux's Theorem we know that $f'(x) > 0$ or $f'(x) < 0$ for all $x \in I$ and Exercise 5.4.3 indicates that f is either strictly increasing or strictly decreasing on I. Furthermore, Exercise 4.3.5(a) then assures us that f is injective on I which is all that is necessary to conclude that f^{-1} exists and is continuous on $f(I)$ by part (b) of the aforementioned exercise; f is of course continuous since it is differentiable.

To identify the derivative of f^{-1}, let $d \in f(I)$. Then there is a point $c \in I$ with $f(c) = d$. Also, let (y_n) be a sequence in $f(I)$ with $y_n \neq d$ for all n and $\lim y_n = d$. For each n, there is a unique point $x_n \in I$ with $f(x_n) = y_n$. Since f^{-1} is continuous, we also see that

$$\lim x_n = \lim f^{-1}(y_n) = f^{-1}(d) = c.$$

Notice also that $x_n \neq c$ for all n since f is injective and $y_n \neq d$ for all n. Now, computing the difference quotient along the sequence (y_n), we have

$$\lim \frac{f^{-1}(y_n) - f^{-1}(d)}{y_n - d} = \lim \frac{x_n - c}{f(x_n) - f(c)} = \lim \frac{1}{\frac{f(x_n) - f(n)}{x_n - c}} = \frac{1}{f'(c)} = \frac{1}{f'(f^{-1}(d))}$$

where the third inequality is valid since $f'(c) \neq 0$ by hypothesis and the fact that (x_n) converges to c. By the definition of functional limit and the derivative we then conclude that

$$(f^{-1})'(d) = \frac{1}{f'(f^{-1}(d))},$$

completing the proof. ☐

Example 5.4.8. To see how the previous theorem is applied, take $g(x) = x^3$ on $(0, \infty)$. This function maps onto $(0, \infty)$, satisfies $g'(x) \neq 0$ on the given domain, and we have shown that it is differentiable with $g'(x) = 3x^2$. We define the function $g^{-1} : (0, \infty) \to (0, \infty)$ by the symbol $g^{-1}(x) = \sqrt[3]{x}$. Bear in mind that this is a formal definition of the cube root whose existence is asserted by the theorem. We can then calculate its derivative by

$$(g^{-1})'(x) = \frac{1}{g'(g^{-1}(x))} = \frac{1}{3(\sqrt[3]{x})^2};$$

we refrain from writing this as $(1/3)x^{-2/3}$ for the moment as we have not discussed algebraic manipulation of exponents. We will prove the most general form of the power rule $(x^r)' = rx^{r-1}$ in Chap. 7.

Notice that the derivative expression also holds for g defined on the interval $(-\infty, 0)$ by exactly the same argument, but we cannot reason in this same manner so as to include the point $x = 0$. Further investigation shows that g^{-1} is not differentiable at 0. To see this, notice that by definition of inverse we have $(\sqrt[3]{x})^3 = x$. Moreover, $g^{-1}(0) = 0$ since $g(0) = 0$. Appealing to the definition of derivative we have

$$(g^{-1})'(0) = \lim_{x \to 0} \frac{\sqrt[3]{x} - 0}{x - 0} = \lim_{x \to 0} \frac{1}{(\sqrt[3]{x})^2};$$

the fact that g^{-1} is continuous (Exercise 4.3.5) shows that this limit does not exist since $\lim_{x \to 0}(\sqrt[3]{x})^2 = 0$.

Exercises

Exercise 5.4.1. Prove part (b) of Theorem 5.4.1.

Exercise 5.4.2. (a) Suppose $f : (a, b) \to \mathbb{R}$ satisfies the conditions of the Mean Value Theorem and suppose $f'(x) = 1$ for all $x \in (a, b)$. Show that $f(x) = x + C$ for some $C \in \mathbb{R}$. Work directly with the Mean Value Theorem and do not simply reference Theorem 5.4.1.
(b) Extend part (a) by showing that $f(x) = mx + C$ if $f'(x) = m$ for all $x \in (a, b)$.
(c) Suppose f is twice continuously differentiable, i.e. f'' exists and is continuous on $[a, b]$. If $f''(x) = 0$ on (a, b) show that f is a linear function.

Exercise 5.4.3. (a) Let $f : (a, b) \to \mathbb{R}$ be differentiable on (a, b). If $f'(x) > 0$ on (a, b), show that f is strictly increasing on (a, b). Prove a similar result for strictly decreasing functions.
(b) Show that the converse of the above statement does not hold by producing a function that is strictly increasing on an interval but which does not satisfy $f'(x) > 0$ on this interval.

Exercise 5.4.4. Use repeated applications of the RaceTrack Principle to prove each of the following inequalities.

(a) Show that $\sin x \leq x$ on $[0, \infty)$.
(b) Show that $\cos x \geq 1 - x^2/2$ on $[0, \infty)$.
(c) Show that $\sin x \geq x - x^3/6$ on $[0, \infty)$.

Exercise 5.4.5. (a) Use the RaceTrack Principle to show that $\ln(1 + x) \leq x$ on $[0, \infty)$. You may assume that $\ln x$ is differentiable on $(0, \infty)$ with $(\ln x)' = 1/x$.
(b) Use the information in part (a) to show that the series $\sum \ln(1 + 1/n^2)$ converges.

Exercise 5.4.6. Prove the following version of the RaceTrack Principle: Suppose I is an open interval containing $[a, b]$. If f and g are differentiable on I and $f'(x) \leq g'(x)$ for every $x \in [a, b]$, then $f(x) - f(a) \leq g(x) - g(a)$ for every $x \in [a, b]$.

Exercise 5.4.7. Show that the root function $\sqrt[n]{x}$ exists (on an appropriate domain) for all $n \in \mathbb{N}$ and find its derivative.

5.5 Taylor Polynomials

One of the main applications discussed in the first-year calculus sequence is that of linear approximation. Simply put, given a differentiable function f and a base point c in the domain of f, we use the values on the tangent line to approximate the value of f for points near c. A simple illustration is given in Fig. 5.5. In explicit form, the line equation is translated into a function $L(x) = f(c) + f'(c)(x - c)$ and then for values of x close to c, it is reasonable to expect that $L(x) \approx f(x)$. What defines

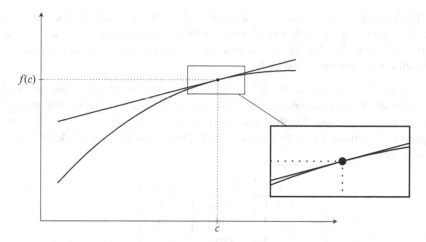

Fig. 5.5 Linear approximation

"near" is up for grabs in the first-year course, whereas we will explore this in more depth. The benefit of such an approximating scheme is that we can always compute values for a linear function whereas computing something like $\ln(2)$ is more difficult without the aid of a calculator.

It is also natural to consider extensions of such a mechanism to improve not only accuracy but also the range of values for the which the approximation scheme is valid. But how do we extend this appropriately? In other words, what are the features of the line which we can replicate? And what type of function should we attempt to construct? For the second question, the answer is a polynomial since we understand these in a more in-depth fashion than other functions. Moreover, we can also easily evaluate polynomials. With this answer, we are left to consider the features of the linear approximation and then try to extend those to higher order polynomials.

The linear approximation is defined by two quantities, $f(c)$ and $f'(c)$. The first guarantees that the approximating function agrees with the given function at the base point while the second ensures us that the first derivatives agree, or that the linear function has the same rate of change as the original at the base point. It seems reasonable therefore that if we were to construct a quadratic polynomial approximation, then we would also want to make sure that the new function and the original have the same second derivative at the base point, and so on for higher order derivatives. The following definition captures this construction for any order.

Definition 5.5.1. Let I be an open interval containing $[a, b]$ and let $c \in (a, b)$. Further suppose $f, f', f'', \ldots, f^n$ exist and are continuous on I. We then define the nth-*degree Taylor polynomial of* f at $c \in (a, b)$ by

$$T_{c,n}(x) = f(c) + f'(c)(x - c) + \frac{f''(c)}{2!}(x - c)^2 + \ldots + \frac{f^{(n)}(c)}{n!}(x - c)^n.$$

The factorial factors may seem unclear at first, but they are what ensures that the derivatives of the polynomial agree with the derivatives of f at the base point c. Notice also that the first degree Taylor polynomial is simply the tangent line approximation discussed above.

Example 5.5.2. To explore this let $f(x) = \sin x$ and $c = 0$. The following table provides all the information necessary to compute the first five Taylor polynomials. Notice the second and fourth degree polynomials will match the first and third degree polynomials, respectively, since both the second and fourth derivatives of f at 0 are 0.

j	$f^{(j)}(x)$	$f^{(j)}(0)$
$j = 1$	$\cos x$	1
$j = 2$	$-\sin x$	0
$j = 3$	$-\cos x$	-1
$j = 4$	$\sin x$	0
$j = 5$	$\cos x$	1

We have

$$T_{0,1}(x) = x, \quad T_{0,3}(x) = x - \frac{1}{6}x^3, \quad T_{0,5}(x) = x - \frac{1}{6}x^3 + \frac{1}{120}x^5.$$

Figure 5.6 shows the graph of each of these polynomials with f. Notice that each polynomial appears to agree with f on a slightly larger set than its predecessor. Using this to approximate, we find that $\sin(1) \approx T_{0,5}(1) = 0.8416666667$. Your calculator is likely to return $\sin(1) \approx 0.8414709848$ which is obtained via the 13th degree Taylor polynomial.

With the example above in mind, we next explore how well the Taylor polynomials approximate a general function f. We will do this by considering the difference between f and $T_{c,n}$. It may seem surprising, but the relationship between these two functions is encoded in the $(n + 1)$st derivative. The statement is an analogue of the Mean Value Theorem for higher order derivatives and reduces to the Mean Value Theorem when $n = 1$.

Theorem 5.5.3 (Taylor's Theorem). *Let I be an open interval containing $[a, b]$ and let $c \in (a, b)$. Further suppose $f, f', f'', \ldots, f^n$ exist and are continuous on I, and that $f^{(n+1)}$ exists on (a, b). If $x \in (a, b)$ with $x \neq c$, then there is a point p strictly between x and c such that*

$$f(x) - T_{c,n}(x) = \frac{f^{(n+1)}(p)}{(n + 1)!}(x - c)^{n+1}.$$

If $x = c$, then $p = c$ satisfies the equality above.

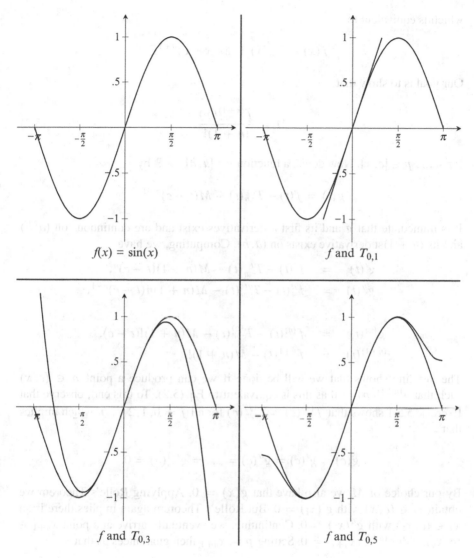

$f(x) = \sin(x)$ f and $T_{0,1}$

f and $T_{0,3}$ f and $T_{0,5}$

Fig. 5.6 Taylor polynomials for $f(x) = \sin x$

Proof. The claim is obviously true when $x = c$ and $p = c$ since both sides of the equation above are 0. We assume that $x > c$ in (a, b) and will not change for the remainder of the proof; the proof for $x < c$ follows analogously. We then set

$$M = \frac{f(x) - T_{c,n}(x)}{(x - c)^{n+1}}$$

which is equivalent to

$$f(x) - T_{c,n}(x) = M(x - c)^{n+1}.$$

Our goal is to show that

$$M = \frac{f^{(n+1)}(p)}{(n+1)!} \tag{5.2}$$

for some $p \in [c, x]$. Now define a function $g : [a, b] \to \mathbb{R}$ by

$$g(t) = f(t) - T_{c,n}(t) - M(t - c)^{n+1}.$$

It is immediate that g and its first n derivatives exist and are continuous on (a, b) and its $(n + 1)$st derivative exists on (a, b). Computing, we have

$$
\begin{aligned}
g'(t) &= f'(t) - T'_{c,n}(t) - M(n+1)(t-c)^n, \\
g''(t) &= f''(t) - T''_{c,n}(t) - M(n+1)n(t-c)^{n-1}, \\
&\vdots \\
g^{(n)}(t) &= f^{(n)}(t) - T^{(n)}_{c,n}(t) - M(n+1)!(t-c), \\
g^{(n+1)}(t) &= f^{n+1}(t) - M(n+1)!.
\end{aligned}
$$

The last line shows that we will be done if we can produce a point $p \in (c, x)$ such that $g^{(n+1)}(p) = 0$ as this is equivalent to Eq. (5.2). To this end, observe that Exercise 5.5.1 shows that $f^{(j)}(c) = T^{(j)}_{c,n}(c)$ for all $j = 0, 1, 2, \dots n$ which implies that

$$g(c) = g'(c) = g''(c) = \dots = g^{(n)}(c) = 0.$$

By our choice of M we also have that $g(x) = 0$. Applying Rolle's Theorem we obtain $x_1 \in (c, x)$ with $g'(x_1) = 0$. But Rolle's Theorem again implies there is an $x_2 \in (c, x_1)$ with $g''(x_2) = 0$. Continuing, we eventually arrive at a point $x_{n+1} \in (c, x_n)$ with $g^{(n+1)}(x_{n+1}) = 0$. Setting $p = x_{n+1}$ then guarantees us that

$$f^{n+1}(p) - M(n+1)! = g^{(n+1)}(p) = g^{(n+1)}(x_{n+1}) = 0$$

as desired.   □

This theorem is very useful in theoretical practice but not so much in practical application for general functions. This is tied to understanding the $(n + 1)$st derivative. For situations where this derivative is not continuous there is little that can be done. However, if the $(n + 1)$st derivative is continuous on the interval $[a, b]$ then it is also bounded there by the Extreme Value Theorem; with this, we can be more precise about how well a Taylor Polynomial approximates a given function. The final result of this section works toward this not by assuming that $f^{(n+1)}$ is continuous but simply that it is bounded.

Theorem 5.5.4 (Taylor Error Bound). *Let I be an open interval containing $[a,b]$ and let $c \in (a,b)$. Further suppose $f, f', f'', \ldots, f^n$ exist and are continuous on I, and that $f^{(n+1)}$ exists on (a,b) with $m \le f^{(n+1)} \le M$ on $[c,b]$. Then*

$$m \frac{(x-c)^{n+1}}{(n+1)!} \le f(x) - T_{c,n}(x) \le M \frac{(x-c)^{n+1}}{(n+1)!}$$

for all $x \in [c,b]$.

Proof. Here we will repeatedly apply the version of the RaceTrack Principle given in Exercise 5.4.6 to the inequality $f^{(n+1)}(x) \le M$ for all $x \in [c,b]$. A first application yields

$$f^{(n)}(x) - f^{(n)}(c) \le Mx - Mc = M(x-c).$$

Applying the principle again to this new inequality yields

$$f^{(n-1)}(x) - f^{(n)}(c)x - \left(f^{(n-1)}(c) - f^{(n)}(c)c \right) \le \frac{M}{2}(x-c)^2$$

which reduces to

$$f^{(n-1)}(x) - \left(f^{(n-1)}(c) + f^{(n)}(c)(x-c) \right) \le \frac{M}{2}(x-c)^2$$

upon gathering like terms on the left. Here we begin to see the construction of the Taylor polynomial in parentheses on the left. After a third application of the RaceTrack Principle and some simplification we arrive at

$$f^{(n-2)}(x) - \left(f^{(n-2)}(c) + f^{(n-1)}(c)(x-c) + \frac{f^{(n)}(c)}{2}(x-c)^2 \right) \le \frac{M}{2 \cdot 3}(x-c)^3.$$

Continuing, after $n+1$ applications we come to the estimate

$$f(x) - T_{c,n}(x) \le M \frac{(x-c)^{n+1}}{(n+1)!}.$$

We leave the lower estimate as an exercise. □

Example 5.5.5. Returning to our previous discussion of $f(x) = \sin x$, we have

$$T_{0,5}(x) = x - \frac{1}{6}x^3 + \frac{1}{120}x^5.$$

With the previous error bound together with the fact that $-1 \le f^{(6)}(x) \le 1$ for all $x \in \mathbb{R}$ we see that

$$\frac{-x^6}{6!} \le f(x) - T_{0,5}(x) \le \frac{x^6}{6!}.$$

In order to determine how accurate out estimate truly is, we must also restrict our focus in the domain. In particular, on the interval $[-1, 1]$ we have that

$$-0.0013\overline{8} \leq f(x) - T_{0.5}(x) \leq 0.0013\overline{8}$$

since $1/6! = 0.0013\overline{8}$. If we move to the interval $[-2, 2]$ the error increases and we have

$$-0.0\overline{8} \leq f(x) - T_{0.5}(x) \leq 0.0\overline{8}.$$

It should be apparent here that a higher order Taylor polynomial not only increases the accuracy of the estimate but also allows us to increase the set of domain points without sacrificing drastically with regard to the error of the approximation.

Exercises

Exercise 5.5.1. Let I be an open interval containing $[a, b]$ and let $c \in (a, b)$. Further suppose $f, f', f'', \ldots, f^n$ exist and are continuous on I. Show that $f^{(j)}(c) = T_{c,n}^{(j)}(c)$ for all $j = 0, 1, 2, \ldots n$.

Exercise 5.5.2. (a) Let p be a polynomial of degree 2. Show that $p = T_{0,2}$.
(b) Extend part (a) by showing that for a polynomial p of degree n, we have $p = T_{0,n}$.

Exercise 5.5.3. Find a Taylor polynomial $T_{0,n}$ for $f(x) = e^x$ which approximates f with an error of no more than 0.0001 on the interval $[-1, 1]$. You may use the fact that f is increasing and estimates from Sect. 3.2 on the value of e.

Exercise 5.5.4. Let $f(x) = 1/(1 - x)$ on the interval $(0, 1)$. Also let $c = \frac{1}{3}$ and $x = \frac{2}{3}$. Find p as specified in the statement of Taylor's Theorem for $n = 1, 2, 3, 4$.

5.6   Eigenvalues and the Invariant Subspace Problem

The goal of this section is to investigate the topic of eigenvalues and two generalizations, the spectrum of an operator and invariant subspaces of an operator. While the concept of differentiation does not play a substantial role in either of these ideas, the derivative does allow us to consider some interesting examples of spaces and operators.

Definition 5.6.1. Let X be a vector space and let $T : X \to X$ be a linear operator. We say λ is an *eigenvalue* for T if there is a nonzero vector $x \in X$ such that

$$Tx = \lambda x.$$

In this case we call x an *eigenvector* corresponding to λ.

Example 5.6.2. For a first example, consider the 2×2 matrix A defined by

$$A = \begin{pmatrix} 2 & 3 \\ 0 & 5 \end{pmatrix}$$

and the corresponding matrix operator $T : \mathbb{R}^2 \to \mathbb{R}^2$ defined by $Tx = Ax$ for $x \in \mathbb{R}^2$. The vector $(1, 0)$ is an eigenvector corresponding to the eigenvalue 2 since

$$\begin{pmatrix} 2 & 3 \\ 0 & 5 \end{pmatrix} \begin{pmatrix} 1 \\ 0 \end{pmatrix} = \begin{pmatrix} 2 \\ 0 \end{pmatrix} = 2 \begin{pmatrix} 1 \\ 0 \end{pmatrix}.$$

It is also easy to see that $(0, 1)$ is an eigenvector corresponding to the eigenvalue 5.

Next take P_n, the space of all polynomials of degree n or less, and consider the composition operator $C_\varphi : P_n \to P_n$ with symbol $\varphi(x) = 2x$; if $p \in P_n$, then

$$(C_\varphi p)(x) = (p \circ \varphi)(x).$$

For $k \leq n$, we find that x^k is an eigenvector corresponding to eigenvalue 2^k since

$$(C_\varphi x^k) = (2x)^k = 2^k x^k.$$

The examples above are simply present to demonstrate the concept, and we now focus on finding eigenvalues. To begin, consider an $n \times n$ matrix A and the operator $T : \mathbb{R}^n \to \mathbb{R}^n$ given by $Tx = Ax$. Our goal is then to find all λ such that

$$Ax = \lambda x$$

for some nonzero $x \in \mathbb{R}^n$. We also let I denote the $n \times n$ identity matrix. We see then that the previous equation is equivalent to

$$(A - \lambda I)x = 0$$

which means that a nonzero vector x is an eigenvalue for λ if and only if $x \in \ker(A - \lambda I)$. Recall from linear algebra that an operator is one-to-one if and only if its kernel contains only the zero vector. Thus we conclude that λ is an eigenvalue for the matrix A (or the operator T) if and only if the matrix $A - \lambda I$ is not one-to-one, which in turn is equivalent to saying that $A - \lambda I$ is not an invertible matrix. Finally, in the finite dimensional setting, this last statement is equivalent to saying that

$$\det(A - \lambda I) = 0.$$

For notation, we set $p(\lambda) = \det(A - \lambda I)$ and call this the *characteristic polynomial* of A. To find the eigenvalues of A, it then suffices to find the zeros of the polynomial $p(\lambda)$. By the definition of determinant it follows quickly that p will be a polynomial of degree n, and hence there are n (possibly complex) eigenvalues (counting multiplicities) for the matrix A.

Example 5.6.3. Consider again the matrix

$$A = \begin{pmatrix} 2 & 3 \\ 0 & 5 \end{pmatrix}.$$

Computing $p(\lambda)$ we find

$$p(\lambda) = \det \begin{pmatrix} 2 - \lambda & 3 \\ 0 & 5 - \lambda \end{pmatrix} = (2 - \lambda)(5 - \lambda)$$

which has zeros at 2 and 5. We conclude that these are the only eigenvalues of A. Finding the eigenvectors (all of them) for each eigenvalue is now simply a matter of solving the homogeneous system

$$(A - \lambda I)x = 0$$

upon substituting $\lambda = 2$ and $\lambda = 5$.

Next consider P, the space of *all* polynomials, and the composition operator C_φ with symbol $\varphi(x) = 2x$. With this space we may consider any $n \in \mathbb{N}$ and, as in the previous example, we see that

$$C_\varphi x^n = (2x)^n = 2^n x^n.$$

This shows that 2^n is an eigenvalue for C_φ acting on this space for every $n \in \mathbb{N}$. Are there more? The technique above does not apply here as we have no concept of determinant in the infinite dimensional setting and thus we cannot conclude that we have found all of the eigenvalues.

The question comes to mind as to whether or not we can quantify how many eigenvalues we should expect an operator to have. For the finite dimensional case we have already mentioned that the number of (real) eigenvalues (counting multiplicities) is no more than the dimension of the domain space. We also saw that for an infinite dimensional space it is possible for an operator to have infinitely many eigenvalues. In that example we computed countably many eigenvalues, but there may be more. The next example shows that it is possible to maximize the number of eigenvalues. Later we will see an example showing that it is possible for an operator to have no eigenvalues.

Example 5.6.4. Define the space $C^1([0, 1])$ to be the set of all functions on $[0, 1]$ which are continuous and have a continuous first derivative. If we define addition and scalar multiplication in the standard pointwise fashion, then Theorems 4.2.6 and 5.2.1 show this is a vector space. Though we are only interested in the vector space structure here, we equip this space with a norm $\| \cdot \|_{C^1} : C^1([0,1]) \to [0, \infty)$ by

$$\|f\|_{C^1} = \|f\|_\infty + \|f'\|_\infty.$$

The norm properties follow from the fact that the sup-norm is a norm and the algebraic properties of the derivative. In Sect. 6.5 we will show that this space is complete and discuss why both terms on the right-hand side of the norm definition are necessary.

We also set $C_\infty([0,1])$ to be the subspace of $C_1([0,1])$ consisting of all functions which have derivatives of all orders. Now consider the operator $T : C_\infty \to C_\infty$ given by

$$Tf = f';$$

we leave it as an exercise to show that T is linear. If we now take $f(x) = e^{\lambda x}$, where $\lambda \in \mathbb{R}$, we find that

$$Tf = (e^{\lambda x})' = \lambda e^{\lambda x} = \lambda f.$$

Thus every real number is an eigenvalue for T.

The previous examples demonstrate that the determinant technique works well in the finite dimensional setting but falls short for larger spaces. This is only one of the drawbacks of using the concept of determinant for computing eigenvalues. Another problem concerns the fact that we must be able to factor the characteristic polynomial in order to find the eigenvalues. However, for large n this task may be extremely difficult. The advance of modern technology has alleviated this problem; however, there is still the simple fact that the polynomial may not factor completely over the real numbers. In other words, even if a matrix has real values, or an operator acts on a real vector space, there is nothing to prohibit complex numbers from entering into the picture, and there is no remedy to this issue.

In the infinite dimensional setting, the properties that an operator is one-to-one and onto are no longer equivalent and thus a given operator can be one-to-one and still fails to be invertible. The concept of eigenvalue is still equivalent to the injectivity of the operator $T - \lambda I$ but to truly extend the notion of eigenvalues to the infinite dimensional setting we have the more general notion of the *spectrum* of an operator: if X is a complete normed linear space and $T : X \to X$ is bounded, then we define the spectrum of T, denoted $\sigma(T)$, to be the set

$$\sigma(T) = \{\lambda : T - \lambda I \text{ is not invertible}\};$$

here λ may be a complex number (though we will restrict our attention to real values) and I is the identity operator on X.

If X is finite dimensional, then the spectrum is simply the set of eigenvalues. If X is infinite dimensional, λ is in $\sigma(T)$ if $T - \lambda I$ is not one-to-one or if $T - \lambda I$ is not onto. If $T - \lambda I$ is not one-to-one, then it is still the case that λ is an eigenvalue, but this is not necessarily true if $T - \lambda I$ is not onto. Thus we see that there are possibly non-eigenvalues in the spectrum; to distinguish between these points we call the set of eigenvalues of T the *point spectrum* of T.

Example 5.6.5. For an example of an operator with no eigenvalues we return to the forward shift operator T on ℓ^2 from Example 4.5.4. We work by contradiction and assume that T has an eigenvalue λ with nonzero eigenvector $a = (a_1, a_2, a_3, \ldots)$. If $\lambda = 0$, the equation $Ta = \lambda a$ implies that $(0, a_1, a_2, a_3, \ldots) = (0, 0, 0, 0, \ldots)$. But this says $a_1 = a_2 = a_3 = \ldots = 0$, a contradiction since $a \neq 0$. If $\lambda \neq 0$, we have the equation $(0, a_1, a_2, a_3, \ldots) = (\lambda a_1, \lambda a_2, \lambda a_3, \ldots)$. This implies that $0 = \lambda a_1$ from which it follows that a_1 must be zero since $\lambda \neq 0$. But then $a_1 = \lambda a_2$ implying that $a_2 = 0$. Continuing we again arrive at the conclusion that $a_1 = a_2 = a_3 = \ldots = 0$, a contradiction. Thus T has no eigenvalues.

On the other hand, notice that $0 \in \sigma(T)$ since the operator $T - 0I = T$ is not onto. Why? In fact, $[-1, 1] \subseteq \sigma(T)$ and these are all of the real numbers in the spectrum. To see this for $\lambda = -1$, let $a = (a_n)$ in ℓ^2. Then we have

$$(T - \lambda I)a = (a_1, a_1 + a_2, a_2 + a_3, \cdots).$$

To show that this operator is not onto, we will produce an element of ℓ^2 which has no pre-image in ℓ^2. Take $b = (1, 0, 0, \ldots)$, which is certainly in ℓ^2. Solving the equation $(T - \lambda I)a = b$ yields

$$(a_1, a_1 + a_2, a_2 + a_3, \cdots) = (1, 0, 0, \ldots)$$

from which it follows that $(a_n) = (1, -1, 1, -1, \ldots)$ is the only sequence of real numbers which maps to b under $T - \lambda I$, but this sequence is not in ℓ^2. Therefore we conclude that there is no sequence in ℓ^2 which maps to b and hence the operator $T - \lambda I$ is not onto; thus, $-1 \in \sigma(T)$. Exercise 5.6.6 will ask you to mimic this procedure to show that $[-1, 1] \subseteq \sigma(T)$.

The study of spectra is one of the most important areas of operator theory and has close connections to the study of the spectra of atoms in theoretical physics. Though we won't devote more time to this topic here we point the interested reader to Sect. 6.6 in [28] and Chaps. 4–6 in [18].

Returning to the determinant discussion from above, there is no analog for infinite dimensional spaces and we present now a concept which unifies the search for eigenvalues and eigenvectors in both the finite and infinite dimensional settings. This approach follows the text [2] which is an extension of the article [3]. Both of these works present a determinant free approach to linear algebra and argue that such an approach is "simpler, clearer, [and] provides more insight" than the standard determinant-driven undergraduate experience.

Definition 5.6.6. Let $(X, \|\cdot\|)$ be a normed linear space and $T : X \to X$ a bounded linear operator. We say that a closed subspace $K \subseteq X$ is an *invariant subspace* for T if $Tk \in K$ for every $k \in K$. Some authors simply write $TK \subseteq K$.

For the most part our discussion will revolve around finite dimensional spaces in which case all subspaces are closed. There are several exercises in the infinite dimensional setting, but rest assured that all given subspaces are closed.

Example 5.6.7. Let $(X, \|\cdot\|)$ be a normed linear space and $T : X \to X$ a bounded linear operator. Then there are always two invariant subspaces, namely $\{0\}$ and X. These will be called trivial invariant subspaces and thus we seek examples of nontrivial invariant subspaces. In this same general context, $\ker(T)$ is always an invariant subspace, and may be trivial or nontrivial depending on the particular operator.

For a more concrete example, let $X = \mathbb{R}^2$, with the Euclidean norm, and define

$$Tx = -x.$$

We claim that every one-dimensional subspace of $\mathbb{R}^2$ is an invariant subspace for T. To see this, let K be a one-dimensional subspace, i.e. $K = \{cy : c \in \mathbb{R}\}$ for some nonzero $y \in \mathbb{R}^2$. Now if $z \in K$, then $z = cy$ for some $c \in \mathbb{R}$. It follows that

$$Tz = -z = -cy$$

which shows that Tz is also in K. Thus K is invariant for K.

Geometrically T is a rotation of $180°$. Furthermore, any one-dimensional subspace is a line through the origin. It makes sense then that a $180°$ rotation maps the line to itself. For a visual, see the left-hand image in Fig. 5.7.

The following proposition demonstrates the connection between eigenvalues and invariant subspaces. We also remark that the statement is true in the infinite dimensional setting though we restrict our attention to the finite dimensional case.

Proposition 5.6.8. *Let $(X, \|\cdot\|)$ be a finite dimensional normed linear space and $T : X \to X$ a bounded linear operator. Then T has an eigenvalue if and only if T has a one-dimensional invariant subspace.*

Proof. First suppose that T has an eigenvalue λ and let x be a nonzero vector in X with $Tx = \lambda x$. The linearity of T implies that

$$T(cx) = cTx = c(\lambda x) = (\lambda c)x$$

for every $c \in \mathbb{R}$, i.e. the image of a scalar multiple of x is also a scalar multiple of x. This shows that the one-dimensional subspace $K = \{cx : c \in \mathbb{R}\}$ is invariant for T.

Conversely, suppose that T has a one-dimensional invariant subspace K. Then $K = \{cy : c \in \mathbb{R}\}$ for some nonzero $y \in K$. Since K is invariant and $y \in K$, we know that $Ty \in K$. Thus there is a $c \in \mathbb{R}$ such that $Ty = cy$ and T has an eigenvalue. $\square$

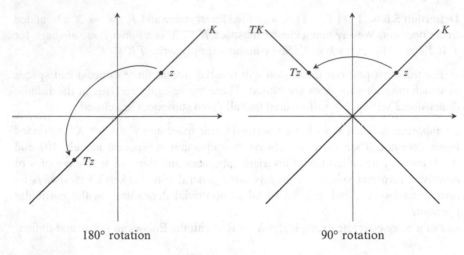

180° rotation   90° rotation

Fig. 5.7 Plane rotations

The proposition tells us that in order to find the eigenvalues of an operator, it suffices to find all the one-dimensional invariant subspaces. We then sift through these and determine which corresponds to eigenvalues. To understand this discrepancy, the operator T from Example 5.6.7 has many one-dimensional invariant subspaces, but only one eigenvalue $\lambda = -1$.

Example 5.6.9. Let $X = \mathbb{R}^2$ and define

$$T(x_1, x_2) = (-x_2, x_1).$$

This operator is a 90° rotation of the plane and it therefore has no invariant one-dimensional subspaces (Fig. 5.7). To see why this is true, we suppose

$$K = \{cy : c \in \mathbb{R}\},$$

for some nonzero $y = (y_1, y_2) \in \mathbb{R}^2$, is a one-dimensional subspace of $\mathbb{R}^2$. To see that it is impossible for K to be invariant for T, suppose $z \in K$ is a nonzero vector. We can then write $z = (cy_1, cy_2)$ and we have

$$Tz = T(cy_1, cy_2) = (-cy_2, cy_1) = c(-y_2, y_1).$$

The vector Tz is then seen to be perpendicular to z, but a one-dimensional subspace cannot have two perpendicular nonzero vectors and thus Tz is not in K. Therefore K is not invariant.

The previous theorem then indicates that T has no real eigenvalues. This approach is somewhat more satisfying than the determinant approach. For if we represent the action of T by multiplication by the matrix

$$A = \begin{pmatrix} 0 & -1 \\ 1 & 0 \end{pmatrix},$$

we find that the characteristic polynomial $p(\lambda)$ of A is

$$p(\lambda) = \det(A - \lambda I) = \lambda^2 + 1$$

which has two complex zeros. With the invariant subspace approach we are able to restrict ourselves only to real eigenvalues with no stipulation that we exclude complex numbers.

Returning to the infinite dimensional setting, we have seen an example that shows that there are spaces and operators which have no eigenvalues and hence no one-dimensional invariant subspaces. This does not say that such an operator has *no* nontrivial invariant subspaces; the forward shift on ℓ^2 has no eigenvalues and hence no one-dimensional invariant subspaces, but Exercise 5.6.9 shows that it has nontrivial invariant subspaces.

Nonetheless, one technique for finding eigenvalues is captured in Proposition 5.6.8, locate all one-dimensional invariant subspaces. This is not necessarily an easy task but it leads to a broader question. Given a bounded operator T acting on an infinite dimensional normed linear space, does T necessarily have a nontrivial invariant subspace? This problem is referred to as the Invariant Subspace Problem and is arguably the most famous unsolved problem in the area of operator theory. Understanding the scope of such a problem is difficult at the moment as we are limited not only by the types of operators that we can consider but also by the spaces on which to consider them. The true nature of such a problem is only fully understood when considering complex Hilbert and Banach spaces of infinite dimension; operators on finite dimensional complex normed linear spaces always have invariant subspaces.

For an infinite dimensional complex Banach space, an example of a space X and a bounded operator T with no nontrivial invariant subspaces was constructed by Per Enflo in a paper published in 1987. In Enflo's work the space X is extremely complicated but the operator is a shift operator. Another example was later given by Charles Read in which the space is simply ℓ^1 and the operator T is more complicated. For more of an introduction to this problem we point the reader to [27].

In the infinite dimensional complex Hilbert space setting, the question is still unanswered though the approach to solving it here has been focused on understanding the problem for particular classes of operators, e.g. compact, normal, etc. It has been shown that operators in these classes all have nontrivial invariant subspaces; this statement, though simple, represents years of work dating back to the 1930s, and the hope is that understanding operators in these classes, and others, will provide insight into the behavior of arbitrary operators. For more general information and some results on the subject, we recommend Sect. 5.4 in [29] and Sect. 4.4 in [18]. For a more thorough treatment, see the text [6].

Finally, if an operator has nontrivial invariant subspaces, can we find them? This too is a difficult question and is still open. In Chap. 2 of [6] it is shown how to use the spectrum to produce invariant subspaces, much like we have used eigenvalues to produce one-dimensional invariant subspaces. There are also examples where the invariant subspaces of an operator have been more completely characterized. An example here is provided by the forward shift operator T acting on ℓ^2. Exercise 5.6.9 will ask you to show that every subspace of the form

$$K_N = \{(a_n)_{n=1}^{\infty} : a_j = 0 \text{ for } 1 \le j \le N\},$$

where $N \in \mathbb{N}$, is invariant under T. Thinking naively about the nature of the operator itself may lead one to believe that these are the only such subspaces, but there are more. To completely characterize the invariant subspaces for this operator, it is helpful to first characterize the invariant subspaces of a certain multiplication operator acting on a particular space of functions. This characterization is outlined in Chap. 17 of [12] and is covered in more detail in [19].

Exercises

Exercise 5.6.1. Let X be a vector space and let $T : X \to X$ be a linear operator. If λ is an eigenvalue for T, let E_λ denote the collection of all eigenvectors corresponding to λ. Show that $E_\lambda \cup \{0\}$ is a subspace of X.

Exercise 5.6.2. Let X be a vector space and let $T : X \to X$ be a linear operator. Show that T is one-to-one if and only if $\ker(T) = \{0\}$.

Exercise 5.6.3. Let P be the space of all polynomials. If $p(x) = a_n x^n + a_{n-1} x^{n-1} + \cdots + a_2 x^2 + a_1 x + a_0$ is in P, we define a norm on the space by

$$\|p\| = \sup\{|a_0|, |a_1|, \ldots, |a_n|\}.$$

(a) Show that P with this norm is a normed linear space.
(b) Show that the operator $T : P \to P$ defined by $Tp = p'$ is not bounded.
(c) Show that 0 is the only eigenvalue for T. Compare this to Example 5.6.4.

Exercise 5.6.4. (a) For $\theta \in [0, 2\pi)$, define the 2×2 matrix A by

$$A = \begin{pmatrix} \cos \theta & \sin \theta \\ -\sin \theta & \cos \theta \end{pmatrix}.$$

The operator $T : \mathbb{R}^2 \to \mathbb{R}^2$ defined by $Tx = Ax$ is then a rotation about the origin of θ radians. Find conditions on θ such that A has a real eigenvalue.

(b) Does your result from (a) make sense from the geometric viewpoint.

Exercise 5.6.5. Show that $C^1([0, 1])$ with the norm from Example 5.6.4 is a normed linear space.

Exercise 5.6.6. Let T be the forward shift on ℓ^2.

(a) Confirm that $0 \in \sigma(T)$ by showing that $T - 0I = T$ is not onto.
(b) Complete Example 5.6.5 by showing that $[-1, 1] \subseteq \sigma(T)$. We have already shown that $-1, 0 \in \sigma(T)$.

Exercise 5.6.7. Consider the normed linear space $(C([0, 1]), \| \cdot \|_\infty)$.

(a) Show that the multiplication operator $M_x : C([0, 1]) \to C([0, 1])$ defined by $(M_x f)(x) = xf(x)$ has no eigenvalues.
(b) Follow the steps below to show that the operator M_x satisfies $\sigma(M_x) = [0, 1]$.

   (i) If $\lambda \notin [0, 1]$, find a function h which is continuous on $[0, 1]$ such that the multiplication operator M_h is the inverse of $M_x - \lambda I$. This shows that no point outside the interval $[0, 1]$ is in $\sigma(M_x)$.

  (ii) Since M_x has no eigenvalues, it follows that for $\lambda \in \sigma(M_x)$, it must be the case that $M_x - \lambda I$ is not onto. Knowing this, for $\lambda \in [0, 1]$, choose a function $h \in C([0, 1])$ such that there is no $f \in C([0, 1])$ for which $(M_x - \lambda I)f = h$. The constant function $h(x) = 1$ is perhaps a good place to start.

 (iii) Conclude that $\sigma(M_x) = [0, 1]$.

(c) Mimic parts (a) and (b) to show that $\sigma(M_g)$ is equal to the range of g when g is a continuous function and $M_g : C([0, 1]) \to C([0, 1])$.

Exercise 5.6.8. Let $(X, \| \cdot \|)$ be a normed linear space and let $T : X \to X$ be a bounded linear operator. Show that $\ker(T)$ is an invariant subspace.

Exercise 5.6.9. Show that each of the following subspaces is invariant for the given operator. You may assume that each of the following subspaces is closed.

(a) Let $X = C([0, 1])$ with the sup-norm, $K = \{f : f(0) = 0\}$, and let M_x be the multiplication operator from Exercise 5.6.7(a).
(b) Let $X = C([0, 1])$ with the sup-norm and let $\{c_1, c_2, \ldots, c_n\}$ be any finite subset of $[0, 1]$. Take $K = \{f : f(c_i) = 0, 1 \le i \le n\}$ and M_x the multiplication operator from Exercise 5.6.7.
(c) Let $X = C([0, 1])$ with the sup-norm, $K = \{f : f(0) = 0\}$, and let C_φ be the composition operator with symbol $\varphi(x) = \sin(x)$.
(d) Let $X = \ell^2$, $K = \{(a_n)_{n=1}^\infty : a_1 = 0\}$, and let T be the forward shift operator.
(e) Let $X = \ell^2$, $K_N = \{(a_n)_{n=1}^\infty : a_j = 0 \text{ for } 1 \le j \le N\}$, for $N \in \mathbb{N}$, and let T be the forward shift operator.

Chapter 6
Sequences and Series of Functions

Sequences and series of functions are natural extensions of the topics discussed in Chaps. 2 and 3 and many of their properties are derived in a straightforward manner from the results covered there. The investigation of series of functions in the late eighteenth and early nineteenth centuries uncovered problems with the foundations of calculus which led to the restructuring of the subject and the birth of analysis as we know it today. Here we investigate the basic forms of convergence for sequences and series of functions with specific attention payed to the representation of functions as power series. This is propagated by our brief study of Taylor polynomials in Chap. 5 and we also provide a construction of a function which is continuous on $\mathbb{R}$ but is not differentiable at any point.

6.1 Sequences of Functions

The concept of a sequence has served us well in the previous chapters providing a common theme unifying the concepts of series, continuity, and differentiation. Here we discuss what is arguably the most natural generalization of a sequence of real numbers. The basic notion of a sequence is to use behavior of the terms in order to identify a limit. Here we will take a collection of functions which are related in some fashion, the most typical relation being a natural number index such as $x, x^2, x^3, \ldots$; the natural concept of a limit here will take the form of a function. With such a definition in place, we hope to determine when properties of the sequence functions, e.g. continuity, differentiability, etc., pass on to the limit function.

Definition 6.1.1. Let $A \subseteq \mathbb{R}$. A *sequence of functions* defined on the common domain A is a countable collection of functions ordered by the natural numbers, i.e. for each natural number n, f_n is a function defined on A.

We will use the symbol (f_n) to denote a sequence of functions. Simple examples include $f_n(x) = x^n$ on $[0, 1]$ and $g_n(x) = \sin(nx)$ on $\mathbb{R}$. Typically the common

M.A. Pons, *Real Analysis for the Undergraduate: With an Invitation to Functional Analysis*, 227
DOI 10.1007/978-1-4614-9638-0_6, © Springer Science+Business Media New York 2014

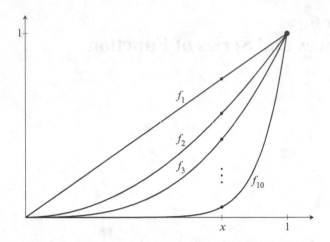

Fig. 6.1 Pointwise convergence of (x^n)

domain is key in investigating properties of a sequence of functions and we will always specify a domain. Just as with our study of sequences of real numbers, we will seek to understand the convergence behavior of this more general form of a sequence. The difference here will center around the fact that a sequence of functions can converge in several distinct manners. We begin with the weakest form of convergence.

Definition 6.1.2. Let (f_n) be a sequence of functions defined on a common domain $A \subseteq \mathbb{R}$. We say (f_n) *converges pointwise* on A to a function f if for each $x \in A$, the sequence of real numbers $(f_n(x))$ converges to $f(x)$. We use the symbol $\lim f_n(x) = f(x)$ to indicate this behavior.

This definition considers each domain point individually reducing the question of convergence to a sequence of real numbers. The fact that a sequence of real numbers has a unique limit implies that the limit function f, if it exists, is well defined and unique. The notation above is meant to reflect the fact that a domain point x is specified. We will use the symbol $\lim f_n = f$ to denote a stronger form of convergence later in this section. To be more explicit, the above definition states that a sequence of functions (f_n) converges pointwise to f if for each $x \in A$ and $\varepsilon > 0$, there is an $N \in \mathbb{N}$ such that $|f_n(x) - f(x)| < \varepsilon$ for all $n \geq N$. In this form, it is clear that a choice for N is allowed to depend on both x and ε.

Example 6.1.3. As a first example take $f_n(x) = x^n$ on $[0, 1]$. To identify a limit function, if it exists, we consider an arbitrary $x \in [0, 1]$ and then attempt to find the limit of the sequence $(f_n(x))$ so as to determine the value f assumes at x (Fig. 6.1). First, if $x = 0$, then $f_n(0) = 0$ for every n and thus $(f_n(0)) \to 0$ as $n \to \infty$. Similarly, if $x = 1$, then $f_n(1) = 1$ for all n and $(f_n(1)) \to 1$. Now take $x \in (0, 1)$. Exercise 2.3.7(a) shows that $(x^n) \to 0$ and hence $(f_n(x)) \to 0$. These three computations show that the sequence $(f_n(x))$ converges for every $x \in [0, 1]$ and the limit function f is defined by

$$f(x) = \lim_{n \to \infty} f_n(x) = \begin{cases} 0, & 0 \le x < 1; \\ 1, & x = 1. \end{cases}$$

Notice that while f_n is continuous on $[0, 1]$ for each n, the limit f has a discontinuity at 1.

Now consider the sequence of functions defined by $g_n(x) = (\sin(nx))/(nx)$ on $[1, \infty)$. For the moment, take $x = 1$. This leaves us to find the limit of the sequence $((\sin(n))/n)$. Can we find the limit of such a sequence? Since we don't necessarily have a limit candidate, though it shouldn't be hard to guess that it should be 0, it will often help to consider convergence statements which provide information about convergence and limits rather than always trying to work directly with definition of convergence; The Algebraic Limit Theorem, Squeeze Theorem, Monotone Convergence Theorem, and Cauchy Criterion are examples of such statements. Here we will use the Squeeze Theorem. The reason being that $\sin x$ is bounded above by 1 and below by -1 setting up an initial inequality

$$\frac{-1}{n} \le \frac{\sin(n)}{n} \le \frac{1}{n}.$$

Since both $(1/n)$ and $(-1/n)$ converge to zero, we conclude that $((\sin(n))/n) \to 0$ as $n \to \infty$.

Moving on to the more general case, let $x \in [1, \infty)$. Then we can use the same strategy. We know that $-1 \le \sin x \le 1$ for all $x \in \mathbb{R}$. Moreover, since $x \in [1, \infty)$, it must be the case that $0 < 1/x \le 1$ and $-1 \le -1/x < 0$. With this information we see that

$$\frac{-1}{n} \le \frac{-1}{nx} \le \frac{\sin(nx)}{nx} \le \frac{1}{nx} \le \frac{1}{n}.$$

The Squeeze Theorem applies again showing that $((\sin(nx))/(nx)) \to 0$ as $n \to \infty$. Our conclusion is that the limit function g is defined on $[1, \infty)$ by $g(x) = 0$.

One conclusion that this example brings to light is the fact that convergence often depends on the domain. We needed to guarantee that $0 < 1/x \le 1$ in order to apply the Squeeze Theorem. However, this does not mean that the sequence does not converge outside this domain, it only means that our technique was specific to the domain. For instance, if $x = 1/2$, the sequence $((2 \sin(n/2))/n)$ does converge to 0. Exercise 6.1.2 asks you extend this example to the interval $(0, 1)$.

Our main goal is to understand when properties of the sequence functions pass successfully to the limit function. For example, if each f_n is continuous on A, is it necessarily the case that f is continuous on A? Unfortunately, the sequence (f_n) from the previous example indicates that this property is not preserved under the limit process. This example also has immediate consequences for a sequence of differentiable functions.

For the moment, suppose (f_n) is a sequence of continuous functions defined on a common domain $A \subseteq \mathbb{R}$. The continuity of f is summarized by the statement

$$\lim_{x \to c} f(x) = f(c)$$

where $c \in A$ is a limit point. We can translate this into a question involving the sequence (f_n) by the statement

$$\lim_{x \to c} \left(\lim_{n \to \infty} f_n(x) \right) = \lim_{n \to \infty} \left(\lim_{x \to c} f_n(x) \right). \qquad (6.1)$$

Pointwise convergence isn't strong enough to ensure that any such outcome always occurs. The reason for this is the fact that pointwise convergence focuses on a single domain point while continuity, though centered around a fixed domain point c, also considers the function at values close to c. Our next point of interest will be to refine our definition of convergence so as to guarantee that this situation is remedied. As a final comment here, Eq. (6.1) shows that the question of continuity of the limit function is equivalent to asking if the two limits commute. This gets at a broader theme, commutativity of operations. Though we won't touch on this in depth here, questions of this type are predominant in mathematics and you have experienced situations where commutativity occurs, real number addition and multiplication, and where it does not, multiplication of matrices, for example.

For a sequence of differentiable functions, the differentiability of the pointwise limit is summarized by the equation

$$\lim_{n \to \infty} \left(\lim_{x \to c} \frac{f_n(x) - f_n(c)}{x - c} \right) = \lim_{x \to c} \left(\lim_{n \to \infty} \frac{f_n(x) - f_n(c)}{x - c} \right)$$

and again we see that the question comes down to a statement about the commutativity of limits. The situation in this case is more delicate than for continuity, which should be expected. In order to guarantee an affirmative solution, we will be required to impose an extremely strict set of hypotheses on the sequence (f_n).

Uniform Convergence

Pointwise convergence takes a local approach to convergence. In order to guarantee that a limit function inherits certain properties from the sequence functions, we will need a global solution.

Definition 6.1.4. Let (f_n) be a sequence of functions defined on a common domain $A \subseteq \mathbb{R}$. We say (f_n) *converges uniformly* on A to a function f if for every $\varepsilon > 0$, there is an $N \in \mathbb{N}$ such that $|f_n(x) - f(x)| < \varepsilon$ for all $n \geq N$ and all $x \in A$. We use the notation $\lim f_n = f$ on A or $(f_n) \to f$ uniformly on A.

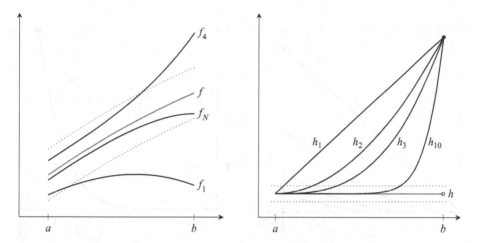

Fig. 6.2 Example of uniform and nonuniform convergence

The notation we are adopting for this definition is meant to convey the fact that we have removed the dependence on specific domain points. While this may seem like a place for potential confusion, rest assured that we will always be explicit about the type of convergence in question when there is a possibility for confusion. One comforting fact is that when a function converges uniformly to a limit f, then f also serves as the pointwise limit.

One way to visualize uniform convergence is demonstrated in Fig. 6.2. If f is the proposed limit function and $\varepsilon > 0$, we can think of this as setting up an ε-neighborhood around the graph of f. The convergence is then uniform if there is an $N \in \mathbb{N}$ such that f_n is inside this ε-neighborhood for all $n \geq N$. The left-hand image in the mentioned figure has this property, i.e. f_N is completely inside the set of dashed lines which represent the neighborhood. The right-hand figure represents a sequence which does not converge uniformly since for each $n \in \mathbb{N}$, some part of h_n is outside the neighborhood.

Example 6.1.5. Let $f_n(x) = x^n$ on $[0, 1/2]$. We know that the pointwise limit of this sequence is $f(x) = 0$ and we now aim to show that the convergence is uniform on this interval. Let $\varepsilon > 0$. We must now produce an $N \in \mathbb{N}$ such that $|f_n(x) - f(x)| < \varepsilon$ for all $n \geq N$ and all $x \in [0, 1/2]$. We know that each f_n is an increasing function and thus $f_n(x) = x^n \leq 1/2^n$ for all $x \in [0, 1/2]$. Also, we know that the sequence $(1/2^n) \to 0$ as $n \to \infty$. Thus if we choose $N \in \mathbb{N}$ such that $1/2^N < \varepsilon$, we see that

$$|f_n(x) - f(x)| = x^n \leq \frac{1}{2^n} \leq \frac{1}{2^N} < \varepsilon$$

for all $n \geq N$ and all $x \in [0, 1/2]$.

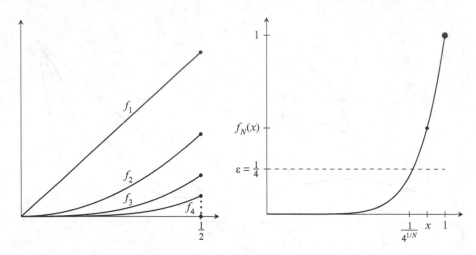

Fig. 6.3 Uniform and nonuniform convergence of (x_n)

The above proof depends on the restricted domain $[0, 1/2]$ together with the fact that $(1/2^n) \to 0$. If we attempt to extend this argument to the larger domain $[0, 1]$, we lose this last component. It turns out that the convergence is not uniform. This is due to the fact that the domain point 1 takes on the value 1 in each function in the sequence while all other domain values produce sequences which tend to zero. To see that the convergence is not uniform, we negate the definition above. Thus we must find an $\varepsilon > 0$ such that for all $N \in \mathbb{N}$ there is an $n \geq N$ and $x \in [0, 1]$ such that

$$|f_n(x) - f(x)| \geq \varepsilon.$$

Let $\varepsilon = 1/4$ and fix $N \in \mathbb{N}$. If we then choose $n = N$ and any $x \in (1/4^{1/N}, 1)$, we see that

$$|f_N(x) - f(x)| = x^N \geq (1/4^{1/N})^N = 1/4$$

and we therefore conclude that the sequence does not converge uniformly on $[0, 1]$ (see Fig. 6.3).

There is something subtle to note here. While the rate of convergence to the zero function slows down as we choose domain points closer to 1, the sequence *does* converge to 0. This observation leads us to believe that perhaps uniform convergence is always possible as long as we only consider domains which are bounded away from 1. Exercise 6.1.3 encapsulates this idea.

The next theorem relates the question of uniform convergence to convergence of a numerical sequence.

Proposition 6.1.6 (M-Test). *Suppose (f_n) is a sequence of functions which converges pointwise on $A \subseteq \mathbb{R}$ to a function f and set*

$$M_n = \sup\{|f_n(x) - f(x)| : x \in A\}.$$

Then (f_n) converges uniformly to f on A if and only if $\lim M_n = 0$.

Proof. We will prove one direction of the theorem and leave the reverse implication as an exercise. Suppose that M_n is defined as in the statement and assume that $\lim M_n = 0$. To show that (f_n) converges uniformly to f on A, let $\varepsilon > 0$, and choose $N \in \mathbb{N}$ such that $M_n < \varepsilon$ for all $n \geq N$. Then we have that

$$|f_n(x) - f(x)| \leq \sup\{|f_n(x) - f(x)| : x \in A\} = M_n < \varepsilon$$

for all $x \in A$ and all $n \geq N$. Thus (f_n) converges to f uniformly on A.   □

Example 6.1.7. The sequence (f_n) from Example 6.1.5 on the domain $[0, 1/2]$ essentially employed this theorem. There we found that $M_n = 1/2^n$ which converges to 0.

For another example, consider $h_n(x) = x/(1 + nx)$ on $(0, 1)$. With this example we show that it is not always necessary to calculate M_n explicitly. By definition $M_n \geq 0$. Thus if we can find a sequence (a_n) such that $M_n \leq a_n$ for all n and $\lim a_n = 0$, then we can apply the Squeeze Theorem to conclude that $\lim M_n = 0$. The theorem above does specify that we know the pointwise limit beforehand. A quick analysis of (h_n) indicates that, for a fixed $x \in (0, 1)$, the sequence $(h_n(x)) \to 0$ since the denominator diverges to ∞ as $n \to \infty$ while the numerator is constant. Thus we claim that (h_n) converges uniformly on $(0, 1)$ to $h(x) = 0$. To apply the theorem, fix $x \in (0, 1)$ and observe that

$$|h_n(x) - h(x)| = \left| \frac{x}{1 + nx} - 0 \right| = \frac{x}{1 + nx} \leq \frac{x}{nx} = \frac{1}{n}.$$

This computation holds for any $x \in (0, 1)$ and thus M_n must satisfy $0 \leq M_n \leq 1/n$. By the Squeeze Theorem, we conclude that $\lim M_n = 0$ and thus (h_n) converges uniformly on $(0, 1)$ to h.

Theorem 6.1.8 (Cauchy Criterion for Uniform Convergence of a Sequence of Functions). *A sequence of functions (f_n) converges uniformly on $A \subseteq \mathbb{R}$ if and only if for every $\varepsilon > 0$, there is an $N \in \mathbb{N}$ such that $|f_n(x) - f_m(x)| < \varepsilon$ for all $n, m \geq N$ and all $x \in A$.*

As was the case with numerical sequences, the advantage of the Cauchy Criterion is that it removes the need to know the limit beforehand. This will be particularly useful when considering series of functions. We will use the phrase *uniformly Cauchy* to signify the above condition.

Proof. First suppose that the sequence (f_n) is uniformly Cauchy on A. To show that (f_n) converges uniformly on A, we must first construct a function f which will serve as the limit. We will do this in a pointwise manner first and then show that the pointwise limit is also a uniform limit. To construct f, we first suppose $x_0 \in A$ and let $\varepsilon > 0$. Then there is an $N \in \mathbb{N}$ such that

$$|f_n(x) - f_m(x)| < \varepsilon$$

for all $x \in A$ and all $n, m \geq N$. Then it must be the case that

$$|f_n(x_0) - f_m(x_0)| < \varepsilon$$

for all $n, m \geq N$. This shows that the sequence $(f_n(x_0))$ is Cauchy, and by the Cauchy Criterion for sequences it must converge. We set $f(x_0) = \lim f_n(x_0)$. The fact that x_0 is an arbitrary point in A shows that we have constructed a function $f : A \to \mathbb{R}$ which satisfies $\lim f_n(x) = f(x)$ for all $x \in A$. This shows that f is the pointwise limit of (f_n).

To complete the proof, we must also show that (f_n) converges uniformly to f on A. To this end, let $\varepsilon > 0$ and choose $N \in \mathbb{N}$ such that $|f_n(x) - f_m(x)| < \varepsilon/2$ for all $x \in A$ and all $n, m \geq N$. Now fix $x_0 \in A$. Using the fact that (f_n) converges pointwise to f, we can choose $N_1 \in \mathbb{N}$ such that $|f_n(x_0) - f(x_0)| < \varepsilon/2$ for all $n \geq N_1$. Now set $N_2 = \max\{N, N_1\}$. If $n \geq N$, we have

$$|f_n(x_0) - f(x_0)| \leq |f_n(x_0) - f_{N_2}(x_0)| + |f_{N_2}(x_0) - f(x_0)| < \varepsilon/2 + \varepsilon/2 = \varepsilon.$$

The above inequality holds for any $x \in A$ if we vary N_1, and hence N_2, as x varies. This guarantees us that

$$|f_n(x) - f(x)| < \varepsilon$$

for all $x \in A$ and all $n \geq N$. Therefore (f_n) converges uniformly on A to f.

We leave the remaining implication as Exercise 6.1.11.   □

The proof above may seem unsettling. Indeed, the convergence seems to be dependent on the choice of N_2 which *does* depend on the domain value x_0 under consideration since we are appealing to the pointwise convergence of (f_n) to obtain it, i.e. as the domain point varies we may be required to add/subtract a different sequence function evaluated at the domain point in question. However, to each x_0 *there is* an associated N_2, and, even though N_2 depends upon the domain point, the choice for N *does not*. The independence of N guarantees us that the convergence is uniform.

We now show that uniform convergence preserves continuity.

Theorem 6.1.9. *Let (f_n) be a sequence of functions defined on a common domain A and suppose each f_n is continuous. If (f_n) converges uniformly on A to a function f, then f is also continuous on A.*

For $c \in A$ and $\varepsilon > 0$, we must identify a $\delta > 0$ such that $|x - c| < \delta$ implies $|f(x) - f(c)| < \varepsilon$. To do this, we must also employ the fact that (f_n) converges uniformly to f. Convergence guarantees us that $f_n(x)$ will be close to $f(x)$ (this also holds for c) if we choose n large enough. Thus it makes sense to approximate f by f_n. This idea and two applications of the triangle inequality lead to the inequality,

$$|f(x) - f(c)| = |f(x) - f_n(x) + f_n(x) - f_n(c) + f_n(c) - f(c)|$$
$$\leq |f(x) - f_n(x)| + |f_n(x) - f_n(c)| + |f_n(c) f(c)|.$$

The first and third term on the right can be controlled simultaneously since the convergence is uniform. The middle term can also be controlled since f_n is continuous for all n. The key here is deciding which property to apply first, convergence or continuity. In order to apply continuity, however, we must first know which function in the sequence is under consideration. Thus we should apply convergence first, specifying a function f_n, and then continuity, providing $\delta > 0$.

Proof. Let (f_n) and f be as in the statement of the theorem. Let $\varepsilon > 0$ and $c \in A$ a limit point; the case for isolated points is trivial. Choose $N \in \mathbb{N}$ such that $|f_n(x) - f(x)| < \varepsilon/3$ for all $n \geq N$ and all $x \in A$. We will work specifically with the function f_N since we only need one sequence function which is a viable approximation of f. With this in mind, since f_N is continuous at c, there is a $\delta > 0$ such that if $x \in A$ with $|x - c| < \delta$, then $|f_N(x) - f_N(c)| < \varepsilon/3$. If we then consider $x \in A$ with $|x - c| < \delta$, we see that

$$|f(x) - f(c)| = |f(x) - f_N(x) + f_N(x) - f_N(c) + f_N(c) - f(c)|$$
$$\leq |f(x) - f_N(x)| + |f_N(x) - f_N(c)| + |f_N(c) - f(c)|$$
$$< \varepsilon/3 + \varepsilon/3 + \varepsilon/3 = \varepsilon$$

and hence f is continuous at c. The fact that c was arbitrary in A implies that f is continuous on A. □

We turn our attention now to the preservation of differentiation. Specifically we seek to determine when a sequence of differentiable functions (f_n) converges to a differentiable function f, and to do this we will be led to consider the convergence of the sequence (f_n'). We mentioned earlier that we will need a strict set of hypothesis in this case; we first consider several examples to motivate these hypotheses and to determine how well our current notions of convergence handle this scenario.

Example 6.1.10. The fact that pointwise convergence does not preserve continuity immediately implies that it is not strong enough to imply differentiability of the limit function. Thus from the outset we will only consider sequences which converge uniformly. First consider the sequence defined by $f_n(x) = (\sin(nx))/n$ on the interval $[0, 1]$. It should be apparent from our previous work that this sequence converges uniformly on $[0, 1]$ to $f(x) = 0$. Notice also that $f_n'(x) = \cos(nx)$ and $f'(x) = 0$. This sequence does not converge uniformly on $[0, 1]$ to f'. Even more concerning is the fact that there are points $x \in [0, 1]$ for which the sequence $f_n'(x)$ diverges; take $x = 1$ for instance. This demonstrates that uniform convergence of a sequence does not imply convergence of any sort for the differentiated sequence.

Now consider the sequence (g_n) defined on $[0, 1]$ by $g_n(x) = (\sin(nx))/n^2$. In this case we see that (g_n) converges uniformly on $[0, 1]$ to $g(x) = 0$, and the sequence of derivatives g_n' satisfies $g_n'(x) = (\cos(nx))/n$ which does converge uniformly to g' on $[0, 1]$. While this example doesn't explain the relationship between the uniform convergence of (g_n) and (g_n'), it does lead us to believe that uniform convergence of the later potentially implies uniform convergence of the former. A third example provides a bit more insight. Take $h_n(x) = ((\sin(nx))/n^2) + n$. This sequence of functions does not converge for *any* $x \in [0, 1]$, but the sequence of derivatives is defined by $h_n'(x) = (\cos(nx))/n$ which does converge uniformly on $[0, 1]$ to the zero function.

These examples motivate our next result which states that uniform convergence of the sequence of derivatives implies uniform convergence of the original sequence as long as the sequence of functions converges for at least one point in the common domain. The need for this extra hypothesis is provided by the last example and the fact that two functions which have the same derivative differ by at most a constant. Requiring convergence of the original sequence for at least one domain point acts as an anchor and alleviates the problematic situation above.

Lemma 6.1.11. *Let (f_n) be a sequence of functions defined on the common domain $[a, b]$ and assume that (f_n') converges uniformly on $[a, b]$. If there is a point $c \in [a, b]$ such that $(f_n(c))$ converges, then (f_n) converges uniformly on $[a, b]$.*

Proof. Our aim is to show that (f_n) is uniformly Cauchy on $[a, b]$ in order to avoid having to construct a limit function, but this will imply that (f_n) converges uniformly on $[a, b]$ by Theorem 6.1.8. We will also use the Mean Value Theorem to provide the connection between f_n and f_n'. To begin, let $\varepsilon > 0$. Using the fact that (f_n') converges uniformly on $[a, b]$, we may choose $N_1 \in \mathbb{N}$ such that

$$|f_n'(x) - f_m'(x)| < \frac{\varepsilon}{2(b-a)} \tag{6.2}$$

for all $x \in [a, b]$ and all $n, m \geq N_1$. Also, choose $N_2 \in \mathbb{N}$ such that $|f_n(c) - f_m(c)| < \varepsilon/2$ for all $n, m \geq N_2$, and set $N = \max\{N_1, N_2\}$. Before applying our choice of N, observe that, for $x \in [a, b]$,

$$|f_n(x) - f_m(x)| = |f_n(x) - f_m(x) - f_n(c) + f_m(c) + f_n(c) - f_m(c)|$$
$$\leq |(f_n(x) - f_m(x)) - (f_n(c) - f_m(c))| + |f_n(c) - f_m(c)|.$$

Focusing on the first term on the right, the Mean Value Theorem applied to the function $f_n - f_m$ guarantees us that there is a point d, dependent on m, n, and x, between x and c such that

$$|(f_n(x) - f_m(x)) - (f_n(c) - f_m(c))| = |x - c| |f_n'(d) - f_m'(d)| \leq (b-a)|f_n'(d) - f_m'(d)|.$$

As with our proof of the Cauchy Criterion, the dependence of d on x is acceptable since there is a choice of d for each $x \in [a, b]$ and the fact that the estimate from Eq. (6.2) holds on $[a, b]$. If we now consider $n, m \geq N$, we have

$$|f_n(x) - f_m(x)| \leq |(f_n(x) - f_m(x)) - (f_n(c) - f_m(c))| + |f_n(c) - f_m(c)|$$
$$\leq (b - a)|f_n'(d) - f_m'(d)| + |f_n(c) - f_m(c)|$$
$$\leq (b - a)\left(\frac{\varepsilon}{2(b - a)}\right) + \varepsilon/2$$
$$= \varepsilon$$

for all $x \in [a, b]$. With this we conclude that (f_n) is uniformly Cauchy on $[a, b]$ and hence uniformly convergent.   □

The final result of this section provides a sufficient condition for a sequence of differentiable functions to converge to a differentiable function.

Theorem 6.1.12. *Let (f_n) be a sequence of functions defined on the common domain $[a, b]$ and assume that (f_n') converges uniformly on $[a, b]$. Further suppose there is a point $c_0 \in [a, b]$ such that $(f_n(c_0))$ converges. Then (f_n) converges uniformly on $[a, b]$ to a function f which satisfies $\lim f_n' = f'$.*

The proof is rather involved and we will dispense with a few technical details and observations here to streamline the flow of the argument. By the previous lemma and the hypothesis on (f_n') we know there exist functions $f, g : [a, b] \to \mathbb{R}$ which are the uniform limits of (f_n) and (f_n'), respectively. Our goal is then to show that $f' = g$. Using the definition of the derivative and the $\varepsilon - \delta$ definition of a functional limit, this equates to showing for $c \in [a, b]$ and $\varepsilon > 0$, there is a $\delta > 0$ such that $0 < |x - c| < \delta$ implies

$$\left| \frac{f(x) - f(c)}{x - c} - g(c) \right| < \varepsilon.$$

To determine a choice for δ and prove this implication we will work with three distinct terms which will appear momentarily. One of these terms will demonstrate that we can approximate slopes of secant lines of f with slopes of secant lines of f_n (for some n to be specified). A second term will use the fact that each f_n is differentiable, and a third term will rely on the fact that (f_n') converges uniformly to g. These three terms are obtained via addition/subtraction and the triangle inequality,

$$\left| \frac{f(x) - f(c)}{x - c} - g(c) \right| \leq \left| \frac{f(x) - f(c)}{x - c} - \frac{f_n(x) - f_n(c)}{x - c} \right|$$
$$+ \left| \frac{f_n(x) - f_n(c)}{x - c} - f_n'(c) \right| + |f_n'(c) - g(c)|.$$

Notice that controlling the middle term here is possible since f_n is differentiable, but its behavior depends on the index. We will therefore work with the first and third terms at the outset to identify which sequence function we will use. We used a similar strategy in the proof of Theorem 6.1.9. For the first term we will use the Mean Value Theorem and a strategy similar to that of Lemma 6.1.11.

Proof. From Lemma 6.1.11, we know that there is a function $f : [a, b] \to \mathbb{R}$ such that (f_n) converges uniformly to f on $[a, b]$. By hypothesis, there is a function $g : [a, b] \to \mathbb{R}$ such that (f'_n) converges uniformly to g on $[a, b]$. Our goal is to show that f is differentiable with $f' = g$ on $[a, b]$. To this end, let $\varepsilon > 0$ and $c \in [a, b]$. Now we must produce $\delta > 0$ such that $0 < |x - c| < \delta$ implies

$$\left| \frac{f(x) - f(c)}{x - c} - g(c) \right| < \varepsilon.$$

To do this we will show that each term on the right of the inequality

$$\left| \frac{f(x) - f(c)}{x - c} - g(c) \right| \leq \left| \frac{f(x) - f(c)}{x - c} - \frac{f_n(x) - f_n(c)}{x - c} \right|$$

$$+ \left| \frac{f_n(x) - f_n(c)}{x - c} - f'_n(c) \right| + |f'_n(c) - g(c)|$$

can be bounded above by $\varepsilon/3$. To deal with the third term here, choose $N_1 \in N$ such that $|f'_n(x) - g(x)| < \varepsilon/3$ for all $x \in [a, b]$ and all $n \geq N_1$.

For the first term, since (f'_n) converges uniformly, and is hence uniformly Cauchy, choose $N_2 \in \mathbb{N}$ such that $|f'_n(x) - f'_m(x)| < \varepsilon/3$ for all $x \in [a, b]$ and all $n, m \geq N_2$. Now, for $m, n \in \mathbb{N}$ and $x \in [a, b]$ with $x \neq c$, we apply the Mean Value Theorem to the function $f_m - f_n$ on the interval $[c, x]$ (or $[x, c]$ though the argument is identical) to obtain a point $d \in (c, x)$ with

$$f'_m(d) - f'_n(d) = \frac{(f_m(x) - f_n(x)) - (f_m(c) - f_n(c))}{x - c};$$

notice that the choice of d does depend on m, n, x, and c. Thus, by varying d with m, n, and x, we have

$$\left| \frac{f_m(x) - f_m(c)}{x - c} - \frac{f_n(x) - f_n(c)}{x - c} \right| = \left| \frac{(f_m(x) - f_m(x)) - (f_n(c) - f_n(c))}{x - c} \right|$$

$$= |f'_m(d) - f'_n(d)| < \varepsilon/3$$

for all $n, m \geq N_2$ and $x \in [a, b]$ with $x \neq c$. Here we apply the Order Limit Theorem to the terms above indexed by m but keeping $n \geq N_2$ fixed to obtain

$$\left| \frac{f(x) - f(c)}{x - c} - \frac{f_n(x) - f_n(c)}{x - c} \right| \leq \varepsilon/3$$

for all $n \geq N_2$ and $x \in [a, b]$ with $x \neq c$.

To complete the proof, we set $N = \max\{N_1, N_2\}$ which is independent of x and c. From the assumption that f_N is differentiable on $[a, b]$, we now choose $\delta > 0$ such that $0 < |x - c| < \delta$ implies

$$\left| \frac{f_N(x) - f_N(c)}{x - c} - f_N'(c) \right| < \varepsilon/3;$$

notice that δ does depend on c, which is permissible. Thus if $0 < |x - c| < \delta$, the fact that $N \geq N_1, N_2$ implies

$$\left| \frac{f(x) - f(c)}{x - c} - g(c) \right| \leq \left| \frac{f(x) - f(c)}{x - c} - \frac{f_N(x) - f_N(c)}{x - c} \right|$$

$$+ \left| \frac{f_N(x) - f_N(c)}{x - c} - f_N'(c) \right| + |f_N'(c) - g(c)|$$

$$< \varepsilon/3 + \varepsilon/3 + \varepsilon/3$$

$$= \varepsilon$$

completing the proof.   □

Exercises

Exercise 6.1.1. Suppose (f_n) converges uniformly on $A \subseteq \mathbb{R}$ to a function f. Show that (f_n) converges pointwise on A to f.

Exercise 6.1.2. (a) Show that the sequence of functions defined on $(0, 1)$ by $g_n(x) = (\sin(nx)/(nx))$ converges pointwise to $g(x) = 0$.
(b) Show that (g_n) converges uniformly to g on $[1, \infty)$.

Exercise 6.1.3. Show that the sequence of functions defined by $f_n(x) = x^n$ converges uniformly to $f(x) = 0$ on any compact subset of $(0, 1)$.

Exercise 6.1.4. Find the pointwise limit of the sequence of functions defined on $\mathbb{R}$ by

$$f_n(x) = \frac{x^2 + nx}{n}.$$

Show that the convergence is uniform on any bounded subset of $\mathbb{R}$.

Exercise 6.1.5. For the sequence (h_n) in Example 6.1.7, discuss the role played by the domain $(0, 1)$. Can you show that the convergence is uniform on $(0, 1]$? On $(0, 2]$? On $(0, \infty)$?

Exercise 6.1.6. Suppose the sequence of functions (f_n) converges uniformly on the interval (a, b). Suppose further that (f_n) converges pointwise at $x = a$ and $x = b$. Show that the sequence converges uniformly on $[a, b]$.

Exercise 6.1.7. Show that the sequence of functions defined by

$$f_n(x) = \begin{cases} \dfrac{x}{x^2 - (1 - nx)^2}, & x \neq \frac{1}{n+1}, \frac{1}{n-1}; \\ 0, & x = \frac{1}{n+1}, \frac{1}{n-1} \end{cases}$$

converges pointwise on $[0, 1]$ but that this convergence is not uniform. To show the convergence is not uniform, consider the vertical asymptotes of f_n and consider the M-test.

Exercise 6.1.8. Show that the sequence of functions defined by

$$f_n(x) = \frac{x^n}{1 + x^n}$$

converges pointwise on $[0, 1]$. Argue that the convergence is not uniform.

Exercise 6.1.9. Let (f_n) and (g_n) be two sequences defined on the common domain A.

(a) If (f_n) converges uniformly to f and (g_n) converges uniformly to g on A, show that $(f_n + g_n)$ converges uniformly to $f + g$ on A.
(b) If (f_n) converges pointwise to f and (g_n) converges pointwise to g on A, show that $(f_n g_n)$ converges pointwise to fg on A.
(c) Use the functions $f_n(x) = x + 1/n$ and $g_n(x) = x$, on $\mathbb{R}$, to show that the above statement does not hold if we replace pointwise convergence with uniform convergence.
(d) Suppose that (f_n) converges uniformly to f and (g_n) converges uniformly to g on A. If we further suppose that there is a constant $M > 0$ such that $|f_n(x)| \leq M$ and $|g_n(x)| \leq M$ on A, show that $(f_n g_n)$ does converge uniformly on A.

Exercise 6.1.10. Complete the proof of Proposition 6.1.6.

Exercise 6.1.11. Complete the proof of Theorem 6.1.8.

Exercise 6.1.12. Theorem 6.1.9 shows that the condition that (f_n) converges uniformly to f is sufficient to conclude that f is also continuous whenever each f_n is continuous. Is this condition also necessary? In other words, can you provide an example of a sequence of functions (f_n) defined on a common domain A for which each f_n is continuous and which converges pointwise but not uniformly to a continuous function f?

Exercise 6.1.13. Let (f_n) be a sequence of functions defined on a common domain A and suppose each f_n is uniformly continuous. If (f_n) converges uniformly on A to a function f, show that f is also uniformly continuous on A.

Exercise 6.1.14 (Dini's Theorem). Suppose (f_n) is a sequence of continuous functions defined on the common domain $[a, b]$ with pointwise limit f which is also continuous. Suppose also that for each $x \in [a, b]$, we know that $f_{n+1}(x) \le f_n(x)$ for all $n \in \mathbb{N}$.

(a) Follow the steps below to show that (f_n) converges uniformly to f.

    (i) Define a new sequence (g_n) by $g_n = f_n - f$. Explain why (g_n) is a decreasing sequence of continuous functions converging pointwise to the zero function on $[a, b]$. Further, interpret the statement "(f_n) converges uniformly to f" into a statement about the sequence (g_n).

    (ii) For $\varepsilon > 0$, define a sequence of sets $E_n = \{x \in [a, b] : g_n(x) < \varepsilon\}$. Show that this sequence of sets satisfies $E_1 \subseteq E_2 \subseteq E_3 \subseteq \dots$. Explain why each E_n is open relative to $[a, b]$, i.e. for each $n \in \mathbb{N}$ there is an open set A_n such that $E_n = A_n \cap [a, b]$.

    (iii) Use the fact that $[a, b]$ is compact to find a finite open cover for $[a, b]$ which guarantees the existence of $N \in \mathbb{N}$ such that $|g_n(x)| < \varepsilon$ for all $x \in [a, b]$ and all $n \ge N$.

(b) Modify your argument to show that the statement from part (a) holds if we assume that for each $x \in [a, b]$, we have $f_n(x) \le f_{n+1}(x)$ for all $n \in \mathbb{N}$.

(c) Show that the conclusion to Dini's Theorem fails if the hypothesis of continuity of the limit function is removed.

Exercise 6.1.15. Suppose (f_n) is a sequence of functions defined on $[0, 1]$ each of which is Lipschitz (see Exercises 4.4.8 and 5.3.5) with the common factor of M.

(a) If (f_n) converges pointwise to f on $[0, 1]$, follow the steps below to show that the convergence is uniform.

    (i) For $\varepsilon > 0$, explain why there is a finite collection of points $\{x_1, x_2, \dots x_K\}$ such that $[0, 1] \subseteq \cup_{k=1}^{K} B_{\varepsilon/3M}(x_k)$.

    (ii) For each $k \in \{1, 2, \dots, K\}$, choose $N_k \in \mathbb{N}$ such that $|f_n(x_k) - f_m(x_k)| < \varepsilon/3$ for all $n, m \ge N_k$.

    (iii) Choosing $N = \max\{N_k : 1 \le k \le K\}$, show that $|f_n(x) - f_m(x)| < \varepsilon$ for all $n, m \ge N$ and all $x \in [0, 1]$.

(b) If (f_n) converges pointwise to f on $[0, 1]$, show that f is also Lipschitz.

6.2 Series of Functions

Many of the results of this section will follow immediately from our study of numerical series and sequences of functions. This may leave the reader feeling as though series are of secondary importance. This is not the case, however, as will be seen in the next section where we consider a particular class of series.

Definition 6.2.1. Let (f_n) be a sequence of functions defined on a common domain $A \subseteq \mathbb{R}$. The *series of functions*

$$\sum_{j=1}^{\infty} f_j(x) = f_1(x) + f_2(x) + f_3(x) + \cdots$$

converges pointwise on A to a function f if the sequence of partial sums (s_n) defined by

$$s_n(x) = f_1(x) + f_2(x) + f_3(x) + \cdots + f_n(x)$$

converges pointwise on A to f. In this case we write $\sum_{j=1}^{\infty} f_j(x) = f(x)$ on A.

If the sequence of partial sums converges uniformly on A to f then we say the series *converges uniformly* on A to f and write $\sum_{j=1}^{\infty} f_n = f$.

As is typical, we will often suppress the notation and write $\sum f_n$ unless the first term of the series is relevant to our work.

Example 6.2.2. As with our study of numerical series, it is often hard to identify the limit for a series of functions and we will typically be satisfied to know if a given series converges. There is, however, one very special example which we have already encountered. Take $f_0(x) = 1$ and $f_n(x) = x^n$ for $n \in \mathbb{N}$ on the interval $(-1, 1)$. It is conventional to assume that $x^0 = 1$ even though this isn't exactly true at $x = 0$; we will assume this convention throughout this chapter. This gives rise to the geometric series of Sect. 3.2 and we have

$$\sum_{n=0}^{\infty} f_n(x) = \sum_{n=0}^{\infty} x^n = \frac{1}{1-x}$$

on $(-1, 1)$. Though this is a pointwise result, we will see later that the convergence is uniform on any compact subset of $(-1, 1)$.

For a second example, for $n \in \{0\} \cup \mathbb{N}$, take $g_n(x) = x^n/n!$ where $0! = 1$, and consider the series $\sum_{n=0}^{\infty} x^n/n!$. We will show that this series converges pointwise on $\mathbb{R}$, but we will not identify the limit function at this time. Choose $x \in \mathbb{R}$. To show that $\sum_{n=0}^{\infty} x^n/n!$ converges, we apply the Ratio Test,

$$\lim \left| \frac{x^{n+1}/(n+1)!}{x^n/n!} \right| = \lim \frac{|x|}{n+1} = 0$$

which shows that the numerical series $\sum_{n=0}^{\infty} x^n/n!$ converges. The fact that x was an arbitrary real number guarantees us that the series converges pointwise on $\mathbb{R}$. We will identify the limit function in the next chapter.

The next two theorems follow almost immediately from their sequence counterparts the details of which are requested in the exercises.

Theorem 6.2.3. *Let $\sum_{n=1}^{\infty} f_n$ be a series which converges uniformly on $A \subseteq \mathbb{R}$ to f. If each f_n is continuous on A, then f is continuous on A.*

Theorem 6.2.4 (Cauchy Criterion for Uniform Convergence of Series). *The series $\sum_{n=1}^{\infty} f_n$ converges uniformly on $A \subseteq \mathbb{R}$ to a function f if and only if for every $\varepsilon > 0$, there is an $N \in \mathbb{N}$ such that*

$$|s_n(x) - s_m(x)| = |f_{m+1}(x) + f_{m+2}(x) + \cdots + f_n(x)| < \varepsilon$$

for all $n > m \geq N$ and all $x \in A$.

We close this section with a variant of Proposition 6.1.6 which relates the question of convergence for a series of functions to a numerical series. The difference here is that the condition is sufficient but not necessary. We will explore an example detailing this following the proof.

Proposition 6.2.5 (Weierstrass M-Test). *Let (f_n) be a sequence of functions defined on a common domain $A \subseteq \mathbb{R}$ and suppose for each $n \in \mathbb{N}$ there is an $M_n > 0$ such that $|f_n(x)| \leq M_n$ for all $x \in A$. If $\sum_{n=1}^{\infty} M_n$ converges, then $\sum_{n=1}^{\infty} f_n$ converges uniformly on A.*

Proof. Suppose (f_n) and M_n are as stated above. Our goal is to relate the two series using the bound provided and the Cauchy Criterion in order to conclude that $\sum f_n$ is uniformly Cauchy. Let $\varepsilon > 0$. Since $\sum M_n$ converges, it is Cauchy, and hence we may choose $N \in \mathbb{N}$ such that

$$|M_{m+1} + M_{m+2} + \cdots + M_n| < \varepsilon$$

for all $n > m \geq N$. The fact that $M_n > 0$ for all n indicates that we do not need the absolute value in the above inequality and we have

$$M_{m+1} + M_{m+2} + \cdots + M_n < \varepsilon$$

for all $n > m \geq N$. Now, applying the given bound relating f_n and M_n and the triangle inequality, we have

$$|f_{m+1}(x) + f_{m+2}(x) + \cdots + f_n(x)| \leq |f_{m+1}(x)| + |f_{m+2}(x)| + \cdots + |f_n(x)|$$
$$\leq M_{m+1} + M_{m+2} + \cdots + M_n$$
$$< \varepsilon$$

for all $x \in A$ and all $n > m \geq N$. With this we conclude that $\sum f_n$ is uniformly Cauchy and hence uniformly convergent. $\qquad\square$

Example 6.2.6. Returning now to the sequence defined by $f_n(x) = x^n$ for $n \in \{0\} \cup \mathbb{N}$ (see Example 6.2.2), we will show that the series of functions converges uniformly on any compact subset $K \subseteq (-1, 1)$. The argument is trivial if $K = \{0\}$ and so we assume that K contains at least one nonzero real number. We know immediately that $\sum f_n$ converges pointwise on K, and we know that K is closed and bounded. Thus there is a $M > 0$ such that $K \subseteq [-M, M] \subseteq (-1, 1)$. With this, we see that $|f_n(x)| \leq M^n$ for all $x \in [-M, M]$, specifically for all $x \in K$. Moreover, since $0 < M < 1$, we know that $\sum M^n$ converges as it is a geometric series, and thus the Weierstrass M-Test indicates that $\sum f_n$ converges uniformly on K.

As a final example we investigate a sequence of functions whose series converges uniformly but for which there is no sequence of bounds (M_n) such that $\sum M_n$ converges. This will explain why the Weierstrass M-test is not a biconditional statement. For $n \in \mathbb{N}$, we define

$$h_n(x) = \begin{cases} 0, & 0 \leq x < 1/(n+1); \\ 1/n, & 1/(n+1) \leq x < 1/n; \\ 0, & 1/n \leq x \leq 1. \end{cases}$$

To see that this series converges uniformly, we first identify a pointwise limit. Notice that $h_n(0) = 0 = h_n(1)$ for all n and hence $\sum h_n(0) = \sum h_n(1) = 0$. Now, if $x \in (0, 1)$ then $x \in [1/(j+1), 1/j)$ for some $j \in \mathbb{N}$. In this case, $h_j(x) = 1/j$ but $h_n(x) = 0$ for all $n \neq j$. Thus $\sum h_n(x) = 1/j$. This shows that the limit function h is a step function taking the value $1/j$ on the interval $[1/(j+1), 1/j)$ and the value zero at 0 and 1. If we now consider the partial sums of the series, we see that $s_n(x)$ is equal to h for all x with $1/(n+1) \leq x \leq 1$ and $x = 0$; furthermore, s_n is zero on the interval $(0, 1/(n+1))$ while h is at most $1/(n+2)$ on this interval. For $\varepsilon > 0$, if we choose $N \in \mathbb{N}$ such that $1/N < \varepsilon$, we see that

$$|s_n(x) - h(x)| \leq \frac{1}{n+2} \leq \frac{1}{N+2} < \frac{1}{N} < \varepsilon$$

for all $n \geq N$ and $x \in [0, 1]$ showing that the convergence is uniform.

Now we must show that there is no sequence of bounds (M_n) such that $\sum M_n$ converges. The most obvious sequence is $M_n = 1/n$ for it is the case that $|h_n(x)| \leq 1/n$ for all $x \in [0, 1]$ and $\sum 1/n$ diverges. However, this does not show that there is no sequence of bounds with the prescribed behavior. We therefore suppose that (M_n) is such a sequence, i.e. $|h_n(x)| \leq M_n$ for all $x \in [0, 1]$. Our aim is now to show that the series $\sum M_n$ cannot converge. This is a simple matter since given any $n \in \mathbb{N}$, we know that $h_n(x) = 1/n$ for some $x \in [0, 1]$ and thus $1/n \leq M_n$. By the Comparison Test we conclude that $\sum M_n$ does not converge.

Exercises

Exercise 6.2.1. If $\sum f_n$ converges uniformly on A, show that the sequence (f_n) converges uniformly to zero on A.

Exercise 6.2.2. Show that the series $\sum g_n$ from Example 6.2.2 converges uniformly on any bounded subset of $\mathbb{R}$.

Exercise 6.2.3. Consider the series

$$\frac{x^2}{1+x^2} + \left(\frac{x^4}{1+x^4} - \frac{x^2}{1+x^2}\right) + \left(\frac{x^6}{1+x^6} - \frac{x^4}{1+x^4}\right) + \cdots .$$

Show that this series converges on all of $\mathbb{R}$ and find its pointwise limit by considering the partial sums of the series.

Exercise 6.2.4. Find the largest subset of $\mathbb{R}$ on which the series

$$\sum_{n=1}^{\infty} \frac{x^n}{1+x^{2n}}$$

converges pointwise.

Exercise 6.2.5. Suppose the partial sums of a series of functions are given by

$$s_n = \frac{nx}{1+n^2x^2}.$$

Show that the series converges uniformly on the interval $[1, 2]$ but only pointwise on the interval $[0, 1]$.

Exercise 6.2.6. Show that the series of functions defined by

$$\sum_{n=1}^{\infty} f_n(x) = (1-x) + x(1-x) + x^2(1-x) + \cdots$$

converges pointwise on $[0, 1]$. Argue that the convergence is not uniform.

Exercise 6.2.7. (a) Prove Theorem 6.2.3.
(b) State and prove a version of Theorem 6.1.12 for series of functions. Be sure to give an explicit form for the derivative.

Exercise 6.2.8. Show that the derivative of the function

$$f(x) = \sum_{n=1}^{\infty} \frac{x^n}{n(n+1)}$$

is given by

$$\sum_{n=0}^{\infty} \frac{x^n}{n+2}$$

for $x \in (-1, 1)$.

Exercise 6.2.9. Show that the function

$$f(x) = \sum_{n=1}^{\infty} \frac{\sin(nx)}{n^4}$$

is twice-differentiable on $\mathbb{R}$. Identify f' and f''.

Exercise 6.2.10. Show that the given series defines a continuous function on the specified domain.

(a) $\displaystyle\sum_{n=1}^{\infty} \frac{x^n}{n^4}$ on $[-1, 1]$ (c) $\displaystyle\sum_{n=1}^{\infty} nx^n$ on $(-1, 1)$

(b) $\displaystyle\sum_{n=1}^{\infty} \frac{\sin(nx)}{n^2}$ on $\mathbb{R}$

Exercise 6.2.11. Supply a proof of Theorem 6.2.4.

Exercise 6.2.12. Suppose $\sum f_n$ is a uniformly convergent series defined on an interval (a, b) and let $g : (a, b) \to \mathbb{R}$ be a bounded function. Show that the series $\sum g(x) f_n(x)$ also converges uniformly.

6.3   Power Series

Definition 6.3.1. Let $(a_n)_{n=0}^{\infty}$ be a sequence of real numbers. A series of the form

$$\sum_{n=0}^{\infty} a_n x^n = a_0 + a_1 x + a_2 x^2 + \cdots$$

is called a *power series centered at $c = 0$* and the a_n's are called the *coefficients* of the series.

Power series are a special class of series of functions. The beauty of this type of series is that all of the relevant information regarding the convergence behavior of the series is encoded in the defining sequence (a_n). The Ratio and Root tests are all that is required to extract this information as we will see in the first theorem below; this is quite important as it tells us where the power series is a well-defined function. Once we have this information in hand we will seek to understand when these series are continuous and differentiable. As a first observation, we see that the series always converges for $x = 0$. Thus we will be interested in cases where the series converges on larger sets. Our first result will show that the set on which a power series converges is always an interval centered around 0, hence the phrasing "power series centered at 0."

More generally, for $c \in \mathbb{R}$, we can define a power series centered at c by

$$\sum_{n=0}^{\infty} a_n(x - c)^n = a_0 + a_1(x - c) + a_2(x - c)^2 + \cdots .$$

All of the results in this section hold in this more general setting. We will work with power series centered at 0 exclusively to simplify computations and will discuss how to move to the more general framework later on. We will also always assume that a power series begins at $n = 0$ unless otherwise stated and will therefore suppress the notation on summations.

Example 6.3.2. We saw several examples of power series in Example 6.2.2. For the sequence $(1)_{n=0}^{\infty}$ we saw that the resulting power series $\sum x^n$ converges for all x in the interval $(-1, 1)$ while the power series defined by the sequence $(1/n!)_{n=0}^{\infty}$ converges for all real numbers.

As a more extreme example, consider the sequence $(n!)_{n=0}^{\infty}$ and let $x \in \mathbb{R}$. To determine where the power series $\sum n! x^n$ converges, we apply the Ratio Test. Fix $x \in \mathbb{R}$ with $x \neq 0$ and observe that

$$\lim \left| \frac{(n + 1)! x^{n+1}}{n! x^n} \right| = |x| \lim n$$

which diverges to infinity. Thus the series diverges except when $x = 0$. These three examples are simple but they give a complete characterization of the types of sets on which a power series can converge: an interval of finite length centered around 0, on $\mathbb{R}$, or only at 0. The following theorem provides a rigorous argument for this characterization.

Theorem 6.3.3. *Let $\sum a_n x^n$ be a power series with $a_n \neq 0$ and suppose*

$$L = \lim \left| \frac{a_{n+1}}{a_n} \right|$$

exists or is infinite.

(a) If $0 < L < \infty$, we set $R = 1/L$ and the series converges absolutely if $|x| < R$ and diverges if $|x| > R$.
(b) If $L = 0$, we set $R = \infty$ and the series converges absolutely for all $x \in \mathbb{R}$.
(c) If $L = \infty$, we set $R = 0$ and the series converges only at $x = 0$.

Proof. Our proof here will rely on the Ratio Test (Theorem 3.3.7) and we will apply it to each point of $\mathbb{R}$ as we did in the previous example. For the first statement, assume $0 < L < 1$ and let $x \in \mathbb{R}$. With an eye toward the Ratio test we first compute

$$\lim \left| \frac{a_{n+1} x^{n+1}}{a_n x^n} \right| = |x| \lim \left| \frac{a_{n+1}}{a_n} \right| = L|x|.$$

This quantity is less than 1 exactly when $|x| < 1/L$ and greater than 1 when $|x| > 1/L$. Setting $R = 1/L$, we see by the Ratio Test that the series $\sum a_n x^n$ converges absolutely when $|x| < R$ and diverges when $|x| > R$.

We leave the remaining cases as Exercise 6.3.2.   □

We call R the *radius of convergence* of the series and $(-R, R)$ is called the *interval of convergence*. These concepts describe how far away from the center we are allowed to choose values for which the series will converge. Notice that the conclusions from part (a) provide no information concerning convergence at the points R and $-R$. This is due to the fact that the Ratio Test is inconclusive when the limit above takes on the value 1. This theorem and the one that follows indicate that the interval of convergence is symmetric about the center, however, this behavior does not carry over to the endpoints. In fact, there are power series which converge at both endpoints, neither endpoint, or one endpoint and not the other.

Example 6.3.4. The power series $\sum x^n$ is an example of a power series which converges at neither endpoint. The sequence $(1/n)_{n=1}^{\infty}$ provides an instance where we have convergence at one endpoint and not the other. Notice first that $R = 1$. We then test the series at $x = -1$ and $x = 1$ which results in the two series

$$\sum_{n=1}^{\infty} \frac{(-1)^n}{n} \quad \text{and} \quad \sum_{n=1}^{\infty} \frac{1}{n},$$

respectively. The first series is the alternating harmonic series which converges while the second is the harmonic series which diverges. Hence the power series $\sum_{n=1}^{\infty} x^n/n$ converges on the set $[-1, 1)$. Exercise 6.3.1 asks you to show that the power series $\sum_{n=0}^{\infty} x^n/n^2$ converges on the interval $[-1, 1]$.

As mentioned earlier, the Root Test can also be used to compute the radius of convergence. In this case let $\sum a_n x^n$ be a power series and suppose

$$L = \lim |a_n|^{1/n}$$

exists or is infinite. We could then draw the same conclusions as Theorem 6.3.3 by applying the Root Test. This approach has several advantages as the Root Test removes the requirement that $a_n \neq 0$ for all n and there are circumstances where the limit used in the Root Test exists while the limit employed in the Ratio Test does not. However, it is often more difficult to apply the Root Test in concrete situations.

The inquisitive mind will now ask what happens in instances where both tests fail. Can we still compute the radius of convergence? The answer is yes. Recall that while limits may fail to exist for a given sequence, the limit superior always exists if the sequence is bounded and is infinite otherwise. In this context, an infinite limit is a valid option as exhibited in part (c) of Theorem 6.3.3. Though we won't go into the proof here, the radius of convergence for a series can always be calculated by computing the limit superior

$$L = \limsup |a_n|^{1/n}$$

which will always satisfy $0 \le L \le \infty$. The conclusion is then the same as the previous theorem.

We now turn our attention to uniform convergence of power series as this will lead us to understanding continuity and differentiation of such functions.

Theorem 6.3.5. *Let $\sum a_n x^n$ be a power series with radius of convergence $R > 0$. If $c \in (-R, R)$, then the series converges uniformly on the interval $[-|c|, |c|]$.*

Proof. Let $\sum a_n x^n$ be a power series with radius of convergence $R > 0$ and let $c \in (-R, R)$. We will apply the Weierstrass M-test on the interval $[-|c|, |c|]$. To do this notice that $|x^n| = |x|^n \le |c|^n = |c^n|$ for all $x \in [-|c|, |c|]$. Thus we have $|a_n x^n| \le |a_n c^n|$ for all $x \in [-|c|, |c|]$. Also, since $c \in (-R, R)$ we know that the series $\sum a_n c^n$ converges absolutely, i.e. $\sum |a_n c^n|$ converges. Thus if we take $M_n = |a_n c^n|$, the M-test guarantees us that $\sum a_n x^n$ converges uniformly on $[-|c|, |c|]$. $\square$

The proof actually shows that we could include the endpoints. Specifically, if a power series converges absolutely at R or $-R$, then the series converges uniformly on $[-R, R]$. This provides some information about the endpoints but leaves open the question as to what happens if we have conditional convergence at one endpoint or the other. Again, the examples above show that conditional convergence at R does not necessarily imply convergence at $-R$ and vice versa. For further investigation of behavior at the endpoints, see Exercise 6.3.7.

Corollary 6.3.6. *Let $\sum a_n x^n$ be a power series with radius of convergence $R > 0$ and suppose $K \subset (-R, R)$ is a compact set. Then the power series is continuous on K. In particular, the power series is continuous at any point of $(-R, R)$.*

The corollary follows from Theorems 6.1.9 and 6.3.5.

The true advantage of power series lies in their connection to differentiation. We will first explore applying the process of differentiation to power series and then return to ask what power series can tell us about differentiation. In trying to take the derivative of a function given by a power series it is tempting to treat the function as a polynomial and differentiate term-by-term. The fact that this is correct is one of most useful properties possessed by power series; what's more important is the fact that the differentiated series is still a power series with the same radius of convergence.

Theorem 6.3.7. *Let $f(x) = \sum_{n=0}^{\infty} a_n x^n$ be a power series with radius of convergence $R > 0$. Then the series $\sum_{n=1}^{\infty} n a_n x^{n-1}$ converges on $(-R, R)$. Moreover,*

$$f'(x) = \sum_{n=1}^{\infty} n a_n x^{n-1}.$$

Proof. For the first statement, let $x \in (-R, R)$. To show that $\sum na_n x^{n-1}$ converges we will apply the Comparison Test. To this end, choose $t \in (0, R)$ such that $|x| < t < R$. We then have

$$\left| na_n x^{n-1} \right| = \frac{n}{t} \left| \frac{x^{n-1}}{t^{n-1}} \right| |a_n t^n| = \frac{n}{t} \left| \frac{x}{t} \right|^{n-1} |a_n t^n|.$$

By Exercise 6.3.5, we know that the series $\sum n|x/t|^{n-1}$ converges since $|x/t| < 1$; thus the sequence $(n|x/t|^{n-1})$ converges to 0 and is therefore bounded. With this it follows that there is an $M > 0$ such that

$$|na_n x^{n-1}| \le \frac{M}{t} |a_n t^n|.$$

Furthermore, the series $\sum |a_n t^n|$ converges by Theorem 6.3.3. By the Algebraic Limit Theorem for series $\sum (M/t)|a_n t^n|$ also converges and the Comparison Test assures us that $\sum_{n=1}^{\infty} na_n x^{n-1}$ also converges.

To verify the second statement, we again fix $x \in (-R, R)$ and choose $t \in (0, R)$ such that $|x| < t < R$. To calculate $f'(x)$ we will appeal to Exercise 6.2.7(b). To apply this exercise, we must verify that $\sum a_n x^n$ converges for some point in $[-t, t]$ and that $\sum na_n x^{n-1}$ converges uniformly on $[-t, t]$. The first of these claims is obviously satisfied for any $x \in [-t, t]$ by our hypothesis. The second claim follows from Theorem 6.3.5 since we have just shown that the radius of convergence of the series $\sum na_n x^{n-1}$ is also R. Exercise 6.2.7 now implies that

$$f'(x) = \sum_{n=0}^{\infty} (a_n x^n)' = \sum_{n=1}^{\infty} na_n x^{n-1}$$

where we are thinking of x^0 as a constant function.   □

The statement has a more general interpretation upon closer inspection which brings us back to the question of what power series can tell us about differentiation. Since the radius of convergence of the differentiated series is the same as the original, the true strength of the theorem above is that a power series can be differentiated as many times as we like without losing any domain points in the interval $(-R, R)$; hence if we have a power series representation for a function, we know that all its derivatives exist. There is one small caveat, the endpoints. The original series may converge at one endpoint, or the other, or both, and the derivative may fail to converge there. This is hidden in the first step of the proof by our choice of t, i.e. our proof used the fact that $(-R, R)$ is an open interval and hence provides no information about the endpoints. An example of where this loss occurs is found in the series $\sum_{n=1}^{\infty} x^n/n$ which converges on $[-1, 1)$ while the differentiated series $\sum_{n=0}^{\infty} x^n$ (with indices shifted for a more pleasant expression) converges only on $(-1, 1)$.

For a general power series $\sum a_n(x-c)^n$, the above theorems all translate nicely via substitution of $x - c$. In this case the interval of convergence takes the form $(c - R, c + R)$ and the theorem on differentiation follows since the derivative of $(x-c)^n$ is $n(x-c)^{n-1}$.

Taylor Series

Upon understanding how beneficial power series can be, the natural thing to consider is whether or not we can identify functions which have a power series representation. If we can, then we would also like to be able to identify the coefficients of the series. The second question is easier to answer and is a result of the fact that power series have derivatives of all orders.

Theorem 6.3.8 (Taylor Coefficients). *Assume f is infinitely differentiable and has a power series representation*

$$f(x) = \sum a_n x^n$$

with radius of convergence $R > 0$. Then the coefficients must satisfy

$$a_n = \frac{f^{(n)}(0)}{n!}. \tag{6.3}$$

Proof. The proof utilizes two steps repeatedly: evaluating at 0 and differentiating. We first expand the summation above so that it looks like a polynomial,

$$f(x) = a_0 + a_1 x + a_2 x^2 + \cdots + a_n x^n + \cdots.$$

Evaluating this equation at the origin gives $a_0 = f(0)$ which matches the form given in the statement of the theorem since $0! = 1$. Differentiating this expression, we have

$$f'(x) = a_1 + 2a_2 x + 3a_3 x^2 + \cdots na_n x^{n-1} \ldots.$$

Again, we evaluate this at 0 to obtain $a_1 = f'(0)$. Upon a second differentiation we obtain

$$f''(x) = 2a_2 + 3 \cdot 2a_3 x + \cdots n(n-1)a_n x^{n-2} \ldots.$$

Evaluating, we find $a_2 = f''(0)/2$. Continuing in this process, at the nth derivative we have

$$f^{(n)}(x) = n(n-1) \cdots 3 \cdot 2 \cdot 1 a_n + (n+1)n \cdots 3 \cdot 2a_{n+1} x + \cdots.$$

which yields

$$a_n = \frac{f^{(n)}(0)}{n!}.$$

□

Keep in mind that in the hypothesis of the above theorem we assume that f *has a power series representation*. We are not claiming that we can *always* construct a power series that is equal to a given function by computing the coefficients. The remainder of this section, however, will investigate the truth of such matters.

First, given a function which is infinitely differentiable, we define the *Taylor series centered at* 0 to be the series $\sum a_n x^n$ where a_n is defined by Eq. (6.3) for all n. Some texts call this a Maclaurin series, but we will use the more general terminology.

Example 6.3.9. Let $f(x) = -\ln(1 - x)$. We will assume the derivative of the logarithm for the time being. To find the Taylor series for f, we compute the coefficients by taking successive derivatives. First we have $a_0 = f(0) = 0$. Differentiating, we have

$$f'(x) = \frac{1}{1-x} \qquad f'(0) = 1 \qquad a_1 = 1$$
$$f''(x) = \frac{1}{(1-x)^2} \qquad f''(0) = 1 \qquad a_2 = \frac{1}{2}$$
$$f'''(x) = \frac{2}{(1-x)^3} \qquad f'''(0) = 2 \qquad a_3 = \frac{1}{3}$$
$$f^{(4)}(x) = \frac{2 \cdot 3}{(1-x)^4} \qquad f^{(4)}(0) = 3! \qquad a_4 = \frac{1}{4}$$
$$\vdots \qquad\qquad \vdots \qquad\qquad \vdots$$
$$f^{(n)}(x) = \frac{(n-1)!}{(1-x)^n} \qquad f^{(n)}(0) = (n-1)! \quad a_n = \frac{1}{n}.$$

Thus the Taylor series for f is given by $\sum_{n=1}^{\infty} x^n/n$ and we have already seen that this series converges on $[-1, 1)$. Again, keep in mind that we are not claiming that

$$-\ln(1 - x) = \sum_{n=1}^{\infty} \frac{x^n}{n}.$$

The basic question is to determine if f is equal to its Taylor series,

$$f(x) = \sum_{n=1}^{\infty} a_n x^n,$$

on some nontrivial interval about 0; the equation always holds at 0 by the choice of a_0. We can rephrase this by asking how well the partial sums of the Taylor series approximate the function f. But, the partial sums of the Taylor series are exactly the Taylor polynomials of f and we explored results concerning the difference $f(x) - T_{0,n}(x)$ in Sect. 5.5.

Theorem 6.3.10. *Assume f is infinitely differentiable on an open interval I and has a Taylor series*

$$\sum_{n=0}^{\infty} a_n x^n$$

with radius of convergence $0 < R < \infty$. If $|f^{(n)}(x)| \le M$ for all n and all $x \in [a, b] \subseteq (-R, R)$, then

$$f(x) = \sum_{n=0}^{\infty} a_n x^n$$

for all $x \in [a, b]$.

Proof. The estimate provided by Theorem 5.5.4 can be rephrased as

$$|f(x) - T_{0,n}(x)| \le \max\{|m| |M|\} \frac{|x|^{n+1}}{(n+1)!}.$$

With our particular hypothesis we see that

$$|f(x) - T_{0,n}(x)| \le M \frac{R^{n+1}}{(n+1)!}.$$

By Exercise 6.3.11, we know that the sequence $(R^{n+1}/(n+1)!)$ converges to zero and thus by Proposition 6.1.6, we see that the Taylor polynomials converge uniformly to f on $[a, b]$ as desired.   □

In the proof, we could have employed Theorem 5.5.3. The theorem can also be easily extended to the case where $R = \infty$ by substituting the bound b^{n+1} instead of R^{n+1}. As a final comment, the requirement that $|f^{(n)}(x)| \le M$ for all n and all $x \in [a, b] \subseteq (-R, R)$ may seem rather strict, but typically this is not problematic. Many of the elementary functions, e.g. e^x, $\sin x$, $\cos x$, satisfy this requirement on bounded intervals due to continuity of their derivatives and similarity of successive derivative. We can, however, relax this requirement slightly by assuming instead that there is a constant $M > 0$ such that

$$|f^{(n)}(x)| \le M^n$$

for all n and all $x \in [a, b] \subseteq (-R, R)$.

Example 6.3.11. Return again to the function $f(x) = -\ln(1 - x)$. We identified its Taylor series as $\sum x^n/n$, which converges on $[-1, 1)$. Our ultimate goal is to now show that the Taylor series does in fact converge to f on the interval $[-1, 1)$; in other words, we want to be able to say

$$-\ln(1 - x) = \sum_{n=1}^{\infty} \frac{x^n}{n}$$

for all $x \in [-1, 1)$. At the moment we will only be able to do this on the interval $[-1, 1/2]$. After doing so, we will discuss the difficulty in dealing with convergence on the interval $[1/2, 1)$ but we will return to this example in Chap. 7 and complete the argument.

In order to show convergence of a Taylor series to its function f, in any situation, comes down to understanding the behavior of successive derivatives of f. Here we will appeal to Theorem 5.5.3 since we don't have an obvious bound for the derivatives of f, due to the asymptote at $x = 1$. For a first case, take $x \in [-1, 0)$. Then there is a $p \in (x, 0)$ with

$$f(x) - T_{0,n}(x) = \frac{f^{(n+1)}(p)}{(n+1)!} x^{n+1} = \frac{n!}{(1-p)^{n+1}(n+1)!} x^{n+1}$$

$$= \frac{1}{n+1} \left(\frac{x}{1-p} \right)^{n+1}$$

$$= \frac{1}{n+1} \left(\frac{x}{1+|p|} \right)^{n+1}.$$

Applying absolute value now, we have

$$|f(x) - T_{0,n}(x)| = \frac{1}{n+1} \left(\frac{|x|}{1+|p|} \right)^{n+1} \leq \frac{1}{n+1}.$$

Since

$$\frac{|x|}{1+|p|} \leq 1,$$

which converges to zero as $n \to \infty$.

For the second case, let $x \in (0, 1/2]$. By the aforementioned theorem, we know that there is a $p \in (0, x)$ such that

$$f(x) - T_{0,n}(x) = \frac{f^{(n+1)}(p)}{(n+1)!} x^{n+1} = \frac{1}{n+1} \left(\frac{x}{1-p} \right)^{n+1}.$$

Since x, and hence p, are in $(0, 1/2]$, we see that $1 - p \geq 1/2$. Applying absolute value, we have

$$|f(x) - T_{0,n}(x)| = \frac{1}{n+1}\left(\frac{x}{1-p}\right)^{n+1} \leq \frac{1}{n+1}\left(\frac{1/2}{1/2}\right)^{n+1} = \frac{1}{n+1}$$

for all $x \in (0, 1/2]$. Moreover, the bound on the right tends to zero as $n \to \infty$.

Thus in either case we see that the Taylor polynomials converge pointwise to f; the M-test for sequences provides the stronger conclusions that the convergence is uniform on $[-1, 0]$ and on $[0, 1/2]$. However, if we try to extend this same argument to the interval $(1/2, 1)$, we fall short because in this case we don't have enough information about the location of p in relation to x in order to guarantee that $x/(1 - p) \leq 1$. In fact, we would need to show that $x + p \leq 1$ in order to satisfy the previous bound, but Theorem 5.5.3 doesn't supply any specific information about the relationship between p and x.

We will return to this example and complete the argument once we have discussed the relationship between power series in integration. Before moving on, notice that a quick treat from our work is that

$$\ln 2 = \sum_{n=1}^{\infty} \frac{(-1)^{n+1}}{n},$$

a sum for the alternating harmonic series!

The most ideal situation would be for a function to equal its Taylor series on its interval of convergence. Unfortunately there is no general theorem which provides this connection. What makes the relationship between a function and its Taylor series truly interesting is that there exist functions which exhibit the worst possible behavior with regard to their Taylor series. An example of such a function is given by

$$f(x) = \begin{cases} 0, & x \leq 0; \\ e^{-1/x}, & x > 0. \end{cases}$$

For this function, one can show that $f^{(n)}(0) = 0$ for all n and hence the Taylor series is simply the zero function. What we have then is a function whose Taylor series converges on a nontrivial interval $(-R, R)$, the real line in this case, but which doesn't agree with f on any open interval containing 0. In other words, the Taylor series is converging to a function which is not f.

Exercises

Exercise 6.3.1. Show that the series $\sum_{n=1}^{\infty} x^n/n^2$ converges on the interval $[-1, 1]$ while the series $\sum_{n=1}^{\infty}(-1)^n x^n/n$ converges on the interval $(-1, 1]$.

Exercise 6.3.2. Complete the proof of Theorem 6.3.3.

Exercise 6.3.3. Prove Theorem 6.3.3 with the quantity L replaced by $L = \lim |a_n|^{1/n}$.

Exercise 6.3.4. Complete the proof of Corollary 6.3.6.

Exercise 6.3.5. Let $0 < t < 1$. Show that series $\sum_{n=1}^{\infty} n t^{n-1}$ converges.

Exercise 6.3.6. Show that two power series, both with positive radius of convergence, are equal if and only if their coefficients are equal.

Exercise 6.3.7. (a) Let (b_n) be a decreasing sequence of nonnegative real numbers and let $\sum a_n$ be a series with bounded partial sums, i.e. suppose there is an $A > 0$ with

$$|a_1 + a_2 + \cdots + a_n| \le A$$

for all n. Show that

$$|a_1 b_1 + a_2 b_2 + \cdots + a_n b_n| \le A b_1.$$

As a hint, consider the summation by parts formula from Exercise 3.1.12.

(b) Let $\sum a_n x_n$ be a power series with radius of convergence $R > 0$. If the series converges at $x = R$, show that the series converges uniformly on $[0, R]$.

(c) Show that we can draw a similar conclusion if the power series converges at $x = -R$.

Exercise 6.3.8. A series $\sum_{n=0}^{\infty} a_n$ is said to be *Abel-summable to L* if the power series

$$f(x) = \sum_{n=0}^{\infty} a_n x^n$$

converges for all $x \in [0, 1)$ and $L = \lim_{x \to 1^-} f(x)$.

(a) Show that any series that converges to a limit L is also Abel-summable to L.

(b) Show that $\sum_{n=0}^{\infty} (-1)^n$ is Abel-summable and find the sum.

Exercise 6.3.9. (a) Find the Taylor series at 0 for the function $f(x) = e^x$. You may assume $f'(x) = e^x$.

(b) Show that the Taylor series for f converges uniformly to f on any interval of the form $[-M, M]$. You may assume f is increasing.

(c) Use the Taylor series and part (b) to show that $(e^x)' = e^x$. This is circular reasoning of course, and we will actually justify these statements in the next chapter.

Exercise 6.3.10. (a) Find the Taylor series at 0 for the function $f(x) = \sin x$. You may assume the derivatives of the sine and cosine functions.

(b) Show that the Taylor series for f converges uniformly to f on any interval of the form $[-M, M]$.

Exercise 6.3.11. Let $0 < M < \infty$. Show that series $\sum_{n=0}^{\infty} M^n / n!$ converges.

Exercise 6.3.12. Consider the function f given by

$$f(x) = \begin{cases} 0, & x \le 0; \\ e^{-1/x}, & x > 0. \end{cases}$$

Show that $f^{(n)}(0) = 0$ for $n = 0, 1, 2, 3$.

6.4 A Continuous Nowhere Differentiable Function

As discussed in Chap. 5, most of the elementary functions encountered in the calculus sequence are differentiable on their domain; at the very worst, they are not differentiable on a finite subset of their domain. The absolute value was our first example of a function which has a point where it is not differentiable and Exercise 5.1.7 asked you to extend this idea to finite and countable subsets of $\mathbb{R}$. Here we show that we can modify this behavior further to produce a function which is continuous on $\mathbb{R}$ but not differentiable at any point.

On a historical note, Weierstrass was the first to publish an example of a continuous nowhere differentiable function. His original paper concerned series of the form

$$\sum_{n=0}^{\infty} a^n \cos(b^n x)$$

for certain values of a and b; we will have similar compression terms in our construction below. Weierstrass' example appeared in 1872, though it is rumored that he knew of the result as early as 1861. In fact, Bolzano is also attributed to having such a result as early as 1830. Weierstrass' example was generalized many times in the following years and our construction is a variation on an example given by van der Waerden in 1930.

We define a function $g : \mathbb{R} \to [0, 1]$ by defining $g(x) = |x|$ on $[-1, 1]$ and then extend this to a function on $\mathbb{R}$ by requiring that $g(x) = g(x + 2)$ for all $x \in \mathbb{R}$. This type of construction repeats the same portion of the absolute value function by translation. We can also represent g by the piecewise function

$$g(x) = \begin{cases} x - j, & \text{if } j \le x < j + 1 \text{ for } j \text{ even}; \\ j + 1 - x, & \text{if } j \le x < j + 1 \text{ for } j \text{ odd}, \end{cases}$$

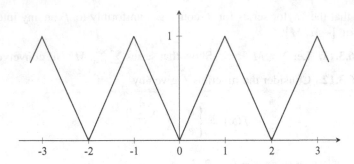

Fig. 6.4 Sawtooth graph

where j is an integer. In either case Fig. 6.4 provides the graph of g which is commonly called a *sawtooth graph*. We will use this function to construct a function which is continuous on $\mathbb{R}$ but which is not differentiable at any point. First we have a lemma.

Lemma 6.4.1. *Let g be defined as above. Then $|g(y) - g(x)| \leq |y - x|$ for all $x, y \in \mathbb{R}$. Moreover, if x, y satisfy $j \leq x, y < j + 1$ for some $j \in \mathbb{N}$, then $|g(y) - g(x)| = |y - x|$.*

Proof. We will break the proof into three cases depending on the distance between x and y, and their orientation with respect to the integers. As a first observation, notice that g maps into $[0, 1]$ which implies that $|g(y) - g(x)| \leq 1$ for all $x, y \in \mathbb{R}$. For our first case, suppose $|y - x| \geq 1$. It immediately follows that $|g(y) - g(x)| \leq 1 \leq |y - x|$.

Next suppose $|y - x| < 1$, i.e. x and y are not more than one unit apart. Thus we have two subcases depending on whether there is an integer between x and y or not. First suppose there is an integer j such that

$$j - 1 < x \leq j \leq y < j + 1.$$

If j is even, then the definition of g shows that

$$|g(y) - g(x)| = |(y - j) - (j - x)| \leq |y - j| + |j - x| = |y - x|$$

since $x \leq j \leq y$. On the other hand, if j is odd, then

$$|g(y) - g(x)| = |(j + 1 - y) - (x - (j - 1))| = |j - y - x + j| \leq |y - j| + |j - x| = |y - x|.$$

For the final case, suppose that there is no integer between x and y. But this means there is a $j \in \mathbb{Z}$ such that $j < x < y < j + 1$. If j is even, then

$$|g(y) - g(x)| = |(y - j) - (x - j)| = |y - x|;$$

if j is odd, we have

$$|g(y) - g(x)| = |(j + 1 - y) - (j + 1 - x)| = |y - x|.$$

Thus in any case we see that $|g(y) - g(x)| \leq |y - x|$ and the equality statement is demonstrated by the final subcase.   □

To construct our desired function we first construct a sequence of functions (f_n). For $n \in \mathbb{N}$ define $f_n : \mathbb{R} \to \mathbb{R}$ by

$$f_n(x) = \left(\frac{3}{4}\right)^n g(4^n x)$$

where g is defined as above. For a fixed n, the graph of the function f_n is a vertical and horizontal compression of the graph of g; the horizontal compression increases the period by 4^n. For a fixed x, this equates to increasing the number of oscillations around x as $n \to \infty$. We then set

$$f(x) = \sum_{n=1}^{\infty} f_n(x) = \sum_{n=1}^{\infty} \left(\frac{3}{4}\right)^n g(4^n x).$$

The vertical compressions of $(3/4)^n$ are present to guarantee that the series converges. The non-differentiability of f is attributed to the fact that adding up the increasingly oscillating sequence functions produces "jagged" behavior in the graph about every domain point; interpreting this as saying that there is a corner at every point in the graph is not entirely correct. Rather, as we will show, adding up these oscillations produces interference in calculating the derivative by permitting secant lines of unbounded slope.

The next lemma asserts that f is defined on $\mathbb{R}$ and is continuous there. We leave its proof as an exercise.

Lemma 6.4.2. *The function f defined above is well defined on $\mathbb{R}$, converges uniformly on $\mathbb{R}$, and is continuous on $\mathbb{R}$.*

The remainder of this section is dedicated to showing that the function f is the function we seek.

Proposition 6.4.3. *There is a function $f : \mathbb{R} \to \mathbb{R}$ which is continuous on $\mathbb{R}$ but is not differentiable at any point of $\mathbb{R}$.*

Proof. The function f defined above is continuous on $\mathbb{R}$. To show that it is not differentiable for any point in $\mathbb{R}$, let $c \in \mathbb{R}$ be fixed. We will construct a sequence (x_n) in $\mathbb{R}$ such that $x_n \neq c$ for all n with $\lim x_n = c$ but for which

$$\lim \frac{f(x_n) - f(c)}{x_n - c}$$

does not exist. We will actually show that

$$\lim \left| \frac{f(x_n) - f(c)}{x_n - c} \right| = \infty$$

meaning that the sequence of difference quotients is unbounded and hence cannot converge.

For $n \in \mathbb{N}$ we define $x_n = c \pm 4^{-n}/2$ where the plus/minus is chosen so that there is no integer between $4^n x_n$ and $4^n c$. This is possible since

$$4^n x_n = 4^n (c \pm 4^{-n}/2) = 4^n c \pm 1/2,$$

i.e. $4^n x_n$ and $4^n c$ differ by at most $1/2$.

To show the sequence of difference quotients is unbounded, we will focus on a fixed n, and then show that the difference quotient corresponding to that index is bounded below by $3^n/2$. The argument is quite technical in nature, so paper and pencil will be handy. Let $n \in \mathbb{N}$. Then, considering the difference quotient for this index, we apply the series representation for f, and the convergence of that series to see that

$$\left| \frac{f(x_n) - f(c)}{x_n - c} \right| = \frac{1}{|x_n - c|} \left| \sum_{j=1}^{\infty} f_j(x_n) - \sum_{j=1}^{\infty} f_j(c) \right|$$

$$= \frac{1}{|x_n - c|} \left| \sum_{j=1}^{\infty} (f_j(x_n) - f_j(c)) \right|.$$

Now we split the last sum above into three terms which we will estimate shortly; the above expression is equal to

$$\frac{1}{|x_n-c|} \left| \left[\sum_{j=1}^{n-1} (f_j(x_n) - f_j(c)) \right] + (f_n(x_n) - f_n(c)) + \left[\sum_{j=n+1}^{\infty} (f_j(x_n) - f_j(c)) \right] \right|.$$

Working with the three pieces individually now, we begin with the third. Keeping in mind that f_j is defined in terms of g, observe that

$$4^j x_n = 4^j (c \pm 4^{-n}/2) = 4^j c + 4^{j-n}/2;$$

also, since $j \geq n + 1$ in the term under consideration, $4^{j-n}/2$ is an even integer. Thus $4^j x_n$ and $4^j c$ differ by an even integer for all $j \geq n + 1$. The fact that g is defined to have period two indicates that

$$g(4^j x_n) = g(4^j c)$$

for all $j \geq n + 1$. It follows then that

$$(f_j(x_n) - f_j(c)) = \left(\frac{3}{4}\right)^j g(4^j x_n) - \left(\frac{3}{4}\right)^j g(4^j c) = 0$$

for all $j \geq n + 1$ and thus the third term above is 0.

For the second term, recall that x_n was chosen so that there is no integer between $4^n x_n$ and $4^n c$ with which we can apply Lemma 6.4.1 in the case of equality and are guaranteed that

$$|f_n(x_n) - f_n(c)| = \left(\frac{3}{4}\right)^n |g(4^n x_n) - g(4^n c)| = \left(\frac{3}{4}\right)^n |4^n x_n - 4^n c| = 3^n |x_n - c|.$$

Finally, for the first term we estimate using the triangle inequality and Lemma 6.4.1,

$$\left| \sum_{j=1}^{n-1} (f_j(x_n) - f_j(c)) \right| \leq \sum_{j=1}^{n-1} |f_j(x_n) - f_j(c)| = \sum_{j=1}^{n-1} \left(\frac{3}{4}\right)^j |g(4^j x_n) - g(4^j c)|$$

$$\leq \sum_{j=1}^{n-1} \left(\frac{3}{4}\right)^j |4^j x_n - 4^j c|$$

$$= \sum_{j=1}^{n-1} 3^j |x_n - c|$$

$$= (4^{-n}/2) \sum_{j=1}^{n-1} 3^j$$

$$= (4^{-n}/2) \left(\frac{3^n - 1}{3 - 1}\right)$$

$$\leq 4^{-n-1} 3^n$$

where the last equality is from the summation identity provided by Eq. (3.4). We will apply this estimate momentarily, but we first multiply by -1 to obtain the lower bound

$$- \left| \sum_{j=1}^{n-1} (f_j(x_n) - f_j(c)) \right| \geq -4^{-n-1} 3^n.$$

Piecing these estimates together and applying a reverse triangle inequality (see Exercise 1.3.2) we have that

$$
\left| \frac{f(x_n) - f(c)}{x_n - c} \right| = \frac{1}{|x_n - c|} \left| \left[\sum_{j=1}^{n-1} (f_j(x_n) - f_j(c)) \right] + (f_n(x_n) - f_n(c)) \right|
$$

$$
\geq \frac{1}{|x_n - c|} \left[|(f_n(x_n) - f_n(c))| - \left| \sum_{j=1}^{n-1} (f_j(x_n) - f_j(c)) \right| \right]
$$

$$
\geq \frac{1}{|x_n - c|} \left[3^n |x_n - c| - 4^{-n-1} 3^n \right]
$$

$$
= 3^n - \frac{1}{|x_n - c|} 4^{-n-1} 3^n
$$

$$
= 3^n - \frac{1}{(4^{-n}/2)} 4^{-n-1} 3^n
$$

$$
= 3^n / 2.
$$

The sequence $(3^n/2)$ diverges to infinity showing the sequence

$$
\frac{f(x_n) - f(c)}{x_n - c}
$$

is unbounded and cannot converge. By Theorem 4.1.4 we conclude that

$$
\lim_{x \to c} \frac{f(x) - f(c)}{x - c}
$$

does not exist and therefore f is not differentiable at c. Moreover, since c was chosen as an arbitrary real number we conclude that f is not differentiable at any point of $\mathbb{R}$.   □

   The function above may seem like an anomaly but it turns out that functions of this sort are more common among the continuous functions than your previous studies have indicated. This misconception can be partially attributed to the adage that a function which is continuous on an interval can be "drawn without lifting your pencil from the paper." We've all heard this, and many of us have used it to give calculus students a visual means of understanding continuity, but you would be hard pressed to sketch the graph of the function f just constructed in this manner. The myth above more accurately describes functions which have a continuous derivative (though some would argue that continuity of the derivative is not necessary).
   So how do we understand the frequency of functions which are not differentiable at any point among the collection of functions which are continuous everywhere

in $\mathbb{R}$? The resolution is a result of the *Baire Category Theorem* which states that a complete metric space is not the union of a countable collection of nowhere dense sets. A nowhere dense subset of a metric space is one whose closure contains no ε-neighborhoods. We then categorize the subsets of a metric space as follows: a set is of the *first category* or *meager* if it can be written as the union of a countable collection of nowhere dense sets and of the *second category* otherwise. The term *residual set* is used to identify a set whose complement is meager. This classification is actually about the size of a subset though in a manner distinct from cardinality, dimension (if in a vector space), or measure (which we will see in the next chapter). First category sets are small, thin sets, e.g. $\mathbb{N}$ and $\mathbb{Z}$ are of the first category in $\mathbb{R}$ while $\mathbb{Q}$ and the irrationals are second category sets. The statement is then that any complete metric space is of the second category.

After digesting these definitions and the Baire Category Theorem, it follows that $C([a, b])$ with any norm is of the second category. To answer our question posed above concerning the frequency of non-differentiable functions, one then shows that the set

$$\{f \in C([a, b]) : f'(x) \text{ exists for some } x \in [a, b]\}$$

is of the first category. See Sect. 8.2 of [1] for more on the Baire Category Theorem and an outline of the proof that the set just defined is of the first category.

Exercises

Exercise 6.4.1. Provide a proof of Lemma 6.4.2

Exercise 6.4.2. Letting g be as defined earlier in this section, show that the function

$$f(x) = \sum_{n=1}^{\infty} \frac{1}{2^n} g(2^n x)$$

also satisfies the conclusion of Theorem 6.4.3.

6.5 Spaces of Continuous Functions

Here we turn our attention to spaces of continuous functions; these were introduced in Example 4.5.6, and Exercise 4.5.5 showed that the space $C([0, 1])$ endowed with the sup-norm is a normed linear space. We return now with an aim toward more fully understanding these collections, completeness in particular. Here we take a slightly more general approach. Let $a, b \in \mathbb{R}$ with $a < b$ and let $C([a,b])$ denote the collection of all functions $f : [a,b] \to \mathbb{R}$ which are continuous. We also define $\|\cdot\|_\infty : C([a,b]) \to [0,\infty)$ by

$$\|f\|_\infty = \sup\{|f(x)| : x \in [a,b]\}.$$

The fact that this collection is a normed linear space follows in a manner identical to that of $C([0,1])$.

Before moving into a proper investigation, it should be noted that, as usual, when we encounter results dealing with closed, bounded intervals, it should seem likely that the same results will hold for general compact sets. In this instance, this makes for a positive conjecture and one can similarly define the space $C(K)$, where $K \subseteq \mathbb{R}$ is a compact set, to consist of all functions $f : K \to \mathbb{R}$ which are continuous. All of the results of this section also hold in this more general context.

Lemma 6.5.1. *Let (f_n) be a sequence in $C([a,b])$. Then (f_n) converges in the sup-norm to $f \in C([a,b])$ if and only if (f_n) converges to f uniformly on $[a,b]$.*

Proof. Let (f_n) be a sequence in $C([a,b])$ and suppose that (f_n) converges uniformly to f on $[a,b]$. This immediately implies that f is continuous on $[a,b]$ and we need to show that (f_n) converges to f in the sup-norm. Let $\varepsilon > 0$ and choose $N \in \mathbb{N}$ such that $|f_n(x) - f(x)| < \varepsilon/2$ for $n \geq N$ and all $x \in [a,b]$. Then, by definition of supremum,

$$\|f_n - f\|_\infty = \sup\{|f_n(x) - f(x)| : x \in [a,b]\} \leq \varepsilon/2 < \varepsilon$$

whenever $n \geq N$. Hence $\|f_n - f\|_\infty \to 0$ and thus (f_n) converges to f in the sup-norm.

We leave the proof of the converse as Exercise 6.5.1. $\qquad\qquad\square$

Proposition 6.5.2. *The space $C([a,b])$ is complete in the sup-norm.*

Proof. Let (f_n) be a Cauchy sequence with respect to the sup-norm in $C([a,b])$. The tactic here is generally the same as previous completeness arguments; we begin by showing that the sequence is necessarily uniformly Cauchy on $[a,b]$. For $\varepsilon > 0$, choose $N \in \mathbb{N}$ such that $\|f_n - f_m\|_\infty < \varepsilon$ for $n, m \geq N$. Now, if $x \in [a,b]$,

$$|f_n(x) - f_m(x)| \leq \|f_n - f_m\|_\infty < \varepsilon$$

for $n, m \geq N$ implying that (f_n) is uniformly Cauchy on $[a,b]$. From our previous work with sequences of functions, this implies that there is a function $f : [a,b] \to \mathbb{R}$ with (f_n) converging to f uniformly on $[a,b]$. Furthermore, this implies that f is continuous on $[a,b]$ and Lemma 6.5.1 shows that (f_n) converges to f in the sup-norm. $\qquad\square$

The previous proposition shows that $(C([0,1]), \|\cdot\|_\infty)$ is a Banach space and the next question to ask is whether or not its norm can be realized as an inner product in

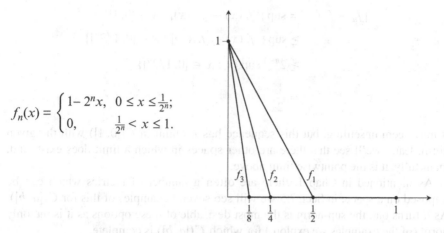

$$f_n(x) = \begin{cases} 1 - 2^n x, & 0 \le x \le \frac{1}{2^n}; \\ 0, & \frac{1}{2^n} < x \le 1. \end{cases}$$

Fig. 6.5 Sequence of functions in $C([0,1])$

the sense of Proposition 1.5.15. Unfortunately the answer is no; we can see this by applying the parallelogram equality to the vectors $f(x) = 1$ and $g(x) = x$. Then $\|f\|_\infty = 1$ and $\|g\|_\infty = 1$. Also, $\|f + g\|_\infty = \|1 + x\|_\infty = 2$ while $\|f - g\| = \|1 - x\|_\infty = 1$. Thus $\|f + g\|_\infty^2 + \|f - g\|_\infty^2 = 5$ and $2\|f\|_\infty^2 + 2\|g\|_\infty^2 = 4$. Therefore $C([0,1])$ cannot be realized as inner product space with respect to the sup-norm.

As in $\mathbb{R}$, the notion of completeness guarantees that all appropriate limits are in a give space. However, a solid understanding of completeness becomes more complicated when there are different modes of convergence. For instance, for a sequence (f_n) in $C([a,b])$, we can currently ask three questions: does the sequence converge pointwise, does it converge uniformly, does it converge in the sup-norm? We have already seen that the last two of these are equivalent, but they do not necessarily coincide with pointwise convergence.

For $C([0,1])$ consider the sequence of functions given in Fig. 6.5. This sequence converges in a pointwise fashion to

$$f(x) = \begin{cases} 1, & x = 0; \\ 0, & 0 < x \le 1, \end{cases}$$

but f is not in $C([0,1])$. This brings the issue to light: if the space is complete, the limit should be there. However, the space $(C([0,1]), \|\cdot\|_\infty)$ is complete with respect to the sup-norm, not with respect to pointwise limits, and the sequence (f_n) is *not* Cauchy in the sup-norm. In fact, if $m, n \in \mathbb{N}$ with $n > m$, then $\|f_n - f_m\|_\infty \ge 1/2$. To see this, consider the difference $|f_n(x) - f_m(x)|$ on $[0, 1/2^n]$,

$$|f_n(x) - f_m(x)| = x(2^n - 2^m) = 2^m x(2^{n-m} - 1) \ge 2^m x(2^{n-m-1}) = 2^{n-1} x.$$

Thus

$$\|f_n - f_m\|_\infty = \sup\{|f_n(x) - f_m(x)| : x \in [0,1]\}$$
$$\geq \sup\{|f_n(x) - f_m(x)| : x \in [0, 1/2^n]\}$$
$$\geq 2^{n-1} \sup\{x : x \in [0, 1/2^n]\}$$
$$= \frac{1}{2}.$$

It may seem unsettling, but this sequence has no limit in $C([0,1])$ with the given norm. Later we'll see that there are other spaces in which a limit does exist, and, thankfully, it is the pointwise limit above.

As mentioned in Chap. 1, there are often a number of metrics which can be imposed on a space; in Sect. 7.5 we will see several examples of this for $C([a,b])$. As it turns out, the sup-norm is the most desirable of these options as it is the only norm (of the examples we explore) for which $C([a,b])$ is complete.

Example 6.5.3. Consider the space $C^1([0,1])$ defined in Example 5.6.4 with norm

$$\|f\|_{C^1} = \|f\|_\infty + \|f'\|_\infty.$$

Here we will show that this space is complete and discuss the structure of the norm given. First suppose we have a sequence (f_n) which is Cauchy in $C^1([0,1])$. Hence for every $\varepsilon > 0$ there is an $N \in \mathbb{N}$ such that

$$\|f_n - f_m\|_\infty + \|f_n' - f_m'\|_\infty < \varepsilon$$

for all $n, m \geq N$. Since both quantities on the left of this inequality are positive, it follows that the sequences (f_n) and (f_n') are both Cauchy with respect to the sup-norm, and Exercise 6.5.1(b) shows that both sequences are uniformly Cauchy on $[0,1]$. This in turn shows that both sequences are uniformly convergent on $[0,1]$. This provides two functions f and g, with (f_n) converging uniformly to f while (f_n') converges uniformly to g. Theorem 6.1.12 shows that $f' = g$ and Theorem 6.1.9 guarantees us that g is also continuous since f_n' is continuous for all n. Thus $f \in C^1([0,1])$ and the sequence (f_n) converges in the C^1 norm to f by applying Lemma 6.5.1.

Let us take a moment now to understand why both terms are needed in the norm given above. Certainly the sup-norm would give a norm on this space but would it guarantee completeness? This is where the second term comes into play and the evidence for this is given by Theorem 6.1.12. The problem is the fact that there are sequences in $C^1[0,1]$ which converge uniformly on $[0,1]$ to non-differentiable functions. Lemmas 6.5.7 and 6.5.8 provide an idea of how to construct such sequences.

So why not take the opposite approach and norm the space with the quantity $\|f'\|_\infty$? The difficulty here is that this is not a norm. Why? The norm above is sufficient and Exercise 6.5.5 asks you to consider whether a less strict version of this norm produces a Banach space structure on the space.

The Weierstrass Approximation Theorem

As the function constructed in Sect. 6.4 demonstrates, general continuous functions can exhibit rather unpleasant behavior with respect to differentiability. Polynomials on the other hand are fairly easy to understand in the greater scheme of things. The final result of this chapter pulls these two classes together in a powerful manner. The Weierstrass Approximation Theorem says that given any continuous function f on an interval $[a, b]$, there is a sequence of polynomials which converges uniformly to f. Thus no matter how badly behaved a function is, we can approximate it to any desired degree of accuracy with a polynomial. In terms of the normed linear space $(C([a, b]), \| \cdot \|_\infty)$, the theorem states that the polynomials form a dense subset of $C([a, b])$.

Weierstrass presented this theorem in 1886, and the proof we supply is that of Lebesgue from 1898. Lebesgue's proof seems to be one of the more constructive which does not require the use of integration; the proof of Bernstein from 1912 also has this flavor. The proof is broken down into a sequence of lemmas so as to keep the argument segmented and digestible.

Theorem 6.5.4 (Weierstrass Approximation Theorem). *Let $f \in C([a, b])$. Then for every $\varepsilon > 0$, there is a polynomial p such that $\| f - p \|_\infty < \varepsilon$.*

The first step in the proof is to show that any continuous function can be approximated with a function whose graph consists of line segments; this is contained in Lemma 6.5.6. Recall that a linear function has the form $f(x) = mx + b$ for some $m, b \in \mathbb{R}$; these functions are not "linear" in the sense of a linear operator, that is, in the algebraic sense. Rather, their classification as "linear" comes from the fact that the graph of such a function is a line.

Definition 6.5.5. A function f on $[a, b]$ is called *piecewise linear* if there exist a finite collection of points $\{x_0, x_1, x_2, \ldots x_n\}$ with $a = x_0 < x_1 < x_2 < \cdots < x_n = b$ such that f is a linear function on each interval (x_{j-1}, x_j) for $j = 1, 2, \ldots, n$. In other words, on the interval (x_{j-1}, x_j), f takes the form

$$f(x) = m_j x + b_j$$

for some $m_j, b_j \in \mathbb{R}$.

For a piecewise linear function f defined on $[a, b]$, we call the set $\{x_0, x_1, x_2, \ldots x_n\}$ the *partition of f*. As an observation, notice that if f is continuous and piecewise linear, then f is linear on the closed subintervals $[x_{j-1}, x_j]$.

Lemma 6.5.6. *Let $f : [a, b] \to \mathbb{R}$ be continuous. Then for every $\varepsilon > 0$, there is a piecewise linear continuous function g such that $|f(x) - g(x)| < \varepsilon$ for all $x \in [a, b]$.*

Proof. Let $f : [a, b] \to \mathbb{R}$ be continuous and let $\varepsilon > 0$. We will construct a piecewise linear continuous function by choosing a suitable partition which depends on the fact that continuous functions on compact sets are uniformly continuous. Thus there is a $\delta > 0$ such that if $x, y \in [a, b]$ with $|x - y| < \delta$, then $|f(x) - f(y)| < \varepsilon$. This dictates how many partition points we need and we choose $N \in \mathbb{N}$ with $\delta > (b - a)/N$. This choice partitions $[a, b]$ into N subintervals of length $(b - a)/N$ with $x_0 = a$ and $x_j = a + j(b - a)/N$ for $j = 1, 2, \ldots N$. Now let $y_j = f(x_j)$ for $j = 0, 1, 2, \ldots N$. The idea is to construct line segments which connect the points (x_{j-1}, y_{j-1}) and (x_j, y_j). On the interval $[x_{j-1}, x_j]$, we define g by

$$g(x) = y_{j-1} + \left(\frac{y_j - y_{j-1}}{x_j - x_{j-1}} \right) (x - x_{j-1}).$$

This shows that g is piecewise linear. It may appear as though we are defining g to take on two different values at x_j for $j = 1, 2, \ldots, N - 1$, but this is not so as $g(x_j) = f(x_j)$ for $j = 0, 1, 2, \ldots N$ by construction. This also shows that g is continuous as it is clearly continuous on the open subintervals dictated by the partition.

To finish the proof we must verify that $|f(x) - g(x)| < \varepsilon$ for every $x \in [a, b]$. Let $x \in [a, b]$. If x is one of the partition points, then $g(x) = f(x)$ by construction and thus the desired inequality clearly holds. Suppose now that x is not a partition point of g. Then there is a $K \in \mathbb{N}$ such that $x \in (x_{K-1}, x_K)$. Moreover, since g is linear on this interval, $g(x)$ satisfies $g(x_{K-1}) \leq g(x) \leq g(x_K)$ (or the reverse inequality from which the conclusion follows identically) which means that $f(x_{K-1}) \leq g(x) \leq f(x_K)$. If g is constant on this interval, then $g(x) = f(y)$ for $y = x_{K-1}$. If g is not constant, then we have $f(x_{K-1}) < g(x) < f(x_K)$ and by the Intermediate Value Theorem, there is a $y \in (x_{K-1}, x_K)$ such that $f(y) = g(x)$. In either case, by our choice of partition we know that $|x - y| < \delta$ and thus $|f(x) - g(x)| = |f(x) - f(y)| < \varepsilon$.   □

The next step in the proof is to show that any piecewise linear continuous function can be approximated uniformly with polynomials. We begin with the simplest such function, the absolute value, and then work to show that we can also approximate translations of the absolute value by polynomials.

Lemma 6.5.7. *Let $f(x) = |x|$. Then there is a sequence of polynomials (p_n) which converges uniformly to f on $[-1, 1]$.*

Proof. The sequence of polynomials is defined in a recursive fashion with $p_0(x) = 1$ on $[-1, 1]$. Then for $x \in [-1, 1]$ we define

$$p_n(x) = \frac{1}{2} \left(x^2 + 2p_{n-1}(x) - p_{n-1}(x)^2 \right)$$

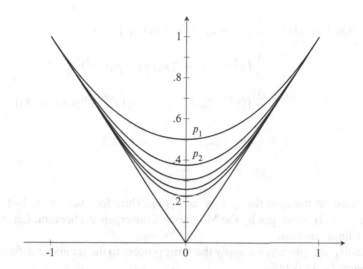

Fig. 6.6 Polynomial approximation of $f(x) = |x|$

for $n \in \mathbb{N}$. The polynomials p_1, p_2, p_3, p_4, and p_5 are given in Fig. 6.6. Our first goal now is to show that this sequence converges pointwise. We will then identify its limit function as $f(x) = |x|$ and confirm that the convergence is uniform.

As is typical with recursively defined sequences, we will use induction and the Monotone Convergence Theorem. For a fixed $x \in [-1, 1]$, it is clear that $|x| \leq p_1(x) \leq 1 = p_0(x)$ since

$$|x| \leq \frac{1}{2}(x^2 + 1) \leq 1.$$

For our inductive hypothesis, suppose that $|x| \leq p_n(x) \leq p_{n-1}(x) \leq 1$ for some $n \in \mathbb{N}$. To show that $|x| \leq p_{n+1}(x) \leq p_n(x)$, notice that

$$p_n(x) - p_{n+1}(x) = p_n - \frac{1}{2}\left(x^2 + 2p_n(x) - p_n(x)^2\right)$$

$$= \frac{1}{2}\left(2p_n(x) - x^2 - 2p_n(x) + p_n(x)^2\right)$$

$$= \frac{1}{2}\left(p_n(x)^2 - x^2\right)$$

$$\geq 0$$

by the inductive hypothesis; this implies that $p_{n+1}(x) \leq p_n(x)$. For the second desired inequality, notice first that our inductive hypothesis $|x| \leq p_n(x) \leq 1$ implies that $1 - |x| \geq 1 - p_n(x) \geq 0$. From this it follows that

$$p_{n+1}(x) - |x| = \frac{1}{2}\left(x^2 + 2p_n(x) - p_n(x)^2\right) - |x|$$

$$= \frac{1}{2}\left(x^2 - 2|x| + 2p_n(x) - p_n(x)^2\right)$$

$$= \frac{1}{2}\left((x^2 - 2|x| + 1) - (p_n(x)^2 - 2p_n(x) + 1)\right)$$

$$= \frac{1}{2}\left((1 - |x|)^2 - (1 - p_n(x))^2\right)$$

$$\geq 0;$$

hence it must be the case that $|x| \leq p_{n+1}(x)$. Thus for every $x \in [-1, 1]$, the sequence $(p_n(x))$ converges by the Monotone Convergence Theorem. Let p be the pointwise limit function.

To identify p explicitly, we apply the limit process to the recursive definition for the sequence (p_n) to obtain

$$p(x) = \frac{1}{2}\left(x^2 + 2p(x) - p(x)^2\right)$$

for all $x \in [-1.1]$. Solving this equation for $p(x)$ results in the equation $p(x)^2 = x^2$. The argument above and the Order Limit Theorem guarantee us that $p(x) \geq |x| \geq 0$ for every x which indicates that $p(x) = x$ if $x \geq 0$ and $p(x) = -x$ if $x < 0$, i.e. $p(x) = |x|$.

Clearly each p_n is a polynomial and is therefore continuous. Moreover, the pointwise limit is also continuous and thus by Dini's Theorem (Exercise 6.1.14), it must be the case that the convergence is uniform completing the proof. $\quad\square$

Lemma 6.5.8. *Let $c \in \mathbb{R}$ and set $g(x) = |x - c|$. Then there is a sequence of polynomials (q_n) which converges uniformly to g on any compact subset of $\mathbb{R}$.*

Proof. The graph of the function g as defined is a translation of the graph of the absolute value function and hence we will appeal to Lemma 6.5.7 to obtain our conclusion. From the lemma we have the sequence of polynomials (p_n) which converges uniformly to $|x|$ on $[-1, 1]$. We begin by choosing an adequate subsequence. For $j \in \mathbb{N}$, choose p_{n_j} such that

$$|p_{n_j}(x) - |x|| < 1/j^2 \tag{6.4}$$

for all $x \in [-1, 1]$ and set $q_j(x) = jp_{n_j}((x - c)/j)$. We then claim that (q_j) converges uniformly to g on any compact subset of $\mathbb{R}$. Before proving this claim, we translate Eq. (6.4) into an inequality concerning q_j. The inequality above is equivalent to the statement

$$|x| - \frac{1}{j^2} < p_{n_j}(x) < |x| + \frac{1}{j^2}$$

for all $x \in [-1, 1]$. Upon substituting $(x - c)/j$ and then multiplying by j, this inequality is equivalent to

$$|x - c| - \frac{1}{j} < j p_{n_j} \left(\frac{x - c}{j} \right) < |x - c| + \frac{1}{j}$$

for all x such that $(x - c)/j \in [-1, 1]$, or for all $x \in [c - j, c + j]$. But this means that

$$|q_j(x) - |x - c|| < 1/j$$

for all $x \in [c - j, c + j]$.

To verify the convergence claim, let $\varepsilon > 0$ and let $K \subseteq \mathbb{R}$ be any compact set. Choose $J_1 \in \mathbb{N}$ such that $1/J_1 < \varepsilon$. Also since K is compact, there is an $M > 0$ such that $K \subseteq [-M, M]$. Choose $J_2 \in \mathbb{N}$ such that $c - J_2 \leq -M < M \leq c + J_2$ which implies $[-M, M] \subseteq [c - J_2, c + J_2]$. We now set $J = \max\{J_1, J_2\}$ and consider $j \geq J$. Then we see that $[-M, M] \subseteq [c - J_2, c + J_2] \subseteq [c - j, c + j]$ and

$$|q_j(x) - |x - c|| < 1/j < 1/J < \varepsilon$$

for all $x \in K \subseteq [-M, M] \subseteq [c - j, c + j]$ proving the claim. $\square$

Before proving the final lemma, we have two more key ideas. First, define the set $P([a, b]) \subseteq C([a, b])$ to be the set of all continuous functions $f : [a, b] \rightarrow \mathbb{R}$ with the property that for every $\varepsilon > 0$, there is a polynomial p which satisfies $|f(x) - p(x)| < \varepsilon$ for all $x \in [a, b]$. It may seem strange to define such a thing as we are actually trying to show that $P([a, b]) = C([a, b])$; the idea is that we will take advantage of the fact that $P([a, b])$ is closed with respect to the addition and scalar multiplication of $C([a, b])$ by showing the preliminary result that $P([a, b])$ is subspace of $C([a, b])$. This is Exercise 6.5.7.

We also define the *positive part* of a real number t by

$$t_+ = \frac{t + |t|}{2} = \max\{t, 0\} = \begin{cases} 0, & t \leq 0; \\ t, & t > 0. \end{cases}$$

If f is a continuous function defined on a set A, then this idea produces a new function $f_+(x)$ which is also continuous by the Algebraic Limit Theorem for continuous functions. Also, Lemma 6.5.8 and Exercise 6.5.7 guarantees us that for each $c \in \mathbb{R}$ the function $(x - c)_+$ is in the set $P([a, b])$ defined above.

Lemma 6.5.9. *Suppose g is a piecewise linear continuous function on $[a, b]$. If g has partition $\{x_0, x_1, x_2, \ldots x_n\}$, then there exist constants $c_0, c_1, c_2, \ldots c_n$ such that*

$$g(x) = c_0 + \sum_{k=1}^{n} c_k (x - x_{k-1})_+.$$

Moreover, any such $g \in P([a, b])$.

Proof. Here we need to show that there exist constants $c_0, c_1, c_2, \ldots c_n$ such that g agrees with the form given in the statement of the theorem. For notation, we have that $\{x_0, x_1, x_2, \ldots x_n\}$ is the partition of g. Let $y_j = g(x_j)$ for $j = 0, 1, 2, \ldots n$. We also know that there are constants $m_j, b_j \in \mathbb{R}$ for $j = 1, 2, \ldots n$ such that $g(x) = m_j x + b_j$ on the interval $[x_{j-1}, x_j]$ since g is continuous; in particular, m_j is a slope given explicitly by

$$m_j = \frac{y_j - y_{j-1}}{x_j - x_{j-1}}.$$

It will be more suitable to have g represented in terms of the partition points, i.e. in a point-slope form rather than a slope-intercept form. To this end, we have that

$$g(x) = y_{j-1} + m_j(x - x_{j-1}) \tag{6.5}$$

on the interval $[x_{j-1}, x_j]$.

Now define $c_0 = y_0$, $c_1 = m_1$ and $c_j = m_j - m_{j-1}$ for $j = 2, \ldots, n$. These constants are identified by solving a system of equations and we omit the details. With these assignments, it is a simple matter to show that the function

$$c_0 + \sum_{k=1}^{n} c_k(x - x_{k-1})_+ \tag{6.6}$$

agrees with Eq. (6.5) on the interval $[x_{j-1}, x_j]$, $j = 1, 2, \ldots n$. One key fact here is that the function $(x - c)_+$ is zero for all $x \le c$; thus on a particular subinterval interval $[x_{j-1}, x_j]$, we can ignore terms from Eq. (6.6) corresponding to index greater than j. We leave this verification as Exercise 6.5.8.

The fact that any piecewise linear continuous function is in $P([a, b])$ follows from Exercise 6.5.7 and the above proof that any such function is a *finite* linear combination of functions in $P([a, b])$.   □

We now piece all of this information together to obtain a proof of Weierstrass' theorem.

Proof (Weierstrass Approximation Theorem). Let $f \in C([a, b])$ and let $\varepsilon > 0$. By Lemma 6.5.6 there is a piecewise linear continuous function $g : [a, b] \to \mathbb{R}$ such that $|f(x) - g(x)| < \varepsilon/3$ for all $x \in [a, b]$. Moreover, by Lemma 6.5.9, $g \in P([a, b])$ and hence there is a polynomial p such that $|g(x) - p(x)| < \varepsilon/3$ for all $x \in [a, b]$. The triangle inequality now guarantees us that

$$|f(x) - p(x)| \le |f(x) - g(x)| + |g(x) - p(x)| < 2\varepsilon/3$$

for all $x \in [a, b]$. Therefore we see that

$$\|f - p\|_\infty \le 2\varepsilon/3 < \varepsilon$$

as desired.   □

Earlier in this section we mentioned that many of our results hold in the more general context of $C(K)$ where K is any compact subset of $\mathbb{R}$. However, our proof of the Weierstrass Approximation Theorem relies on the fact that we are considering functions defined on a compact interval. Rest assured that there are more general versions of the theorem, collected under the title of the Stone–Weierstrass Theorem and credited to Stone in 1937. This statement goes beyond the real number line to general topological spaces X and provides a sufficient condition for a subalgebra to be dense in $C(X)$, the set of all continuous real-valued functions defined on X. The recent article [26] provides an interesting historical review of these results.

Exercises

Exercise 6.5.1. (a) In the proof of Lemma 6.5.1 we showed that a sequence (f_n) in $C([a,b])$ converges in the sup-norm if it is uniformly convergent. Show that the converse also holds.
(b) In the proof of Proposition 6.5.2 we showed that a sequence (f_n) in $C([a,b])$ is uniformly Cauchy on $[a,b]$ if it is Cauchy in the sup-norm. Show that the converse also holds.

Exercise 6.5.2. Show that the sup-norm on $C([a,b])$ does not arise from an inner product in the sense of Proposition 1.5.15.

Exercise 6.5.3. Let $m, n \in \mathbb{N}$ with $n > m$. Show that $2^{n-m} - 1 \geq 2^{n-m-1}$.

Exercise 6.5.4. Show that the sequence of functions (f_n) defined by

$$f_n(x) = \begin{cases} 0, & 0 \leq x \leq \frac{1}{2} - \frac{1}{n}; \\ nx + \frac{2-n}{2}, & \frac{1}{2} - \frac{1}{n} < x \leq 1/2; \\ 1, & 1/2 < x \leq 1 \end{cases}$$

is not Cauchy in $C([0,1])$.

Exercise 6.5.5. (a) Consider the space $C^1([0,1])$ with $\|f\| = \|f'\|_\infty$. Show that this is not a norm.
(b) Consider the space $C^1([0,1])$ with $\|f\| = |f(0)| + \|f'\|_\infty$. Is this a norm? Does it provide a Banach space structure on C^1?

Exercise 6.5.6. Let $L([a,b])$ be the set of all functions which are Lipschitz on $[a,b]$ and, for $f \in L([a,b])$, define the *Lipschitz constant* for f by

$$L(f) = \inf\left\{ M : \left|\frac{f(x) - f(y)}{x - y}\right| \leq M \text{ for all } x, y \in [a,b]\right\}.$$

Then define a norm on $L([a, b])$ by

$$\|f\|_L = \|f\|_\infty + L(f).$$

(a) Explain why $L([a, b]) \subseteq C([a, b])$.
(b) Show that $(L([a, b], \|\cdot\|_L)$ is a normed linear space.
(c) Is this space complete? Explain.

Exercise 6.5.7. (a) Show that the set $P([a, b])$ defined after the proof of Definition 6.5.5 is a subspace of $C([a, b])$.
(b) Show that $P([a, b])$ contains all the polynomials. Can you prove this without using the polynomial in question?

Exercise 6.5.8. Complete the proof of Lemma 6.5.9.

Exercise 6.5.9. Consider the space $(C([0, 1]), \|\cdot\|_\infty)$ and the multiplication operator $M_x : C([0, 1]) \to C([0, 1])$. Further suppose K is a nontrivial invariant subspace for M_x. Show that the function $f(x) = 1$ is not in K. Keep in mind that an invariant subspace is assumed to be closed, meaning it contains all its limit points.

Chapter 7
The Riemann Integral

Here we consider the familiar integral from calculus which is generally attributed to Riemann, though the idea of upper and lower sums for finding areas was used previously by Cauchy; other mathematicians had used such sums before Cauchy for estimating integrals but not for calculating exact values. The method we present is cleaner from the technical point of view and is attributed to Darboux; this method employs the notions of upper and lower integrals. To avoid confusion we will discuss the equivalence of this approach with the familiar Riemann sum approach. We will also devote some time to the famed Fundamental Theorem of Calculus and use this to derive some of the familiar properties of the exponential and logarithmic functions. The final section in this chapter will investigate the dysfunctional relationship between the limit process (with regard to a sequence of functions) and the Riemann integral propelling us toward the more sturdy Lebesgue integral.

7.1 The Riemann Integral

Of the various constructs we have studied, the integral is by far the oldest, dating back to the ancient world, the earliest known reference being of Antiphon the Sophist between 480 and 411 BCE. Other Greeks, including Archimedes, studied this problem as well, but not in terms of antiderivatives as we have come to understand integrals. These ancient mathematicians were interested in calculating the areas of certain regions but it was not until the seventeenth century that the connection with antiderivatives was made. It is safe to say that these early mathematicians influenced the work of Kepler, Cavalieri, Pascal, Fermat, Newton, Leibniz, Fourier, Cauchy, and, finally, Riemann. The influence is seen in that most of the techniques for calculating areas involve breaking a region down into smaller components whose area can be found and then applying some limit process. However, since the formal notion of limit did not enter into the picture until the time

M.A. Pons, *Real Analysis for the Undergraduate: With an Invitation to Functional Analysis*, 275
DOI 10.1007/978-1-4614-9638-0__7, © Springer Science+Business Media New York 2014

of Cauchy, these other mathematicians were forced to resort to other methods, such as the method of exhaustion, which, while clever, is quite tedious.

The process of integration was devised (among other things) to find the area of regions which are rectangular on three sides and bounded by a curve on the fourth; in order to be a true area calculator we necessarily need to assume our curve lies entirely above the x-axis, but we will also be able to integrate functions which take on negative values. As mentioned above, the common theme among many of the techniques used to compute areas involves breaking the given region down into smaller components. As our regions are defined by a function we will accomplish this by breaking our domain interval into smaller components.

Definition 7.1.1. Let $[a, b] \subseteq \mathbb{R}$. A *partition of* $[a, b]$ is a finite ordered set $P = \{x_0, x_1, x_2, \ldots, x_n\}$ such that $a = x_0 < x_1 < x_2 < \cdots < x_n = b$. If P and Q are partitions of $[a, b]$ with $P \subseteq Q$, then we say Q is a *refinement* of P. If $R = P \cup Q$, then we call R the *common refinement* of P and Q.

Example 7.1.2. Consider the interval $[1, 4]$. Each of the sets $P_1 = \{1, 2, 4\}$, $P_2 = \{1, 2, 3, 4\}$, and $P_3 = \{1, 1.5, 2, 2.5, 3, 3.5, 4\}$ are partitions and satisfy $P_1 \subseteq P_2 \subseteq P_3$.

For the interval $[2, 7]$, $P = \{2, 4, 6, 7\}$ and $Q = \{2, 3, 5, 7\}$ are both partitions and $R = \{2, 3, 4, 5, 6, 7\}$ is their common refinement.

Now we wish to make precise the process of computing the area between the graph of a nonnegative function and the x-axis over an interval $[a, b]$. To do this we will use the notion of a partition to break the region down into approximating rectangles, and then apply the concept of refinements to exhaust the area of the given region. We will also only consider functions which are *bounded* (meaning that the range of f is a bounded subset of $\mathbb{R}$), an assumption we will take with us throughout this chapter. The reason for this will be apparent after the following definition.

Definition 7.1.3. Let $P = \{x_0, x_1, x_2, \ldots, x_n\}$ be a partition of $[a, b]$ and suppose $f : [a, b] \to \mathbb{R}$ is bounded. For $i = 1, 2, \ldots n$, we set

$$M_i = \sup\{f(x) : x \in [x_{i-1}, x_i]\} \quad \text{and} \quad m_i = \inf\{f(x) : x \in [x_{i-1}, x_i]\}.$$

The *upper Riemann sum* of f with respect to P is given by

$$U(f, P) = \sum_{i=1}^{n} M_i (x_i - x_{i-1})$$

and the *lower Riemann sum* by

$$L(f, P) = \sum_{i=1}^{n} m_i (x_i - x_{i-1}).$$

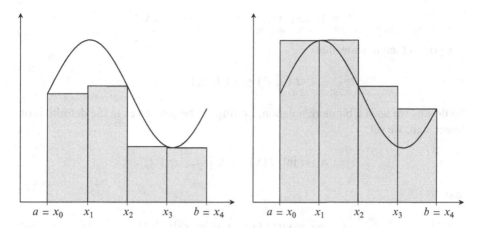

Fig. 7.1 Lower and upper Riemann sums

See Fig. 7.1 for an illustration of the upper and lower Riemann sums. To be clear, the lower sum takes a worst case scenario under approximation of the function over the subinterval $[x_{i-1}, x_i]$ while the upper sum takes a worst case scenario over approximation. Notice that we need the function to be bounded in order to guarantee the existence of m_i and M_i. We then add up the area of the n rectangles, each of which is represented by $m_i(x_i - x_{i-1})$ and $M_i(x_i - x_{i-1})$, respectively, to find worst case scenarios for over and under area estimates of the given region. The next step is to take finer and finer refinements in order to better approximate the area of the region in question. To understand the effect that this has on the upper and lower sums, we first have a lemma.

Proposition 7.1.4. *Let* $f : [a, b] \to \mathbb{R}$ *and let* P, Q *be partitions of* $[a, b]$. *Then*

(a) $L(f, P) \le U(f, P)$;
(b) *if* $P \subseteq Q$, *then* $L(f, P) \le L(f, Q)$ *and* $U(f, P) \ge U(f, Q)$;
(c) $L(f, P) \le U(f, Q)$.

Proof. For (a), let $P = \{x_0, x_1, x_2, \ldots, x_n\}$ be a partition of $[a, b]$. By definition of supremum and infimum, it is immediately clear that $m_i \le M_i$ for each $1 \le i \le n$. With this we see that

$$L(f, P) = \sum_{i=1}^n m_i(x_i - x_{i-1}) \le \sum_{i=1}^n M_i(x_i - x_{i-1}) = U(f, P).$$

For the second conclusion, we begin with the statement for lower sums. As a first case, suppose $P = \{x_0, x_1, x_2, \ldots, x_n\}$ is a partition of $[a, b]$ and let P_1 be a refinement which contains exactly one more point than P. With this assumption we can write

$$P_1 = \{x_0, x_1, x_2, \ldots, x_{j-1}, \alpha, x_j, \ldots, x_n\};$$

our goal is then to show that

$$L(f, P) \le L(f, P_1).$$

To do this we need a bit more notation. Letting m_i be defined as in the definition of lower sum, we set

$$n_1 = \inf\{f(x) : x \in [x_{j-1}, \alpha]\}$$

and

$$n_2 = \inf\{f(x) : x \in [\alpha, x_j]\}.$$

By the properties of infimum it follows that $m_j \le n_1, n_2$. If we consider now the definition of lower sum, we can separate out the term corresponding to the subinterval $[x_{j-1}, x_j]$ and we have

$$L(f, P) = \sum_{i=1}^{n} m_i(x_i - x_{i-1})$$

$$= \left[\sum_{i=1}^{j-1} m_i(x_i - x_{i-1})\right] + m_j(x_j - x_{j-1}) + \left[\sum_{i=j+1}^{n} m_i(x_i - x_{i-1})\right]$$

$$= \left[\sum_{i=1}^{j-1} m_i(x_i - x_{i-1})\right] + m_j(x_j - \alpha + \alpha - x_{j-1})$$

$$+ \left[\sum_{i=j+1}^{n} m_i(x_i - x_{i-1})\right]$$

$$\le \left[\sum_{i=1}^{j-1} m_i(x_i - x_{i-1})\right] + n_2(x_j - \alpha) + n_1(\alpha - x_{j-1})$$

$$+ \left[\sum_{i=j+1}^{n} m_i(x_i - x_{i-1})\right]$$

$$= L(f, P_1).$$

Figure 7.2 shows the area of the rectangle represented by the term $m_j(x_j - x_{j-1})$ as compared to the area of refined rectangle represented by $n_2(x_j - \alpha) + n_1(\alpha - x_{j-1})$.

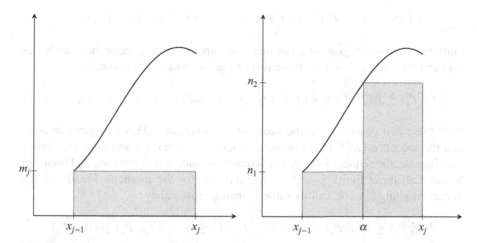

Fig. 7.2 Refinement of lower Riemann sum

Now, if Q is any refinement of P, we know that the number of points in Q exceeds that of P by at most a finite number, say N. Hence we can find refine P one point at a time to obtain Q. We can represent this with the chain

$$P \subseteq P_1 \subseteq P_2 \subseteq \cdots \subseteq P_{N-1} \subseteq Q,$$

and, by the simple case just shown, we have the inequality

$$L(f, P) \leq L(f, P_1) \leq L(f, P_2) \leq \cdots \leq L(f, P_{N-1}) \leq L(f, Q).$$

The statement for upper sums is completed in a similar manner.

Part (c) follows immediately from (a) and (b) if we use the notion of a common refinement. Setting $R = P \cup Q$, we have $P, Q \subseteq R$ and

$$L(f, P) \leq L(f, R) \leq U(f, R) \leq U(f, Q),$$

completing the proof. □

In order to motivate the definition of the integral, observe the following result of the proposition. Let (P_n) be a sequence of partitions of $[a, b]$ with $P_n \subseteq P_{n+1}$, i.e. a sequence of refinements. If we then consider the sequence of upper sums $(U(f, P_n))$, part (b) of the proposition states that this sequence of real numbers is decreasing,

$$\cdots \leq U(f, P_{n+1}) \leq U(f, P_n) \leq \cdots \leq U(f, P_3) \leq U(f, P_2) \leq U(f, P_1).$$

Similarly, the sequence of lower sums $(L(f, P_n))$ forms an increasing sequence of real numbers,

$$L(f, P_1) \leq L(f, P_2) \leq L(f, P_3) \leq \cdots \leq L(f, P_n) \leq L(f, P_{n+1}) \cdots .$$

Furthermore, part (c) guarantees us that *every* upper sum is greater than *any* lower sum and thus we can combine these two strings of inequalities to obtain

$$L(f, P_1) \leq L(f, P_2) \leq L(f, P_3) \leq \cdots \leq U(f, P_3) \leq U(f, P_2) \leq U(f, P_1).$$

With these two observations, the Monotone Convergence Theorem guarantees us that the sequence $(L(f, P_n))$ converges since it is increasing and bounded above, and likewise, the sequence $(U(f, P_n))$ converges since it is decreasing and bounded below. Call these limits L and U, respectively, for the moment. Finally, if the integral is going to exist, call its value I, then it must satisfy

$$L(f, P_1) \leq L(f, P_2) \leq \cdots \leq L \leq I \leq U \leq \cdots \leq U(f, P_2) \leq U(f, P_1)$$

since the lower sums are always underestimates and the upper sums are always overestimates. Ideally we would then prefer the value of the integral to be the common limit of these two sequences. Moreover, we would want this value to remain unchanged should we choose a different sequence of refinements. The following definitions capture this desire.

Definition 7.1.5. We define the *upper Riemann integral* of $f : [a, b] \to \mathbb{R}$ by

$$U(f) = \inf\{U(f, P) : P \text{ is a partition of } [a, b]\}$$

and the *lower Riemann integral* by

$$L(f) = \sup\{L(f, P) : P \text{ is a partition of } [a, b]\}.$$

Notice that the definition of the lower Riemann integral is influenced by the proof of the Monotone Convergence Theorem which states that the limit of an increasing sequence which is bounded above is the supremum, and the infimum for a sequence which is decreasing and bounded below, however, the fact that $U(f)$ and $L(f)$ always exist needs verification; Exercise 7.1.4 asks you to verify this statement and show that $L(f) \leq U(f)$. With these two definitions we are ready for the integral.

Definition 7.1.6. We say that a bounded function $f[a, b] \to \mathbb{R}$ is *Riemann integrable* on $[a, b]$ if $U(f) = L(f)$. In this case we use the symbols

$$\int_a^b f(x)\, dx \quad \text{or} \quad \int_a^b f\, dx$$

to represent this common value.

Often we will simply say the function is integrable; the term Riemann integrable should indicate, however, that there are other techniques of integration. We will explore one of these, the Lebesgue integral, in Chap. 9. Also, as a comment on the notation, we will typically use the second form unless there are multiple variables or parameters in play.

As a final comment, recall that the Riemann integral is designed to return the area of the region between a nonnegative function f and the x-axis over the interval $[a, b]$, but we have *not* required that the function f be nonnegative in our definition. So how do we interpret this when the function takes on negative values, or both negative and positive values? If the function is nonpositive on the interval $[a, b]$ then it is easy to see that m_i and M_i will be nonpositive, and we see that the integral is returning the negative of the area of the region, or the *signed area*. When the function takes on both positive and negative values, then cancelation occurs when we compute the upper and lower sums. In other words, the graph of the function defines a region above the x-axis and a region below. The integral then computes the difference in the area of these two regions, or the *net area* of the region between the graph of the function and the x-axis.

Example 7.1.7. Consider the function $g(x) = 2$ on the interval $[2, 4]$. If our process has truly been a means to calculate the area between the graph of a positive-valued function and the x-axis over an interval, then we should find that this function integrates to 4. Notice that for any partition $P = \{x_0, x_1, x_2, \ldots, x_n\}$ we have

$$L(f, P) = \sum_{i=1}^{n} m_i(x_i - x_{i-1}) = \sum_{i=1}^{n} 2(x_i - x_{i-1}) = 2(4 - 2) = 4.$$

Similarly,

$$U(f, P) = \sum_{i=1}^{n} M_i(x_i - x_{i-1}) = \sum_{i=1}^{n} 2(x_i - x_{i-1}) = 2(4 - 2) = 4.$$

The fact that these two statements hold for any partition indicates that g is integrable and

$$\int_2^4 g \, dx = U(f) = L(f) = 4.$$

Next we take the Dirichlet function from Example 4.1.7,

$$f(x) = \begin{cases} 1, & x \in \mathbb{Q}; \\ 0, & x \notin \mathbb{Q}. \end{cases}$$

This function is relatively simple as it only has two range values, but considering the behavior of the domain sets adds a level of complication when trying to integrate f. For our domain we take the interval $[0, 1]$. If $P = \{x_0, x_1, x_2, \ldots, x_n\}$ is a partition of this interval, then the density of both the rational and irrational numbers indicates that for each $1 \leq i \leq n$, there is a rational r_i and an irrational q_i in the subinterval $[x_{i-1}, x_i]$. This shows that $m_i = 0$ and $M_i = 1$ for all $1 \leq i \leq n$; hence

$$L(f, P) = \sum_{i=1}^{n} m_i (x_i - x_{i-1}) = 0$$

and

$$U(f, P) = \sum_{i=1}^{n} M_i (x_i - x_{i-1}) = \sum_{i=1}^{n} (x_i - x_{i-1}) = 1,$$

where the last equality is due to the fact that $[0, 1]$ has length 1. Since P represents an arbitrary partition of $[0, 1]$, this then implies that

$$L(f) = 0 \qquad \text{and} \qquad U(f) = 1$$

from which we conclude that the function is not integrable.

For a third example, let

$$h(x) = \begin{cases} 1, 0 \leq x < 1; \\ 2, 1 \leq x \leq 2 \end{cases}$$

as a function on the interval $[0, 2]$. The graph of this function consists of two line segments and the area between this graph and the x-axis over $[0, 2]$ is three. To see that the integral can calculate this value, we will show that $L(f) = U(f)$. To this end, let $\varepsilon > 0$ and let $P = \{0, 1 - \varepsilon, 1, 2\}$ be a partition of $[0, 2]$. Calculating the lower and upper sums, we have

$$L(f, P) = \sum_{i=1}^{n} m_i (x_i - x_{i-1}) = 1((1 - \varepsilon) - 0) + 1(1 - (1 - \varepsilon)) + 2(2 - 1) = 3$$

and

$$U(f, P) = \sum_{i=1}^{n} M_i (x_i - x_{i-1}) = 1((1 - \varepsilon) - 0) + 2(1 - (1 - \varepsilon)) + 2(2 - 1) = 3 + \varepsilon.$$

With these two calculations we have

$$3 = L(f, P) \leq L(f) \leq U(f) \leq U(f, P) = 3 + \varepsilon$$

which implies that

$$U(f) - L(f) \leq U(f, P) - L(f, P) = \varepsilon.$$

Since ε was arbitrarily chosen, we conclude that $L(f) = U(f)$ and hence the function is integrable. By varying the partition P with ε and letting ε tend to zero, we can conclude that

$$\int_0^2 h\,dx = U(f) = L(f) = 3.$$

The previous example motivates a means for determining when a function is Riemann integrable. In short a function is Riemann integrable if it is possible to find upper and lower sums which are arbitrarily close, and not Riemann integrable if the upper and lower sums remain bounded away from each other, i.e. if $L(f) < U(f)$.

Theorem 7.1.8 (Riemann's Criterion for Integrability). *Let $f : [a, b] \to \mathbb{R}$ be a bounded function. Then f is integrable if and only if for every $\varepsilon > 0$ there is a partition P_ε of $[a, b]$ such that*

$$U(f, P_\varepsilon) - L(f, P_\varepsilon) < \varepsilon.$$

Proof. For the first direction, suppose f is integrable on $[a, b]$ and let $\varepsilon > 0$. We must now produce a partition P_ε which satisfies the inequality in the statement. Using the fact that $L(f)$ is defined as a supremum and Lemma 1.2.10, we can find a partition P_1 of $[a, b]$ such that $L(f) - \varepsilon/2 < L(f, P_1)$. Likewise, we can find a partition P_2 such that $U(f) + \varepsilon/2 > U(f, P_2)$. Now we set $P_\varepsilon = P_1 \cup P_2$, the common refinement. By Proposition 7.1.4 we have

$$U(f, P_\varepsilon) \leq U(f, P_2) < U(f) + \varepsilon/2$$

and

$$L(f, P_\varepsilon) \geq L(f, P_1) > L(f) - \varepsilon/2.$$

Rearranging this last inequality,

$$-L(f, P_\varepsilon) < -(L(f) - \varepsilon/2).$$

The integrability of f now implies that $U(f) = L(f)$; combining the above inequalities then implies

$$U(f, P_\varepsilon) - L(f, P_\varepsilon) < (U(f) + \varepsilon/2) - (L(f) - \varepsilon/2) = U(f) - L(f) + \varepsilon = \varepsilon.$$

The converse follows similarly to our work from the third example above. Suppose $\varepsilon > 0$ and let P_ε be a partition of $[a, b]$ such that

$$U(f, P_\varepsilon) - L(f, P_\varepsilon) < \varepsilon.$$

The fact that it is always the case that

$$L(f, P_\varepsilon) \leq L(f) \leq U(f) \leq U(f, P_\varepsilon)$$

implies that

$$U(f) - L(f) \leq U(f, P_\varepsilon) - L(f, P_\varepsilon) < \varepsilon$$

and we conclude that $U(f) = L(f)$ as desired.   □

Keep in mind that although this theorem will tell us when a function is integrable, it does not necessarily give us any information about the value of the integral. Exercise 7.1.5 will explore extensions of this exercise and provide a sequential criterion for the Riemann integral. The previous example shows that there are integrable and non-integrable functions and the previous theorem provides a means for determining integrability, but it does so in a manner that is somewhat disconnected from the natural attributes of the function. Our next theorem assures us that this process is valid for a large class of familiar functions.

Theorem 7.1.9. *If f is continuous on $[a, b]$, then f is integrable on $[a, b]$.*

To prove the theorem we will appeal to the previous theorem. Notice that if $P = \{x_0, x_1, x_2, \ldots, x_n\}$ is a partition of $[a, b]$, then we can combine the difference of the upper and lower Riemann sums as

$$U(f, P) - L(f, P) = \sum_{i=1}^{n} (M_i - m_i)(x_i - x_{i-1}).$$

With this observation it is clear that we can control the size of the quantity $U(f, P) - L(f, P)$ by controlling the size of the $M_i - m_i$, which represents the greatest variation of f over the subinterval $[x_{i-1}, x_i]$. In particular if we can make these differences small for each i, then we can guarantee that $U(f, P) - L(f, P)$ will also be small. But this is exactly what continuity gives us, that is, the ability to make variations in the range small by only considering small variations in the domain. The key here is that we will need to do this for each subinterval of the partition simultaneously, i.e. in a *uniform* manner.

Proof. Using the fact that f is continuous on $[a, b]$, Theorem 4.4.10 assures us that f is in fact uniformly continuous there. Let $\varepsilon > 0$. Then there is a $\delta > 0$ such that $|f(x) - f(y)| < \varepsilon/(b - a)$ whenever $|x - y| < \delta$ in $[a, b]$. To show that f is integrable, let P_ε be any partition of $[a, b]$ such that $x_i - x_{i-1} < \delta$ for each $i = 1, 2, \ldots, n$. In particular, one can choose $n \in \mathbb{N}$ such that $(b - a)/n < \delta$ and let $x_1 = a + (b - a)/n$, $x_2 = x_1 + (b - a)/n$, etc.

Now consider a particular subinterval from this partition $[x_{i-1}, x_i]$. Using the continuity of the function again, the Extreme Value Theorem (Theorem 4.2.12) guarantees the existence of points $y_i, z_i \in [x_{i-1}, x_i]$ such that $f(y_i) = M_i$ and

$f(z_i) = m_i$. The fact that these two points are both in the same subinterval also guarantees us, by choice of P_ε, that $|y_i - z_i| < \delta$. Finally, the choice of δ further implies that

$$M_i - m_i = |f(y_i) - f(z_i)| < \frac{\varepsilon}{b-a}.$$

With this estimate in hand for each $i = 1, 2, \ldots, n$, it follows then that

$$U(f, P_\varepsilon) - L(f, P_\varepsilon) = \sum_{i=1}^{n}(M_i - m_i)(x_i - x_{i-1}) < \left(\frac{\varepsilon}{b-a}\right)\sum_{i=1}^{n}(x_i - x_{i-1})$$

$$= \left(\frac{\varepsilon}{b-a}\right)(b-a) = \varepsilon$$

and thus f is integrable.   □

We encountered a similar theorem when discussing differentiability. There we saw that any function which has a derivative at a point is continuous at a point, but we also discussed examples of continuous functions which are not differentiable at domain points. We have seen a similar pattern here. Continuity implies integrability, but there are integrable functions which are not continuous; see the function h from Example 7.1.7. These two ideas demonstrate that differentiability is a stronger property than continuity, which in turn is stronger than integrability.

The functions f and h from Example 7.1.7 should now challenge us to think about which bounded discontinuous functions are integrable; the function f is discontinuous everywhere and h has exactly one discontinuity. The question that concerns us then is how discontinuous a function can be and still be integrable. It should seem reasonable that a bounded function with a finite number of discontinuities is integrable. Indeed, an induction argument in conjunction with a technique similar to that employed for the function h of the mentioned example will demonstrate this fact. Moving on to infinite sets, can a bounded function have a countable number of discontinuities and still be integrable? An uncountable number of discontinuities? The answer to this question is a deep result and is postponed until Theorem 9.3.9. This theorem is Lebesgue's Criterion for Riemann integrability and explicitly identifies when a function is integrable in terms of the size of the set of discontinuities of the function.

For the sake of clarity we now turn to a discussion of ideas which arise in the calculus sequence and deserve some attention. First, it is typical to define two forms of integration, definite and indefinite. Clearly the definite integral is what we have been discussing thus far. The notion of indefinite integration is simply present to provide formal notation for the process of antidifferentiation. The link between these two is then supplied by the Fundamental Theorem of Calculus, but this was not always the case. Before Cauchy and Riemann developed their theories of integration, it was common to use the indefinite form as the integral. However, this produced a smaller class of integrable functions due in large part to Darboux's

Theorem which states that any derivative must have the intermediate value property. In fact, the function

$$h(x) = \begin{cases} 1, \ 0 \leq x < 1; \\ 2, \ 1 \leq x \leq 2 \end{cases}$$

is not integrable in this sense as it does not have the intermediate value property and hence cannot be the derivative of any function. With Riemann's integral, the Fundamental Theorem still ties the idea of integral and antiderivative, but only for continuous functions which do have the intermediate value property.

For the second point, in the calculus sequence the integral is defined for functions which are continuous, but the process there takes a slightly different form than what we have presented above. It is then natural to wonder if these two methods agree. Rest assured that they do and that the sphere of ideas presented here represents a broader form of integration, in part because we have not limited ourselves to continuous functions. To see this, let $f : [a, b] \to \mathbb{R}$ be continuous. We will now outline briefly the construction given in a standard calculus course. For $n \in \mathbb{N}$, we set $\Delta x = (b - a)/n$ and then divide $[a, b]$ into n subintervals of length Δx by defining $x_1 = a + \Delta x$, $x_2 = x_1 + \Delta x$, and so on, producing a partition $P_n = \{x_0, x_1, x_2, \ldots, x_n\}$; it should be easy to see that $x_n = b$. Then from each subinterval $[x_{i-1}, x_i]$, we choose a sample point x_i^*. The *Riemann sum* R_n is then defined by

$$R_n = \sum_{n=1}^{n} f(x_i^*) \Delta x$$

and we say that f is Riemann integrable if $\lim_{n \to \infty} R_n$ exists. In this case we set

$$\int_a^b f \, dx = \lim_{n \to \infty} R_n.$$

To see how this ties into our construction, notice that for any point $x_i^* \in [x_{i-1}, x_i]$ it must be the case that

$$m_i \leq f(x_i^*) \leq M_i$$

and thus

$$L(f, P_n) \leq R_n \leq U(f, P_n)$$

for every $n \in \mathbb{N}$ (Fig. 7.3). Now, since f is continuous and therefore uniformly continuous on $[a, b]$, for $\varepsilon > 0$, we can choose $\delta > 0$ so that $|x - y| < \delta$ implies $|f(x) - f(y)| < \varepsilon/(b - a)$. We can also choose $N \in \mathbb{N}$ so that $\Delta x < \delta$ for all $n \geq N$. The proof of Theorem 7.1.9 then shows that

$$U(f, P_n) - L(f, P_n) < \varepsilon$$

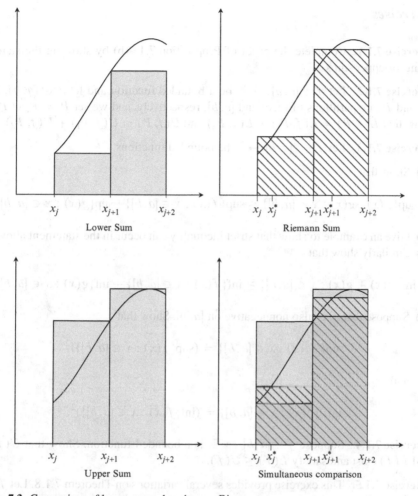

Fig. 7.3 Comparison of lower, general, and upper Riemann sums

for all $n \geq N$ and Exercise 7.1.6 implies

$$\int_a^b f\, dx = \lim_{n \to \infty} U(f, P_n) = \lim_{n \to \infty} L(f, P_n);$$

thus we have

$$\int_a^b f\, dx = \lim_{n \to \infty} R_n$$

by the Squeeze Theorem.

Exercises

Exercise 7.1.1. Complete the proof of Proposition 7.1.4(b) by showing the statement for upper sums.

Exercise 7.1.2. Let $f : [a,b] \to \mathbb{R}$ be a bounded function and let $c \in (a,b)$. If P_1 and P_2 are partitions of $[a,c]$ and $[c,b]$, respectively, and we set $P = P_1 \cup P_2$, show that $L(f, P) = L(f, P_1) + L(f, P_2)$ and $U(f, P) = U(f, P_1) + U(f, P_2)$.

Exercise 7.1.3. Let $f, g : [a,b] \to \mathbb{R}$ be bounded functions.

(a) Show that

$$\sup\{f(x)+g(x) : x \in [a,b]\} \le \sup\{f(x) : x \in [a,b]\}+\sup\{g(x) : x \in [a,b]\}.$$

(b) Give an example to show that strict inequality can occur in the statement above.
(c) Similarly show that

$$\inf\{f(x) + g(x) : x \in [a,b]\} \ge \inf\{f(x) : x \in [a,b]\} + \inf\{g(x) : x \in [a,b]\}.$$

(d) Suppose that f is also nonnegative on $[a,b]$. Show that

$$\sup\{f^2(x) : x \in [a,b]\} = (\sup\{f(x) : x \in [a,b]\})^2$$

and

$$\inf\{f^2(x) : x \in [a,b]\} = (\inf\{f(x) : x \in [a,b]\})^2.$$

Exercise 7.1.4. Suppose $f : [a,b] \to \mathbb{R}$ is a bounded function. Show that $U(f)$ and $L(f)$ exist and satisfy $L(f) \le U(f)$.

Exercise 7.1.5. This exercise provides several variations on Theorem 7.1.8. Let $f : [a,b] \to \mathbb{R}$ be a bounded function.

(a) Show that f is integrable on $[a,b]$ if and only if there is a sequence of partitions (P_n) such that

$$\lim_{n\to\infty} U(f, P_n) - L(f, P_n) = 0.$$

In this situation, show that

$$\int_a^b f(x)\,dx = \lim_{n\to\infty} U(f, P_n) = \lim_{n\to\infty} L(f, P_n).$$

(b) Extend the previous result by showing that the sequence (P_n) can be chosen as a sequence of refinements, i.e. so that $P_1 \subseteq P_2 \subseteq P_3 \subseteq \cdots$.

(c) If $P = \{x_0, x_1, x_2, \ldots, x_n\}$ is a partition of $[a, b]$, we define the *mesh of* P by

$$\text{mesh}(P) = \max\{|x_j - x_{j-1}| : j = 1, 2, \ldots, m\}.$$

Show that we can further refine the sequence of partitions from parts (a) and (b) so that $\text{mesh}(P_n) \to 0$ as $n \to \infty$.

Exercise 7.1.6. Suppose $f : [a, b] \to \mathbb{R}$ is continuous on $[a, b]$. Show that

$$\int_a^b f(x)\, dx = \lim_{n \to \infty} U(f, P_n) = \lim_{n \to \infty} L(f, P_n)$$

whenever (P_n) is a sequence of partitions with $\text{mesh}(P_n) \to 0$ as $n \to \infty$.

Exercise 7.1.7. Suppose $f : [a, b] \to \mathbb{R}$ is bounded and integrable on $[c, b]$ for every $c \in (a, b)$. Show that f is integrable on $[a, b]$. Does the statement remain true if we remove the hypothesis that f is bounded on $[a, b]$? Does the same result hold at the right endpoint?

Exercise 7.1.8. Suppose $f : [a, b] \to \mathbb{R}$ is bounded and increasing on $[a, b]$. Show that f is integrable.

Exercise 7.1.9. Let t be Thomae's function from Exercise 4.1.10. Assuming t is integrable, show that $\int_0^1 t(x)\, dx = 0$.

7.2 Properties of the Riemann Integral

Before proving the Fundamental Theorem of Calculus we investigate the various algebraic properties of the integral. The first result should not be surprising.

Theorem 7.2.1 (Algebraic Limit Theorem for Integrals). *Let $f, g : [a, b] \to \mathbb{R}$ be integrable on $[a, b]$.*

(a) $f + g$ is integrable with $\displaystyle\int_a^b f + g\, dx = \int_a^b f\, dx + \int_a^b g\, dx.$

(b) cf is integrable for every $c \in \mathbb{R}$ with $\displaystyle\int_a^b cf\, dx = c\int_a^b f\, dx.$

Proof. As a first observation, notice that Exercise 7.1.3 shows that

$$U(f + g, P) \leq U(f, P) + U(g, P)$$

and

$$L(f + g, P) \geq L(f, P) + L(g, P)$$

for P any partition of $[a, b]$. Now, to show that $f + g$ is integrable it suffices to find a sequence of partitions (R_n) such that

$$\lim_{n \to \infty} U(f + g, R_n) - L(f + g, R_n) = 0$$

by Exercise 7.1.5(a). Appealing to this same exercise (note the if and only if in the statement) for f and g, respectively, we can find sequences of partitions (P_n) and (Q_n) such that

$$\lim_{n \to \infty} U(f, P_n) - L(f, P_n) = 0$$

and

$$\lim_{n \to \infty} U(g, Q_n) - L(g, Q_n) = 0.$$

Set $R_n = P_n \cup Q_n$. Then, by our initial observation and Proposition 7.1.4,

$$0 \leq U(f + g, R_n) - L(f + g, R_n) \leq U(f, R_n) + U(g, R_n) - L(f, R_n) - L(g, R_n)$$
$$\leq U(f, P_n) + U(g, Q_n) - L(f, P_n) - L(g, Q_n)$$
$$= (U(f, P_n) - L(f, P_n)) + (U(g, Q_n) - L(g, Q_n))$$

and thus

$$\lim_{n \to \infty} U(f + g, R_n) - L(f + g, R_n) = 0$$

by the Squeeze Theorem. Thus $f + g$ is integrable.

To show the integral equality, first notice that the sequences $(U(f + g, R_n))$ and $(L(f + g, R_n))$ both converge by Exercise 7.1.5(a). Working with upper sums for a moment, observe that

$$\int_a^b f + g \, dx = \lim U(f + g, R_n) \leq \lim U(f, R_n) + U(g, R_n)$$
$$\leq \lim U(f, P_n) + U(g, Q_n)$$
$$= \int_a^b f \, dx + \int_a^b g \, dx.$$

Using the lower sums in a similar fashion we can show that

$$\int_a^b f + g \, dx \geq \int_a^b f \, dx + \int_a^b g \, dx.$$

Hence we conclude that the two quantities must be equal.

The proof of (b) is similar and is left as an exercise. $\square$

Notice that we have (as with series) not included a rule for products and quotients. While it is true that a product of integrable functions is integrable, there is no general relationship between the integral of the product and the product of the integrals. Exercise 7.2.3 will touch on this for one particular case. The situation for quotients is more complicated as we are only allowed to integrate bounded functions; thus for a quotient to be integrable we would have to further assume that the denominator is bounded away from zero on $[a, b]$ so that its reciprocal is bounded. For compositions, Exercise 7.2.4 will provide one instance where integration is possible. We will see a more general result in Sect. 9.3 which will allow us to prove more general statements for products and compositions (Exercise 9.3.11).

Before proving a version of the order limit theorems, we first have a lemma concerning the behavior of suprema and infima.

Lemma 7.2.2. *Let $A \subseteq \mathbb{R}$ be nonempty and bounded, and define the set $|A| = \{|a| : a \in A\}$. Then*

$$\sup(|A|) - \inf(|A|) \leq \sup(A) - \inf(A).$$

Proof. The proof here depends on several cases and we recommend a review of Exercises 1.3.10 and 1.3.11. First assume that $A \subseteq [0, \infty)$. In this case $|A| = A$ and the result is immediate.

If $A \subseteq (-\infty, 0]$, then $|A| = -A$, and therefore $\sup(|A|) = \sup(-A) = -\inf(A)$ and $\inf(|A|) = \inf(-A) = -\sup(A)$. Thus

$$\sup(|A|) - \inf(|A|) = -\inf(A) - (-\sup(A)) = \sup(A) - \inf(A).$$

For the third case we suppose A contains both positive and negative values. Then the second of the referenced exercises guarantees us that we have the following string of inequalities,

$$\inf(A) < 0 \leq \inf(|A|) \leq \sup(A) \leq \sup(|A|).$$

Now we have two subcases depending on whether the final inequality on the right is an equality or a strict inequality. If $\sup(A) = \sup(|A|)$, then

$$\sup(|A|) - \inf(|A|) = \sup(A) - \inf(|A|).$$

Furthermore, the fact that $\inf(A) \leq \inf(|A|)$ implies that

$$\sup(A) - \inf(|A|) \leq \sup(A) - \inf(A).$$

Thus we see that

$$\sup(|A|) - \inf(|A|) = \sup(A) - \inf(|A|) \leq \sup(A) - \inf(A).$$

Finally, assume that $\sup(A) < \sup(|A|)$. It should be clear that in this case we have $\sup(|A|) = -\inf(A)$. By the fact that $\inf(|A|)$ and $\sup(A)$ are both positive, we have

$$\sup(|A|) - \inf(|A|) \leq \sup(|A|) = -\inf(A) \leq \sup(A) - \inf(A),$$

completing the proof.   □

Theorem 7.2.3 (Order Limit Theorem for Integrals). *Let* $f, g : [a, b] \to \mathbb{R}$ *be integrable on* $[a, b]$.

(a) If $m \leq f(x) \leq M$ *on* $[a, b]$, *then* $m(b - a) \leq \displaystyle\int_a^b f \, dx \leq M(b - a)$.

(b) If $f(x) \leq g(x)$ *on* $[a, b]$, *then* $\displaystyle\int_a^b f \, dx \leq \int_a^b g \, dx$.

(c) The function $|f|$ *is integrable on* $[a, b]$ *with* $\left| \displaystyle\int_a^b f \, dx \right| \leq \int_a^b |f| \, dx$.

Proof. For (a), let $P = \{a, b\}$. It follows then that $L(f, P) \geq m(b - a)$ and $U(f, P) \leq M(b - a)$. We then have

$$m(b - a) \leq L(f, P) \leq L(f) = \int_a^b f \, dx = U(f) \leq U(f, P) \leq M(b - a)$$

by definition of upper and lower integrals.

The proof of (b) is requested in Exercise 7.2.5.

For the final statement we first show that $|f|$ is integrable. Assuming that f is integrable, for $\varepsilon > 0$ we can choose a partition $P = \{x_0, x_1, x_2, \dots, x_n\}$ so that

$$U(f, P) - L(f, P) = \sum_{i=1}^n (M_i - m_i)(x_i - x_{i-1}) < \varepsilon.$$

If we now set

$$M_i^* = \sup\{|f(x)| : x \in [x_{i-1}, x_i]\} \quad \text{and} \quad m_i^* = \inf\{|f(x)| : x \in [x_{i-1}, x_i]\},$$

Lemma 7.2.2 shows that

$$M_i^* - m_i^* \leq M_i - m_i.$$

Therefore

$$U(|f|, P) - L(|f|, P) = \sum_{i=1}^{n} (M_i^* - m_i^*)(x_i - x_{i-1}) \le \sum_{i=1}^{n} (M_i - m_i)(x_i - x_{i-1}) < \varepsilon$$

and $|f|$ is integrable. For the integral inequality, the fact that

$$-|f(x)| \le f(x) \le |f(x)|$$

for all $x \in [a, b]$ together with (b) and Theorem 7.2.1(b) shows that

$$-\int_a^b |f(x)|\, dx \le \int_a^b f(x)\, dx \le \int_a^b |f(x)|\, dx$$

from whence it follows that

$$\left| \int_a^b f\, dx \right| \le \int_a^b |f|\, dx. \qquad \Box$$

The subdivision property of the Riemann integral has no analog for continuous and differentiable functions because these properties are local, while integration takes place in a more global context. The benefit of such a theorem will be apparent in the next section, though its functionality should be clear momentarily. To summarize, the property allows us to integrate over subintervals of $[a, b]$ and identifies how this relates to integrating over the entire interval.

Theorem 7.2.4 (Subdivision Property). *Let $f : [a, b] \to \mathbb{R}$. If $c \in (a, b)$, then f is integrable on $[a, b]$ if and only if f is integrable on $[a, c]$ and on $[c, b]$. Moreover,*

$$\int_a^c f\, dx + \int_c^b f\, dx = \int_a^b f\, dx.$$

Proof. We first show the equivalence and then the corresponding integral statement. First suppose f is integrable on $[a, b]$ and let $\varepsilon > 0$. By Theorem 7.1.8 there is a partition Q of $[a, b]$ such that

$$U(f, Q) - L(f, Q) < \varepsilon.$$

To generate partitions for $[a, c]$ and $[c, b]$, we first set $P = Q \cup \{c\}$; this is a refinement of Q and hence

$$U(f, P) - L(f, P) < \varepsilon$$

by Proposition 7.1.4. Next, let $P_1 = P \cap [a, c]$ and $P_2 = P \cap [c, b]$. By Exercise 7.1.2 and the previous inequality, we have

$$U(f, P) - L(f, P) = (U(f, P_1) + U(f, P_2)) - (L(f, P_1) + L(f, P_2))$$
$$= (U(f, P_1) - L(f, P_1)) + (U(f, P_2) - L(f, P_2))$$
$$< \varepsilon$$

from which it follows that both quantities $U(f, P_1) - L(f, P_1)$ and $U(f, P_2) - L(f, P_2)$ must be less than ε since both values are positive. Therefore f is integrable on $[a, c]$ and $[c, b]$.

For the converse, we assume f is integrable on $[a, c]$ and $[c, b]$. To show f in integrable on $[a, b]$, let $\varepsilon > 0$ and choose partitions P_1 and P_2 of $[a, c]$ and $[c, b]$, respectively, such that

$$U(f, P_1) - L(f, P_1) < \frac{\varepsilon}{2} \quad \text{and} \quad U(f, P_2) - L(f, P_2) < \frac{\varepsilon}{2}.$$

Setting $P = P_1 \cup P_2$ and applying Exercise 7.1.2 again we have

$$U(f, P) - L(f, P) = (U(f, P_1) - L(f, P_1)) + (U(f, P_2) - L(f, P_2)) < \varepsilon$$

showing that f is integrable on $[a, b]$.

For the integral equality, we will show two inequality statements. To begin, let $\varepsilon > 0$ and let P_1, P_2 and $P = P_1 \cup P_2$ be as above (in either part of the proof as the constructions are equivalent). Notice then that

$$\int_a^b f \, dx \leq U(f, P) < L(f, P) + \varepsilon = L(f, P_1) + L(f, P_2) + \varepsilon$$

$$\leq \int_a^c f \, dx + \int_c^b f \, dx + \varepsilon.$$

The fact that this relationship holds for every positive ε implies that

$$\int_a^b f \, dx \leq \int_a^c f \, dx + \int_c^b f \, dx.$$

In a similar fashion,

$$\int_a^c f \, dx + \int_c^b f \, dx \leq U(f, P_1) + U(f, P_2)$$

$$< L(f, P_1) + L(f, P_2) + \varepsilon = L(f, P) + \varepsilon \leq \int_a^b f \, dx + \varepsilon,$$

and hence

$$\int_a^c f\,dx + \int_c^b f\,dx \le \int_a^b f\,dx.$$

With these two inequalities we see that it must be case that

$$\int_a^c f\,dx + \int_c^b f\,dx = \int_a^b f\,dx. \qquad \square$$

The following definition extends the definition of the integral to degenerate intervals and allows us to integrate in the reverse orientation. We could modify the definition of the integral to cover such situations, but it's equally efficient to state these as facts.

Definition 7.2.5. If $f : [a,b] \to \mathbb{R}$ is integrable on $[a,b]$, we define

$$\int_b^a f\,dx = -\int_a^b f\,dx.$$

If c is any point at which f is defined, we set

$$\int_c^c f\,dx = 0.$$

With this definition we can extend Theorem 7.2.4 to encompass any triple of numbers a, b, c. In particular, if $a, b, c \in \mathbb{R}$ and the three integrals below exist, then

$$\int_a^c f\,dx + \int_c^b f\,dx = \int_a^b f\,dx.$$

For the final result of this section we show that a sequence of integrable functions converges to an integrable function provided the convergence is uniform on the interval $[a,b]$. In the exercises you will show that this result does not hold if we only assume pointwise convergence and we will touch on this again in Sect. 7.5.

Theorem 7.2.6. *Let (f_n) be a sequence defined on the common domain $[a,b]$ and suppose each f_n is integrable on $[a,b]$. If (f_n) converges uniformly on $[a,b]$ to f, then f is integrable and*

$$\lim_{n\to\infty} \int_a^b f_n\,dx = \int_a^b f\,dx. \tag{7.1}$$

Proof. For the proof we will show that f is integrable and leave the limit statement as Exercise 7.2.12. To show that the limit function f is integrable, we will again appeal to Theorem 7.1.8. Let $\varepsilon > 0$. By the definition of uniform convergence, there exists $N \in \mathbb{N}$ such that

$$|f_n(x) - f(x)| < \frac{\varepsilon}{3(b-a)}$$

or

$$f_n(x) - \frac{\varepsilon}{3(b-a)} < f(x) < f_n(x) + \frac{\varepsilon}{3(b-a)} \tag{7.2}$$

for all $n \geq N$ and all $x \in [a, b]$. To generate a suitable partition for f, we will work with the function f_N (any $n \geq N$ will do). Choose a partition $P = \{x_0, x_1, x_2, \ldots, x_n\}$ such that

$$U(f_N, P) - L(f_N, P) < \frac{\varepsilon}{3}.$$

Next we need to obtain a relationship between the upper and lower sums for f and f_N. Considering the upper inequality in Eq. (7.2) together with Exercises 1.3.8 and 1.3.9, we have

$$\sup\{f(x) : x \in [x_{i-1}, x_i]\} \leq \sup\left\{ f_N(x) + \frac{\varepsilon}{3(b-a)} : x \in [x_{i-1}, x_i]\right\}$$

$$= \sup\{f_N(x) : x \in [x_{i-1}, x_i]\} + \frac{\varepsilon}{3(b-a)}$$

for any subinterval $[x_{i-1}, x_i]$. This immediately implies that

$$U(f, P) \leq U\left(f_N + \frac{\varepsilon}{3(b-a)}, P \right) = U(f_N, P) + \frac{\varepsilon}{3}.$$

Using the lower inequality in Eq. (7.2) in a similar fashion, we obtain

$$L(f, P) \geq L(f_N, P) - \frac{\varepsilon}{3}.$$

With these two relationships it follows that

$$U(f, P) - L(f, P) \leq \left(U(f_N, P) + \frac{\varepsilon}{3}\right) - \left(L(f_N, P) - \frac{\varepsilon}{3}\right)$$

$$= U(f_N, P) - L(f_N, P) + \frac{2\varepsilon}{3} = \varepsilon$$

and thus f is integrable.   □

Two corollaries for series follow immediately.

Corollary 7.2.7. *Let* (f_n) *be a sequence defined on the common domain* $[a, b]$ *and suppose each* f_n *is integrable on* $[a, b]$. *If* $\sum f_n$ *converges uniformly on* $[a, b]$ *to* f, *then* f *is integrable and*

$$\sum \int_a^b f_n \, dx = \int_a^b \sum f_n \, dx = \int_a^b f \, dx.$$

Corollary 7.2.8. *Let* $f(x) = \sum_{n=0}^\infty a_n x^n$ *be a power series with radius of convergence* $R > 0$, *and suppose* $[a, b] \subseteq (-R, R)$. *Then*

$$\int_a^b \sum a_n x^n \, dx = \sum a_n \int_a^b x^n \, dx.$$

The first corollary follows immediately from Theorem 7.2.6 while the second also requires the fact that a power series converges uniformly on any compact subset of its interval of convergence.

Exercises

Exercise 7.2.1. Prove part (b) of Theorem 7.2.1.

Exercise 7.2.2. Suppose f is a nonnegative, continuous function on $[a, b]$ with

$$\int_a^b f(x) \, dx = 0.$$

Show that $f(x) = 0$ on $[a, b]$. It may be helpful to consider the contrapositive.

Exercise 7.2.3. Let $f : [a, b] \to \mathbb{R}$ be integrable and nonnegative.

(a) Show that f^2 is integrable on $[a, b]$.
(b) If f integrates to zero, show that f^2 integrates to zero also.
(c) If f^2 integrates to zero, show that f integrates to zero also.
(d) If $g : [a, b] \to \mathbb{R}$ is also integrable and nonnegative, show that fg is also integrable on $[a, b]$. Hint: The identity $ab = \frac{1}{2}((a + b)^2 - a^2 - b^2)$ may be useful.

Exercise 7.2.4. Suppose f is integrable on $[a, b]$ with $m \leq f(x) \leq M$ for all $x \in [a, b]$. Further suppose g is continuous and increasing on $[m, M]$. Show that $g \circ f$ is integrable on $[a, b]$.

Exercise 7.2.5. Prove part (b) of Theorem 7.2.3.

Exercise 7.2.6. Provide an example of a function $f : [a, b] \to \mathbb{R}$ for which $|f|$ is integrable but f is not.

Exercise 7.2.7. Prove the general subdivision property: if $a, b, c \in \mathbb{R}$ and the three integrals below exist, then

$$\int_a^c f \, dx + \int_c^b f \, dx = \int_a^b f \, dx.$$

Exercise 7.2.8. Supply a proof for Corollary 7.2.7.

Exercise 7.2.9. Supply a proof for Corollary 7.2.8.

Exercise 7.2.10. Suppose f is a function defined on an interval $[a, b]$.

(a) If f is zero *except* at one point in $[a, b]$, show that f integrates to zero on $[a, b]$.
(b) Let g be a function which agrees with f *except* at one point. If f is integrable, show that g is integrable with

$$\int_a^b f(x) \, dx = \int_a^b g(x) \, dx.$$

To do this, first consider the function $g - f$.
(c) Use an induction argument to show that the function f remains integrable (and integrates to the same value) if we change the value of f at any finite number of points.
(d) Give an example to show that changing the value of f at a countable number of points can result in function which is not integrable.

Exercise 7.2.11. Consider the sequence of functions (f_n) defined on the interval $[0, 1]$ by

$$f_n(x) = \begin{cases} 0 & x = 0; \\ n & 0 < x \le \frac{1}{n}; \\ 0 & \frac{1}{n} < x \le 1. \end{cases}$$

Compute $\int_0^1 f_n \, dx$ for each n and compare this to $\int_0^1 f \, dx$ where f is the pointwise limit of the sequence of functions.

Exercise 7.2.12. (a) Show that the limit statement in Theorem 7.2.6 holds.
(b) Theorem 7.2.6 does not hold if we only assume pointwise convergence and can fail in three distinct (and exhaustive) ways.

(i) Both sides of Eq. (7.1) exist but are not equal (see Exercise 7.2.11).
(ii) The limit on the left may fail to exist.
(iii) The limit function may not be Riemann integrable.

Give examples demonstrating parts (ii) and (iii).

7.3   The Fundamental Theorem of Calculus

The Fundamental Theorem of Calculus provides the link between derivatives and integrals by relating integrals to antiderivatives. This remarkable theorem provides a tractable means of computing integrals without which the theory of calculus would be far less useful in explaining natural phenomenon. The theorem is stated in two parts which, while logically equivalent, seem to be doing very distinct things:

(1) if we differentiate an integral, then we return to the same function;
(2) if we integrate a derivative, we return to the same function.

In other words, both say that differentiation and integration are inverse processes. But there is a slight problem with this kind of interpretation: differentiation yields a function and integration yields a value. How do we then compare these seemingly different objects, values and functions? The key to understanding this discrepancy for the first statement above is found in an understanding of what we will call area functions.

For the moment consider $f : [a, b] \to \mathbb{R}$, an integrable and nonnegative function. We then form the *area function* F defined on $[a, b]$ by

$$F(x) = \int_a^x f(t)\,dt.$$

This function computes the area (and hence the name) between the graph of f and the x-axis over the interval $[a, x]$; see the left-hand image in Fig. 7.4. If we could differentiate the function F, which is the integral of f, we would like to recover the original function f. To get an idea for why this a reasonable conjecture, consider the following computation where we apply Definition 7.2.5 and Theorem 7.2.4,

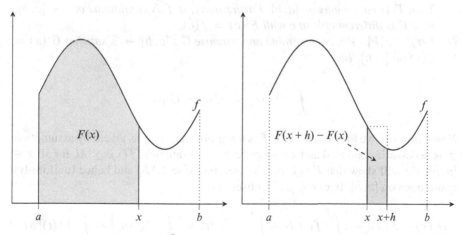

Fig. 7.4 Area function $F(x) = \int_a^x f(t)\,dt$

$$F(x + h) - F(x) = \int_a^{x+h} f(t)\,dt - \int_a^x f(t)\,dt = \int_x^{x+h} f(t)\,dt.$$

Now, this last integral represents the area between f and the x-axis over the interval $[x, x + h]$. And, if the function f is continuous, then the area of this region can be approximated by the rectangle with area $f(x) \cdot h$; for a visual see the right-hand image in Fig. 7.4. Hence

$$F'(x) = \lim_{h \to 0} \frac{F(x + h) - F(x)}{h} \approx \lim_{h \to 0} \frac{f(x) \cdot h}{h} = f(x).$$

This is an intuitive argument and our proof below will present a more rigorous explanation.

The second part of the theorem is more computational in nature and is used to evaluate integrals; it provides a way around the definition just as the product, chain and quotient rules provide shortcuts for taking derivatives. The idea here is to begin with an integrable function g and any antiderivative G. The theorem states that upon integrating g we obtain the value $G(b) - G(a)$. In short, the integral of the derivative returns the net change in the antiderivative over $[a, b]$. So while we haven't exactly recovered the original function G, we do recover some information regarding this function.

Theorem 7.3.1 (Fundamental Theorem of Calculus).

(a) Let $f : [a, b] \to \mathbb{R}$ be integrable and define a function $F : [a, b] \to \mathbb{R}$ by

$$F(x) = \int_a^x f(t)\,dt.$$

Then F is continuous on $[a, b]$. Furthermore, if f is continuous at $c \in [a, b]$, then F is differentiable at c with $F'(c) = f(c)$.
(b) Let $g : [a, b] \to \mathbb{R}$ be integrable and suppose $G : [a, b] \to \mathbb{R}$ satisfies $G'(x) = g(x)$ on $[a, b]$. Then

$$\int_a^b g(x)\,dx = G(b) - G(a).$$

Proof. For (a), we first assume that f is integrable and F is as given; by assumption f is necessarily bounded and we suppose $M > 0$ satisfies $|f(x)| \leq M$ for all $x \in [a, b]$. We will show that F is Lipschitz (see Exercise 4.4.8) and hence (uniformly) continuous on $[a, b]$. If $x, y \in [a, b]$, observe that

$$|F(y) - F(x)| = \left| \int_a^y f(t)\,dt - \int_a^x f(t)\,dt \right| = \left| \int_x^y f(t)\,dt \right| \leq \int_x^y |f(t)|\,dt$$
$$\leq M|y - x|$$

where we have employed Definition 7.2.5, and Theorems 7.2.4 and 7.2.3; the use
of the absolute value in the last term is present to account for the fact that we are
allowing $y < x$. This estimate shows that

$$\left| \frac{F(y) - F(x)}{y - x} \right| \leq M$$

for all $x, y \in [a, b]$. Therefore F is Lipschitz and hence continuous on $[a, b]$.

We next assume that the function f is continuous at c. To conclude that F is
differentiable at c with $F'(c) = f(c)$, for $\varepsilon > 0$ we must produce $\delta > 0$ such that

$$\left| \frac{F(x) - F(c)}{x - c} - f(c) \right| < \varepsilon$$

whenever $|x - c| < \delta$. Two observations are immediate:

$$\frac{F(x) - F(c)}{x - c} = \frac{1}{(x - c)} \int_c^x f(t)\, dt$$

and

$$f(c) = \frac{1}{(x - c)} \int_c^x f(c)\, dt.$$

It follows now that

$$\left| \frac{F(x) - F(c)}{x - c} - f(c) \right| = \frac{1}{|x - c|} \left| \int_c^x f(t)\, dt - \int_c^x f(c)\, dt \right|$$

$$= \frac{1}{|x - c|} \left| \int_c^x f(t) - f(c)\, dt \right|$$

$$\leq \frac{1}{|x - c|} \int_c^x |f(t) - f(c)|\, dt;$$

as a point of clarity, if it is the case that $x < c$, then the bounds on the last integral
would also reverse to preserve the positivity of the quantity. We now appeal to the
continuity of f. For $\varepsilon > 0$, there is a $\delta > 0$ such that $|f(t) - f(c)| < \varepsilon$ whenever
$|t - c| < \delta$. Thus if we only consider $x \in [a, b]$ such that $|x - c| < \delta$, it follows
that $|t - c| < \delta$ for every $t \in [c, x]$ (or $t \in [x, c]$). This further implies that $|f(t) -
f(c)| < \varepsilon$ for every such t which forces

$$\left| \frac{F(x) - F(c)}{x - c} - f(c) \right| \leq \frac{1}{|x - c|} \int_c^x |f(t) - f(c)|\, dt \leq \frac{\varepsilon |x - c|}{|x - c|} = \varepsilon$$

by Theorem 7.2.3 and where we have incorporated the case $x < c$ by using $|x - c|$.
Therefore we conclude that $F'(c) = f(c)$ as desired.

For the second statement we assume that g is integrable with antiderivative G.
To reach our conclusion we will first show that

$$L(g) \leq G(b) - G(a) \leq U(g).$$

To this end, let $P = \{a = x_0, x_1, x_2, \ldots, x_n = b\}$ be an arbitrary partition of $[a, b]$.
We can then write

$$G(b) - G(a) = \sum_{k=1}^{n} G(x_k) - G(x_{k-1}). \tag{7.3}$$

Applying the Mean Value Theorem to G on each subinterval $[x_{k-1}, x_k]$, which is
permissible since G is differentiable and hence continuous on $[a, b]$, we choose
$t_k \in (x_{k-1}, x_k)$ such that

$$\frac{G(x_k) - G(x_{k-1})}{x_k - x_{k-1}} = G'(t_k) = g(t_k)$$

which is equivalent to

$$G(x_k) - G(x_{k-1}) = g(t_k)(x_k - x_{k-1}). \tag{7.4}$$

Furthermore,

$$m_k \leq g(t_k) \leq M_k$$

for each k which implies

$$L(g, P) \leq \sum_{k=1}^{n} g(t_k)(x_k - x_{k-1}) \leq U(g, P).$$

Applying Eq. (7.4), it follows that

$$L(g, P) \leq \sum_{k=1}^{n} G(x_k) - G(x_{k-1}) \leq U(g, P)$$

which implies

$$L(g, P) \leq G(b) - G(a) \leq U(g, P)$$

by Eq. (7.3). The fact that P was chosen as an arbitrary partition of $[a, b]$ together
with properties of infimum and supremum shows that

$$L(g) \leq G(b) - G(a) \leq U(g).$$

To complete the proof, the fact that g is integrable assures us that $L(g) = U(g)$ and thus the previous string of inequalities must be a string of equalities. Therefore, by the definition of the integral, we have

$$\int_a^b g(x)\, dx = U(g) = L(g) = G(b) - G(a). \qquad \square$$

Example 7.3.2. The following examples concern the first part of the fundamental theorem. First take the function

$$f(x) = \begin{cases} x, & 0 \leq x \leq 1; \\ 1, & 1 < x \leq 2. \end{cases}$$

This function is continuous on $[0, 2]$ and thus the function

$$F(x) = \int_0^x f(t)\, dt$$

should be differentiable throughout $[0, 2]$ with $F'(x) = f(x)$. To compute $F(x)$ we consider two cases depending on the location of x with respect to the domain components of f. First, if $x \in [0, 1]$, we find that

$$F(x) = \int_0^x f(t)\, dt = \int_0^x t\, dt = \frac{x^2}{2}$$

where you can evaluate this integral using either the formula for the area of a triangle or part (b) of the fundamental theorem. Now, if $x \in (1, 2]$, we first split the integral according to the domain of f and then integrate,

$$F(x) = \int_0^x f(t)\, dt = \int_0^1 t\, dt + \int_1^x 1\, dt = \frac{1}{2} + (x - 1) = x - \frac{1}{2}.$$

Putting these two computations together we find an explicit form for F,

$$F(x) = \begin{cases} \frac{x^2}{2}, & 0 \leq x \leq 1; \\ x - \frac{1}{2}, & 1 < x \leq 2. \end{cases}$$

The relationship between f and F is exhibited in Fig. 7.5.

Now we verify that $F'(x) = f(x)$ for $x \in [0, 2]$. If $x \in [0, 1)$ it easy to see that $F'(x) = x = f(x)$. Similarly, if $x \in (1, 2]$ we see that $F'(x) = 1 = f(x)$.

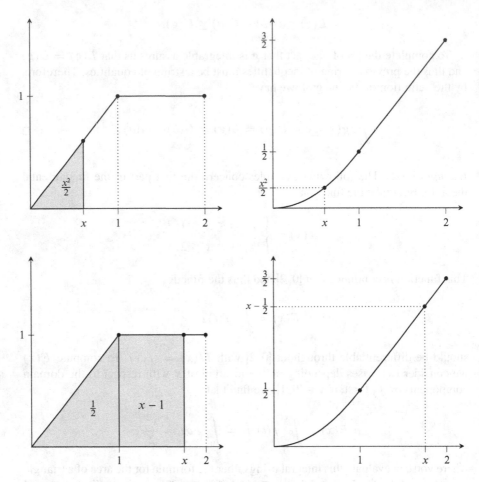

Fig. 7.5 Relationship between f and area function F

If $x = 1$, we must resort to the definition of the derivative and left/right-hand limits to compute $F'(1)$,

$$\lim_{x \to 1+} \frac{F(x) - F(1)}{x - 1} = \lim_{x \to 1+} \frac{(x - \frac{1}{2}) - \frac{1}{2}}{x - 1} = 1$$

and

$$\lim_{x \to 1-} \frac{F(x) - F(1)}{x - 1} = \lim_{x \to 1+} \frac{\frac{x^2}{2} - \frac{1}{2}}{x - 1} = 1.$$

Thus $F'(1) = 1 = f(1)$. Considering the graphs of f and F, notice that although f is continuous on its domain, it is not differentiable at 1. The integrated function

F however is differentiable at one. Though this is a simple example it illustrates an important point, that is, we can interpret the process of integrating a function as a smoothing process.

For a second example we consider what happens when we attempt to apply the fundamental theorem to a function with a discontinuity. The simplest case to consider is a removable discontinuity. Take g to be the function

$$g(x) = \begin{cases} 0, & x \neq 1; \\ 1, & x = 1. \end{cases}$$

With $[0, 2]$ as our domain, we define the function

$$G(x) = \int_0^x g(t)\, dt.$$

By Exercise 7.2.10(a), G is identically 0 on $[0, 2]$. Then it must be the case that $G'(x) = 0$ for all $x \in [0, 2]$, but $G'(1) = 0 \neq 1 = g(1)$. This shows that without the hypothesis of continuity the theorem fails to hold.

For the second part of the fundamental theorem, notice that we assume that the function g is integrable *and* has an antiderivative. From your experience in calculus it is natural to link these two ideas but this is faulty logic. It is possible to construct a function G which is differentiable on an interval $[a, b]$ for which the derivative g is not Riemann integrable; for a reference here we recommend Sect. 7.6 in [1]. It is clear that in this case the equation

$$\int_a^b g(x)\, dx = G(b) - G(a)$$

does not hold since the left-hand side of the previous equation does not exist in this case.

Example 7.3.3 (The Logarithmic Function).   Consider the function $f(x) = 1/x$ on the interval $(0, \infty)$. We can then define a new function

$$\ln x = \int_1^x \frac{1}{t}\, dt,$$

for all $x > 0$, which we call the *(natural) logarithm*; we will typically forgo the adjective as this is the only type of logarithm we will consider. This function is well defined since $1/x$ is continuous (and hence integrable) on every closed, bounded subinterval of $(0, \infty)$ (Fig. 7.6).

To compute the derivative of $\ln x$, notice that for $x \geq 1$, the fundamental theorem immediately implies that $\frac{d}{dx}[\ln x] = \frac{1}{x}$. If $0 < x < 1$, then we cannot apply the theorem directly as stated since $x < 1$ in this case. However, if we choose a point $a \in \mathbb{R}$ such that $0 < a < x < 1$, then we can write

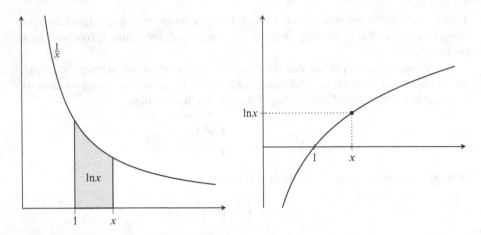

Fig. 7.6 Area function for $\frac{1}{x}$ and the logarithmic function

$$\ln x = \int_1^x \frac{1}{t}\, dt = \int_1^a \frac{1}{t}\, dt + \int_a^x \frac{1}{t}\, dt$$

by Exercise 7.2.7. With this, the fundamental theorem implies that

$$\frac{d}{dx}[\ln x] = \frac{d}{dx}\left[\int_1^a \frac{1}{t}\, dt + \int_a^x \frac{1}{t}\, dt \right] = \frac{1}{x}$$

since the first term in the sum above is a constant. Thus

$$\frac{d}{dx}[\ln x] = \frac{1}{x}$$

for all $x > 0$.

Several other immediate facts follow quickly. First, $\ln(1) = 0$ by Definition 7.2.5. Second, the fact that $1/x > 0$ on $(0, \infty)$ implies that $\ln x$ is strictly increasing. To see that the familiar logarithm property

$$\ln(ab) = \ln a + \ln b$$

holds for every $a, b > 0$, consider two functions

$$f(x) = \ln(ax) \quad \text{and} \quad g(x) = \ln a + \ln x,$$

where a is some fixed positive number and $x \in (0, \infty)$. The desired relationship will hold if we can show that $f(x) = g(x)$ for every $a, x > 0$. Notice first that f and g are differentiable with

$$f'(x) = \frac{a}{ax} = \frac{1}{x} \quad \text{and} \quad g'(x) = \frac{1}{x},$$

where we have used the chain rule for the derivative of f and the fact that a and $\ln a$ are constants. This equality of derivatives indicates that there is a real number C such that $f(x) = g(x) + C$ for all $x > 0$. Evaluating f and g at $x = 1$ then shows that

$$\ln a = f(1) = g(1) + C = \ln a + \ln 1 + C = \ln a + C.$$

This equation indicates that $C = 0$ indicating $f(x) = g(x)$ for all $x > 0$, and thus the desired relationship holds for all $a, b > 0$.

The rules

$$\ln(a/b) = \ln a - \ln b \quad \text{and} \quad \ln(a^r) = r \ln a,$$

where $a, b > 0$ and $r \in \mathbb{R}$, will be explored in the exercises and Sect. 7.4.

Example 7.3.4. We now return to the function $f(x) = -\ln(1 - x)$ from Example 6.3.11. We know that the Taylor series for this function is

$$\sum_{n=1}^{\infty} \frac{x^n}{n},$$

with interval of convergence $[-1, 1)$, and we know that it converges to f on $[-1, 1/2]$. Here we will provide a general argument to show that the Taylor series agrees with f on $[-1, 1)$. First, we define a new function

$$g(x) = \int_0^x \sum_{n=0}^{\infty} t^n \, dt$$

on $(-1, 1)$. If we fix $x \in (-1, 1)$, then there is a $b \in (0, 1)$ such that $|x| < b < 1$. Then we know the power series used to define g is continuous, and hence integrable, on $[-b, b] \subseteq (-1, 1)$ by Theorem 6.3.6. Thus

$$g'(x) = \sum_{n=0}^{\infty} x^n$$

by the Fundamental Theorem. Recognizing this as the power series form for the geometric series, we have

$$g'(x) = \frac{1}{1 - x}$$

for all $x \in (-1, 1)$. Notice also that

$$f'(x) = \frac{1}{1-x}$$

for all $x \in (-1, 1)$ and thus there is a constant C such that

$$f(x) = g(x) + C.$$

Evaluating both of these functions at $x = 0$ shows that $C = 0$. Therefore

$$-\ln(1-x) = \int_0^x \sum_{n=0}^{\infty} t^n \, dt$$

for all $x \in (-1, 1)$. Finally, we can evaluate this integral by Corollary 7.2.8 and we have

$$-\ln(1-x) = \int_0^x \sum_{n=0}^{\infty} t^n \, dt = \sum_{n=0}^{\infty} \int_0^x t^n \, dt$$

$$= \sum_{n=0}^{\infty} \frac{x^{n+1}}{n+1}$$

$$= \sum_{n=1}^{\infty} \frac{x^n}{n}.$$

Thus the Taylor series agrees with f for a $x \in (-1, 1)$. With the previous proof that the Taylor series also converges to f at $x = -1$, we conclude that

$$-\ln(1-x) = \sum_{n=1}^{\infty} \frac{x^n}{n}$$

for all $x \in [-1, 1)$.

Exercises

Exercise 7.3.1. Show that every continuous function g is the derivative of some function G.

Exercise 7.3.2. Use part (a) of the Fundamental Theorem of Calculus to supply a proof for part (b) with the additional assumption that g is continuous on $[a, b]$.

Exercise 7.3.3. Suppose $f : [a, b] \to \mathbb{R}$ is integrable and nonnegative on $[a, b]$. Show that the function $F : [a, b] \to \mathbb{R}$ defined by

$$F(x) = \int_a^x f(t) \, dt$$

is increasing on $[a, b]$.

Exercise 7.3.4. Find all functions $f : [0, 1] \to \mathbb{R}$ such that f is continuous and satisfies

$$\int_0^c f(t) \, dt = \int_c^1 f(t) \, dt$$

for all $c \in [0, 1]$.

Exercise 7.3.5 (Integration by Parts). Suppose $f, g : [a, b] \to \mathbb{R}$ are functions with continuous derivatives. Explain why the functions $f'g$ and fg' are integrable, and show that

$$\int_a^b f(x)g'(x) \, dx = f(b)g(b) - f(a)g(a) - \int_a^b g(x)f'(x) \, dx.$$

Exercise 7.3.6. (a) Show that the logarithmic function defined in Example 7.3.3 is strictly increasing on $(0, \infty)$ directly, that is, without using the fact that its derivative is strictly positive.
(b) Show that $\ln(a/b) = \ln a - \ln b$ for every $a, b > 0$.

Exercise 7.3.7 (Average Value). Suppose $g : [a, b] \to \mathbb{R}$ is continuous.

(a) Show that there is a point $t \in (a, b)$ such that

$$g(t) = \frac{1}{b - a} \int_a^b g \, dx.$$

We call $g(t)$ the *average value* of g on $[a, b]$. Draw a sketch representing this situation.
(b) Compute the average value for $g(x) = \sin x$ on the interval $[0, 2\pi]$. You may use an appropriate antiderivative to compute this integral.
(c) Suppose that g is also a linear function and let $[c, d] \subseteq [a, b]$. Show that the average value of g on the interval $[c, d]$ is assumed at the midpoint of $[c, d]$, that is

$$g\left(\frac{c + d}{2}\right) = \frac{1}{d - c} \int_c^d g \, dx. \tag{7.5}$$

(d) Follow the steps below to show that the converse to part (c) holds: suppose g is twice continuously differentiable, i.e. g'' exists and is continuous on $[a, b]$. If Eq. (7.5) holds for every interval $[c, d] \subseteq [a, b]$, show that g is a linear function.

(i) Form the new function

$$G(x) = \frac{1}{x - c} \int_c^x g(t) \, dt.$$

By our hypothesis,

$$G(x) = g\left(\frac{x + c}{2}\right)$$

for all $x \in (c, b]$. Compute the second derivative of each term in the equation above to obtain relationship between g' and g''.

(ii) Use the Mean Value Theorem to show that there is a point $t \in (\frac{x+c}{2}, x)$ such that $g''\left(\frac{x+c}{2}\right) = 2g''(t)$.

(iii) Apply a continuity argument to show that $g''(c) = 0$.

(iv) Argue that the above statement holds for all $c \in [a, b]$ and use this to conclude that g is linear.

(e) Let $f : [a, b] \to \mathbb{R}$ be nonnegative and integrable. Still assuming that g is continuous, show that there is a point $t \in [a, b]$ such that

$$\int_a^b g(x) f(x) \, dx = g(t) \int_a^b f(x) \, dx.$$

You may assume that the product gf is integrable.

Exercise 7.3.8. Given a function f on $[a, b]$, define the *total variation* of f to be

$$V(f) = \sup \left\{ \sum_{k=1}^n |f(x_k) - f(x_{k-1})| \right\},$$

where the supremum is taken over all possible partitions of $[a, b]$.

(a) Consider the function $f(x) = x$ on the interval $[0, 1]$. Calculate $V(f)$ using the definition above.

(b) Assume f is a continuously differentiable function on $[a, b]$, i.e. f' exists and is continuous on $[a, b]$. Use the steps below to show that

$$V(f) = \int_a^b |f'(t)| \, dt.$$

(i) Use the fundamental theorem to show that $V(f) \leq \int_a^b |f'(t)| \, dt$.
(ii) Use the Mean Value Theorem and the idea of a lower sum to obtain the reverse inequality.

7.4 The Exponential Function

In this section we provide a rigorous treatment of the exponential and logarithmic functions even though we have used these functions from time to time earlier in the text. Though we could have taken all of this information as given it is never the less worthwhile to understand how the familiar properties of these functions arise from the definitions. For the definition of the logarithmic function we refer to Example 7.3.3.

Definition 7.4.1. For each $x \in \mathbb{R}$, define the *exponential function* $\exp : \mathbb{R} \to \mathbb{R}$ by

$$\exp(x) = \sum_{n=0}^{\infty} \frac{x^n}{n!},$$

which is well defined by Example 6.2.2.

The fact that this new function is continuous and differentiable follows from our work with power series.

Proposition 7.4.2. *The function* $\exp(x)$ *is continuous and differentiable on* $\mathbb{R}$. *Moreover,*

$$\frac{d}{dx}[\exp(x)] = \exp(x)$$

for all $x \in \mathbb{R}$.

Proof. By Example 6.2.2, the function converges on all of $\mathbb{R}$. Furthermore, by Theorems 6.3.3 and 6.3.6 we know that the power series defining the function is continuous on $\mathbb{R}$. Theorem 6.3.7 then shows that the function is differentiable on $\mathbb{R}$, and, if $x \in \mathbb{R}$, we have

$$\frac{d}{dx}[\exp(x)] = \sum_{n=1}^{\infty} \frac{n x^{n-1}}{n!} = \sum_{n=0}^{\infty} \frac{x^n}{n!} = \exp(x)$$

as desired. ∎

The statement above justifies our use of the familiar derivative rule previously in the text. We now focus on developing the other important properties of the exponential function and its relationship with the logarithmic function.

Proposition 7.4.3. *The function* $\exp(x)$ *satisfies*

(a) $\exp(0) = 1$;
(b) $\exp(x + y) = \exp(x) \cdot \exp(y)$ *for all* $x, y \in \mathbb{R}$;
(c) $\exp(-x) = 1/\exp(x)$ *for all* $x \in \mathbb{R}$;
(d) $\exp(x) \geq 1$ *for all* $x \geq 0$ *and* $0 < \exp(y) < 1$ *for all* $y < 0$;
(e) $\exp(x) < \exp(y)$ *whenever* $x < y$ *in* $\mathbb{R}$.

Proof. The first statement is obvious from the definition of the exponential function.

For (b), we apply the formula for multiplying series stated at the end of Sect. 3.1. Using this in conjunction with the Binomial Expansion Theorem from Exercise 1.2.9 shows that

$$
\exp(x) \cdot \exp(y) = \left(\sum_{n=0}^{\infty} \frac{x^n}{n!} \right) \left(\sum_{n=0}^{\infty} \frac{y^n}{n!} \right) = \sum_{n=0}^{\infty} \sum_{j=0}^{n} \frac{x^j}{j!} \frac{y^{n-j}}{(n-j)!}
$$

$$
= \sum_{n=0}^{\infty} \frac{1}{n!} \left(\sum_{j=0}^{n} \frac{n!}{j!(n-j)!} x^j y^{n-j} \right)
$$

$$
= \sum_{n=0}^{\infty} \frac{(x+y)^n}{n!}
$$

$$
= \exp(x + y).
$$

Though we have not formally discussed rearrangements of series, the absolute convergence of the three series above justifies the manipulation in the second equality.

Part (c) is requested in Exercise 7.4.1.

Moving on to part (d), notice first that $\exp(x) \geq 1$ for all $x \geq 0$ since all the terms in the power series are positive in this case and the first term is equal to 1. Now, if $y < 0$, then $y = -x$ for some $x > 0$. By (c), $\exp(y) = \exp(-x) = 1/\exp(x)$ and thus $0 < \exp(y) < 1$ since $\exp(x) > 1$ for all $x > 0$.

Part (e) is left as Exercise 7.4.1. $\square$

Next we identify the exponential function in terms of powers of e, but to do this we must first understand the powers of e. The natural number powers are defined in the standard algebraic fashion and we have also shown that the root functions are well defined on their natural domain; see Exercises 4.3.4 and 5.4.7. The intent here is to understand rational and irrational powers of e and while this may seem familiar, keep in mind that this is a definition.

Definition 7.4.4. If $n \in \mathbb{N}$ we define e^n to be the n-fold product

$$
e^n = \underbrace{e \cdot e \cdots \cdot e}_{n}, \quad \text{and} \quad e^{1/n} = \sqrt[n]{e}
$$

where $\sqrt[n]{e}$ denotes the unique positive nth root of e. For a positive rational number p/q, we define

$$e^{p/q} = \left(e^{1/q}\right)^p = \left(\sqrt[q]{e}\right)^p.$$

Furthermore, for a positive irrational t, we define

$$e^t = \sup\{e^r : r \in \mathbb{Q} \text{ with } r < t\}.$$

Notice that this last expression exists by the Axiom of Completeness. Finally, for a negative real number y, we define

$$e^y = \frac{1}{e^{|y|}}.$$

Corollary 7.4.5. *For every $x \in \mathbb{R}$ we have $\exp(x) = e^x$.*

In the proof we will deal exclusively with powers of e, except in the one instance where we use the exponential rule $(x^p)^q = (x^q)^p$ for $p, q \in \mathbb{N}$; this fact follows from the definition of a natural number power of a real number.

Proof. If $x = 1$, then $\exp(1) = e$ by Definition 3.2.8. If $n \in \mathbb{N}$, repeated applications of the previous proposition and the definition of e^n show that

$$\exp(n) = \exp(\underbrace{1 + 1 + \cdots + 1}_{n}) = \underbrace{\exp(1) \cdot \exp(1) \cdots \cdots \exp(1)}_{n} = \underbrace{e \cdot e \cdots \cdot e}_{n} = e^n.$$

Also, in the proof of the previous proposition we saw that $\exp(-x) = 1/\exp(x)$. This shows that

$$e^{-n} = \frac{1}{e^n} = \frac{1}{\exp(n)} = \exp(-n).$$

Next, notice that

$$\exp(1/n)^n = \underbrace{\exp(1/n) \cdot \exp(1/n) \cdots \cdots \exp(1/n)}_{n} = \exp(1) = e$$

and hence

$$\exp(1/n) = e^{1/n}.$$

For a general rational power, first observe that for $p, q \in \mathbb{N}$,

$$\left(\left(e^p\right)^{1/q}\right)^q = e^p = \left(\left(e^{1/q}\right)^q\right)^p = \left(\left(e^{1/q}\right)^p\right)^q;$$

it follows that

$$\left(e^p\right)^{1/q} = \left(e^{1/q}\right)^p$$

since both terms here are positive and the power function x^q is strictly increasing and hence one-to-one on $(0, \infty)$. Therefore, if p/q is a positive rational number, we have

$$\exp(p/q)^q = \underbrace{\exp(p/q) \cdot \exp(p/q) \cdots \exp(p/q)}_{q} = \exp(p) = e^p$$

and hence

$$\exp(p/q) = \left(e^p\right)^{1/q} = \left(e^{1/q}\right)^p = e^{p/q}.$$

For a negative rational number, we have

$$\exp(r) = \frac{1}{\exp(|r|)} = \frac{1}{e^{|r|}} = e^r.$$

For the irrational case, first consider a positive irrational number t and let (r_k) be a sequence of rational numbers which converges to t from below, i.e. an increasing sequence. The monotonicity of the exponential function assures us that $(\exp(r_k))$ is also an increasing sequence and $\exp(r_k) = e^{r_k}$ for every $k \in \mathbb{N}$; it follows that $\exp(r_k) \in \{e^r : r \in \mathbb{Q} \text{ with } r < t\}$ for every $k \in \mathbb{N}$. The continuity of the exponential function with the Order Limit Theorem also shows that

$$\lim_{k \to \infty} \exp(r_k) = \exp(t) \le e^t.$$

If we also take a decreasing sequence of rational numbers (s_k) converging to t, we can apply the same logic to show that

$$\lim_{k \to \infty} \exp(s_k) = \exp(t) \ge e^t$$

since in this case each value e^{s_k} is an upper bound for the set $\{e^r : r \in \mathbb{Q} \text{ with } r < t\}$ by the monotonicity of the exponential function. Thus we conclude that $\exp(t) = e^t$. Finally, if t is a negative irrational number, then

$$\exp(t) = \frac{1}{\exp(|t|)} = \frac{1}{e^{|t|}} = e^t$$

completing the proof. $\square$

With this corollary, we can forgo the more formal $\exp(x)$ notation and use symbol e^x exclusively. Also, it would be easy to extend the previous definition and replace e

with any positive real number. But we don't need to do this, as we will see once we have established the relationship between the exponential and logarithmic functions. In other words, we will use this to supply a definition for exponentiating other positive bases. For nonpositive bases, the situation is somewhat more delicate and we will also discuss this briefly.

Moving now toward a discussion of the logarithm, we make two limit observations about the behavior of the exponential function. First notice that $\lim_{n \to} e^n = \infty$ since $e > 1$; see Example 2.3.5. This immediately implies that

$$\lim_{n \to \infty} e^{-n} = \lim_{n \to \infty} \frac{1}{e^n} = 0.$$

We will use this now to show that the exponential function maps onto $(0, \infty)$.

Proposition 7.4.6. *The function e^x is one-to-one from $\mathbb{R}$ onto $(0, \infty)$. Moreover, the inverse of the exponential function is the natural logarithm defined in Example 7.3.3.*

Proof. The fact that the function is one-to-one is immediate since the function is strictly increasing. It is also clear that the function maps into $(0, \infty)$ by Theorem 7.4.3(d). To show that it maps onto $(0, \infty)$, let $y \in (0, \infty)$. The limit observations above indicate that there exist $N_1, N_2 \in \mathbb{N}$ with

$$e^{-N_1} \leq y \leq e^{N_2}.$$

The Intermediate Value Theorem then implies that there is an $x \in [-N_1, N_2]$ such that $e^x = y$. With these two statements, we conclude e^x is invertible and hence there is a function $g : (0, \infty) \to \mathbb{R}$ with

$$g(e^x) = x \quad \text{and} \quad e^{g(y)} = y$$

for every $x \in \mathbb{R}$ and $y \in (0, \infty)$, respectively. Also, by the Inverse Function Theorem we have that

$$g'(y) = \frac{1}{e^{g(y)}} = \frac{1}{y}$$

where the derivative is taken with respect to y. Recall that the natural logarithm defined from $(0, \infty)$ to $\mathbb{R}$ also satisfies

$$\frac{d}{dy}[\ln y] = \frac{1}{y}$$

and thus by Theorem 5.4.1 there is a constant C such that

$$\ln y - g(y) = C$$

for all $y \in (0, \infty)$. To find C, we evaluate the function $\ln y - g(y)$ at $y = 1$. Keep in mind that $\ln(1) = 0$ by definition and $g(1) = 0$ since $e^0 = 1$. Thus we find that

$$C = \ln(1) - g(1) = 0$$

and therefore we conclude that $g(y) = \ln y$ as desired.   □

We conclude this section with a definition for powers of positive real numbers. With this definition we can prove all of the common exponentiation rules and these are requested as Exercise 7.4.4.

Definition 7.4.7. For $x > 0$ and $a \in \mathbb{R}$, we define $x^a = \exp(a \ln x)$.

For $x = 0$ and $a > 0$ the fact that $x^n = 0$ and $x^{1/n} = 0$ for all $n \in \mathbb{N}$ imply that we should set $x^a = 0$. If $x < 0$ we know that $x^{1/n}$ is not defined (or is a complex number) for even n though it is defined for odd n and is an odd function, i.e $(-a)^{1/n} = -a^{1/n}$. With these considerations, for a rational number p/q where p and q are in lowest terms and q is odd, we define

$$x^{p/q} = \begin{cases} |x|^{p/q} & p \text{ even;} \\ -|x|^{p/q} & p \text{ odd.} \end{cases}$$

For irrational powers, x^r is necessarily a complex number and hence we will take it as undefined.

As a final remark, we make good on our promise from Example 5.4.8 and prove the general power rule $(x^r)' = r x^{r-1}$. Applying the Chain Rule, we have

$$\frac{d}{dx}[x^r] = \frac{d}{dx}[\exp(r \ln x)] = \exp(r \ln x)\left(\frac{r}{x}\right) = x^r \left(\frac{r}{x}\right) = r x^{r-1}$$

for $x > 0$. For the case when $x < 0$ and all r for which the expression x^r is defined, we first write $x^r = (-|x|)^r = (-1)^r |x|^r$. Using this substitution we see that,

$$\frac{d}{dx}[x^r] = (-1)^r r |x|^{r-1}(-1) = r(-1)^{r+1}|x|^{r-1} = r(-1)^{r-1}|x|^{r-1} = r x^{r-1}$$

where the -1 in the second term comes from the derivative of $|x|$ for $x < 0$ and we have also used the fact that $(-1)^{r+1} = (-1)^{r-1}$. For $x = 0$, the definition of derivative shows that the derivative of x^r is 0 for $r > 1$.

The relationship with the logarithmic function together with the chain rule makes quick work of this formula but there are proofs which do not rely on logarithms, for instance only using sequences of functions and Theorem 6.1.12. The approach in this case is to first show the formula for natural number powers, and then to extend this to rational powers. An application of the mentioned theorem then allows one to complete the proof for irrational powers, though this case requires a complicated convergence argument and requires one to work only on bounded intervals; as a reference we point the reader to Example 8–8 in [14].

Exercises

Exercise 7.4.1. 1. Prove part (c) of Proposition 7.4.3.
2. Use the Mean Value Theorem directly to prove part (e) of Proposition 7.4.3.

Exercise 7.4.2. Show that $f(x) = e^x$ is the unique function which satisfies $f' = f$ with $f(0) = 1$. *Hint:* Suppose there is another function g which has these properties and consider the function $h(x) = g(x)/e^x$.

Exercise 7.4.3. Use the relationship between the exponential function and the logarithmic function to show that $\ln(xy) = \ln x + \ln y$ and $\ln(x^r) = r \ln x$ for all $x, y \in (0, \infty)$ and $r \in \mathbb{R}$.

Exercise 7.4.4. Let $x \in (0, \infty)$ and let $a, b \in \mathbb{R}$. Show each of the following:

(a) $x^0 = 1$;
(b) $x^{a+b} = x^a \cdot x^b$;
(c) $x^{-a} = 1/x^a$;
(d) $x^{a-b} = x^a/x^b$;
(e) $x^a \le x^b$ if $0 < a < b$ and $x > 1$;
(f) $x^a \ge x^b$ if $0 < a < b$ and $0 < x < 1$.

Exercise 7.4.5. Repeat parts (b)–(d) for $x \in \mathbb{R}$ and a, b appropriate rational numbers for which the expressions x^a and x^b exist.

Exercise 7.4.6. Show that $(-1)^{r+1} = (-1)^{r-1}$ provided the two values exist in $\mathbb{R}$.

Exercise 7.4.7. Compute the derivative of the function $f : \mathbb{R} \to \mathbb{R}$ given by $f(x) = a^x$ where $a > 0$.

7.5 Spaces of Continuous Functions Revisited

We return now to the collection of functions $C([a, b])$, but here we turn our attention to norms which are defined by the Riemann integral. Though we restrict our focus to two particular norms for the space, there are many others that we could also consider.

Example 7.5.1. For $f \in C([a, b])$, define $\|f\|_1$ by

$$\|f\|_1 = \int_a^b |f(x)| \, dx.$$

We already know that $C([a, b])$ is a vector space and Exercise 7.5.1 asks you to show that the rule above defines a norm.

Example 7.5.2. For $f, g \in C([a, b])$, define $\langle f, g \rangle_2$ by

$$\langle f, g \rangle_2 = \int_a^b f(x) g(x) \, dx.$$

Inner product properties (a), (c), and (d) follow immediately. For property (b), the non-negativity follows from the order properties of the Riemann integral. For the zero condition, one direction is trivial and the second is a consequence of Exercise 7.2.2. Proposition 1.5.15 provides a norm and for $f \in C([a, b])$, we have

$$\|f\|_2 = \sqrt{\langle f, f \rangle_2} = \left(\int_a^b |f(x)|^2 \, dx \right)^{1/2}.$$

This norm looks similar to the norm of the previous example and as vector spaces they contain the same elements. Be aware, however, that they do have different norm structures as Exercise 7.5.2 will demonstrate. Comparatively, keep in mind that while ℓ^1 and ℓ^2 appear similar in the same manner that these spaces seem similar, they are distinct spaces. Specifically, there are vectors in ℓ^2 which are not in ℓ^1.

Our first goal in this section is to show that these spaces are not complete. This may seem like a significant loss, but in actuality it opens the door to a wider world of spaces and also strengthens the impetus for expanding our notion of integration.

Proposition 7.5.3. *The space* $(C([a, b]), \| \cdot \|_1)$ *is not complete.*

Proof. For the proof we will work on $[0, 1]$ and exhibit a sequence of functions which is Cauchy with respect to $\| \cdot \|_1$ but which has no limit in the space. For our sequence, we define f_n for $n \geq 2$ in Fig. 7.7.

For each $n \geq 2$, it's obvious that the function f_n is continuous on $[0, 1]$ and

$$\|f_n\|_1 = \int_0^1 |f_n(x)| \, dx = \frac{1}{2} + \frac{1}{2n}.$$

To see that the sequence is Cauchy, let $m, n \in \mathbb{N}$ with $n > m \geq 2$. Using the linearity of the integral, we have

$$\|f_n - f_m\|_1 = \int_0^1 |f_n(x) - f_m(x)| \, dx = \int_0^1 f_m(x) - f_n(x) \, dx = \frac{1}{2m} - \frac{1}{2n}.$$

Using the fact that the sequence $(1/2n)$ is Cauchy in $\mathbb{R}$, for $\varepsilon > 0$, choose $N \in \mathbb{N}$ such that $|1/2m - 1/2n| < \varepsilon$ whenever $m, n \geq N$. It immediately follows that $\|f_n - f_m\|_1 < \varepsilon$ for $m, n \geq N$ and hence (f_n) is Cauchy in $(C([0, 1]), \| \cdot \|_1)$.

$$f_n(x) = \begin{cases} 0, & 0 \le x \le \frac{1}{2} - \frac{1}{n}; \\ nx + \frac{2-n}{2}, & \frac{1}{2} - \frac{1}{n} < x \le \frac{1}{2}; \\ 1, & \frac{1}{2} < x \le 1. \end{cases}$$

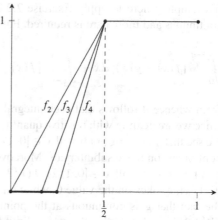

Fig. 7.7 Cauchy sequence in $C([0, 1])$

In order to conclude that this sequence has no limit in the space, we first show that the sequence does converge in the norm $\|\cdot\|_1$ to its pointwise limit

$$f(x) = \begin{cases} 0, & 0 \le x < \frac{1}{2}; \\ 1, & \frac{1}{2} \le x \le 1. \end{cases}$$

A calculation shows that

$$\|f_n - f\|_1 = \int_0^1 |f_n(x) - f(x)|\, dx = \int_{\frac{1}{2} - \frac{1}{n}}^{\frac{1}{2}} |f_n(x)|\, dx = \frac{1}{2n}$$

and thus $\|f_n - f\|_1 \to 0$ as $n \to \infty$. Here we must be careful. The sequence (f_n) is Cauchy and does converge in norm to f, but f is not in our space. This is not enough to conclude that the space under consideration is not complete as there may be some other function in the space which will serve as the limit. To show that this cannot happen we proceed by contradiction. Suppose g is in $C([0, 1])$ and that $\|f_n - g\|_1 \to 0$ as $n \to \infty$. Let $\varepsilon > 0$ and choose $N \in \mathbb{N}$ such that $\|f_n - f\|_1 < \varepsilon/2$ and $\|f_n - g\|_1 < \varepsilon/2$ for all $n \ge N$. In particular, this shows that

$$\int_0^1 |f(x) - g(x)|\, dx \le \int_0^1 |f_N(x) - f(x)|\, dx + \int_0^1 |f_N(x) - g(x)|\, dx < \varepsilon$$

and thus it must be the case that

$$\int_0^1 |f(x) - g(x)|\, dx = 0.$$

It's tempting here to apply Exercise 7.2.2 to conclude that $f = g$, but f is not continuous and more care is required. For $t \in (0, 1/2)$ we can write

$$\int_0^{\frac{1}{2}-t} |f(x) - g(x)|\, dx + \int_{\frac{1}{2}-t}^{\frac{1}{2}+t} |f(x) - g(x)|\, dx + \int_{\frac{1}{2}+t}^1 |f(x) - g(x)|\, dx = 0$$

from whence it follows that each integral on the left-hand side of this equation is 0 since we are dealing with positive quantities. Focusing on the first and third integral, we see that $f(x) = g(x)$ for all $x \in [0, 1/2-t] \cup (1/2+t, 1]$ since both f and g are continuous on these subintervals. Moreover, since t is arbitrary we are assured that $f(x) = g(x)$ for all $x \in [0, 1/2) \cup (1/2, 1]$. The contradiction is apparent now. On $[0, 1/2)$, g takes on the value 0, while on $(1/2, 1]$ it takes the value 1 contradicting the fact that g is continuous at the point $x = 1/2$. Thus there is no function in $C([0, 1])$ which can play the role of limit for this particular Cauchy sequence.   □

At this point it is insightful to think about why this is happening. It might seem natural to place the blame on the integral-norm since the space is complete with respect to the sup-norm of the previous section. For a moment though let us focus on the particular collection of functions under consideration. Is it possible to grow this set in order to "complete" the space? With the above proof in mind, we might take the largest possible set that we can and consider the collection of all functions which are Riemann integrable on the interval $[a, b]$. The space is clearly a vector space thanks to the linear properties of the integral. Also, the function f from the previous proof is Riemann integrable on $[0, 1]$. This seems to alleviate a significant portion of the problem, but only for this particular sequence of functions. However, we are now faced with the fact that $\| \cdot \|_1$ is not a norm on this vector space since it does not satisfy norm axiom (a) bringing us back to the integral. This problem will persist on any space larger than $C([a, b])$. Plainly stated, the Riemann integral produces the same output if we change a function on a "small" set. Assuming we could alleviate this issue, it is still the fact that there are sequences of Riemann integrable functions which are Cauchy with respect to $\| \cdot \|_1$ but which do not converge to any Riemann integrable function. For an example on $[0, 1]$, if $\{r_1, r_2, r_3, \ldots\}$ is an enumeration of $\mathbb{Q} \cap [0, 1]$, define a sequence by

$$f_n(x) = \begin{cases} 1, & x \in \{r_1, r_2, \ldots, r_n\}; \\ 0, & else. \end{cases}$$

The pointwise limit of this sequence of Riemann integrable functions is the Dirichlet function which is not Riemann integrable while each of the sequence functions integrates to zero.

The bigger picture seemingly reveals that the problem is twofold, the collection of functions we're considering and the manner in which we've defined a norm. But both of these are actually statements about the Riemann integral. First, not enough functions are Riemann integrable. Second, when dealing with discontinuous functions, the Riemann integral cannot distinguish between functions which disagree on "small" sets. Neither of these is insurmountable; a stronger form of integration is required for the first issue while a weakening of our notion of function equality will alleviate the second.

To explore the relationship between the integral and the limit process more thoroughly, suppose (f_n) is a sequence of functions defined on $[a, b]$ and let f be the pointwise limit of this sequence. As with sequences of continuous or differentiable functions, we should then consider when the equation

$$\lim_{n \to \infty} \int_a^b f_n \, dx = \int_a^b f \, dx \qquad (7.6)$$

holds true; in other words, how does the integral hold up with respect to the limit process? Recall that the analogous statement for continuous functions is true when the convergence is uniform, and is true for differentiable functions when the sequence of derivatives converges uniformly with the additional hypothesis that the sequence of functions converges to the limit function for at least one point in the common domain. On the other hand, while it is true that the equation above holds true when (f_n) converges uniformly to f (Theorem 7.2.6), the fact that integrability is a weaker condition than continuity and differentiability provides sufficient cause for us to hope that some weaker condition, e.g. pointwise convergence, is enough to guarantee Eq. (7.6). What we actually find is that pointwise convergence guarantees nothing of the sort. This should not be a discouragement, rather, it should be taken as a motivator! The limit process is the foundation of analysis and the fact that the Riemann integral does not behave well with respect to this process has pushed mathematicians to seek out new and sturdier forms of integration.

The remainder of this section demonstrates that with only the hypothesis of pointwise convergence, Eq. (7.6) can fail to hold in three distinct (and exhaustive) cases; this was explored in Exercises 7.2.11 and 7.2.12. First, the left-hand side of the equation may fail to exist. For an example of this consider a sequence of functions (f_n) on $[0, 1]$ which converges pointwise to 0 with the property that $\int_0^1 f_n(x) \, dx$ is 1 when n is odd and 2 when n is even. We leave it as an exercise to come up with an explicit representation for such a sequence. It should be evident though that

$$\lim_{n \to \infty} \int_0^1 f_n(x) \, dx$$

does not exist for such a sequence.

Second, the right-hand side may not exist. This is really saying that the pointwise limit is not Riemann integrable and we encountered such a sequence earlier in this discussion. The third case occurs when both sides of the equation exist but do not share the same value. To provide an example showing that this is possible take the sequence of functions

$$f_n(x) = \begin{cases} 0, \ x = 0; \\ n, \ 0 < x < \frac{1}{n}; \\ 0, \ \frac{1}{n} \le x \le 1. \end{cases}$$

The pointwise limit here is the function $f(x) = 0$ whose integral is 0, but each function in our sequence has integral equal to 1.

In the hopes of alleviating some of these issues, we now move into a study of the Lebesgue integral. With this construction our hope is to find some middle ground concerning limits not present in the case of the Riemann integral. Specifically we seek a hypothesis which is stronger than pointwise convergence, but weaker than uniform convergence that we can place on a sequence of functions and/or the limit function in order to force Eq. (7.6). Section 9.3 will show that this is possible for the Lebesgue integral. However, we also mention two results there for the Riemann integral. These statements are similar to the convergence theorems for the Lebesgue integral, but require the additional hypothesis that the limit function is Riemann integrable.

Exercises

Exercise 7.5.1. Show that $(C([a, b]), \| \cdot \|_1)$ is a normed linear space.

Exercise 7.5.2. (a) Let $f(x) = 1$ and $g(x) = x$. Show that f and g satisfy the parallelogram equality in $(C([0, 1]), \| \cdot \|_1)$.
(b) Use the functions $f(x) = x$ and $g(x) = 1 - x$ to show that $\| \cdot \|_1$ cannot be realized as an inner product on $C([0, 1])$ in the sense of Proposition 1.5.15.
(c) Generalize part (b) to $C([a, b])$ with $\| \cdot \|_1$.

Exercise 7.5.3. (a) Why does the sequence of functions defined in Fig. 6.5 not suffice to show that $(C([0, 1]), \| \cdot \|_1)$ is not complete?
(b) Modify the proof of Proposition 7.5.3 to show that $(C([a, b]), \| \cdot \|_1)$ is not complete.

Exercise 7.5.4. Show that $(C([0, 1]), \| \cdot \|_2)$ is not complete.

Exercise 7.5.5. Let P be the space of all polynomials. Show that the rule

$$\langle p, q \rangle = \int_0^1 p(x)q(x) \cos x \, dx$$

defines an inner product on P.

Exercise 7.5.6. Let P be the space of all polynomials with the norm $\| \cdot \|_2$ defined in Example 7.5.2.

(a) Show that the functional $T : P \to \mathbb{R}$ defined by

$$Tp = \int_0^1 p(x) \cos x \, dx$$

defines a linear functional on P.
(b) Can you conclude that the functional in (b) is bounded? It may help to consider the larger space $(C([0, 1]), \| \cdot \|_2)$ and the Cauchy–Schwarz inequality.

Exercise 7.5.7. Consider the space $(C([0, 1]), \| \cdot \|_1)$ and let $g \in C([0, 1])$. Define the multiplication operator $M_g : C([0, 1]) \to C([0, 1])$.

(a) Show that M_g is a bounded operator.
(b) For $c \in [0, 1]$, show that the set $K_c = \{f : f(x) = 0 \text{ on } [0, c]\}$ is an invariant subspace for M_g.

Exercise 7.5.8. Consider the space $(C([0, 1]), \| \cdot \|_2)$ and let $g \in C([0, 1])$. Show that the multiplication operator $M_g : C([0, 1]) \to C([0, 1])$ is a bounded operator.

Exercise 7.5.9 (Volterra Operator). Consider the space $(C([0, 1]), \| \cdot \|_\infty)$ and define the *Volterra operator* $V : C([0, 1]) \to C([0, 1])$ by

$$Vf(x) = \int_0^x f(t) \, dt.$$

This operator maps a continuous function to the antiderivative F which satisfies $F(0) = 0$.

(a) Show that V is linear and maps into $C([0, 1])$.
(b) Show that V is a bounded operator.

Chapter 8
Lebesgue Measure on $\mathbb{R}$

The goal of the next two chapters is to extend the notion of integration. Though the Riemann integral is the first integral that a typical student encounters and its construction is geometrically beautiful, there are shortcomings which, in the age of modern mathematics, deem it insufficient. Our goal is to develop a new form of integration that, while not perfect, overcomes many of the innate flaws of the Riemann integral. This development will take place in three steps, each aimed at correcting a minor drawback of the Riemann integral. Section 7.5 explored the more severe deficiencies in detail. The first step in this process will be to construct an extension of the notion of length of an interval which will allow us to consider integrating over more complicated sets; intervals are the simplest subsets of $\mathbb{R}$ and the fact that the Riemann integral is restricted to these sets is a minor, but fundamental obstacle. The next step will be to develop a broader class of functions to integrate. The Riemann integral insists that a function be bounded to be integrable, but not all bounded functions are integrable, e.g. the Dirichlet function. Though this function is easy to define, its behavior with respect to domain values is rather problematic. However, its range is profoundly simple, consisting of merely two points. The final step will then be to develop a new integral which focuses on the range of a function rather than on its domain. As a glimpse of what's to come, with our new integral the Dirichlet function will actually be one of the simplest to integrate! This chapter focuses on the first step while the following deals with the latter steps in the construction.

8.1 Length and Measure

As we have seen, subsets of the real number system range from elementary, e.g. singletons, intervals, finite unions and intersections of intervals, etc., to fairly complicated, e.g. the rationals, the irrationals, the Cantor set, etc. Intervals are the simplest subsets of $\mathbb{R}$ and the earliest type of set that young mathematicians encounter, be it as the solution set to an inequality or as the domain or range of

M.A. Pons, *Real Analysis for the Undergraduate: With an Invitation to Functional Analysis*, 325
DOI 10.1007/978-1-4614-9638-0_8, © Springer Science+Business Media New York 2014

an elementary function. Concretely an interval is defined to be the set of all points lying between two given points either including or excluding the endpoints; the inclusion of the endpoints allows us to classify intervals as open, closed, or neither, (a, b), $[a, b]$, $(a, b]$, and $[a, b)$. One of the properties of intervals that make them so attractive is that every interval has a natural length associated with it; in fact, these two ideas go hand in hand. Whether including or excluding endpoints, the length of the interval (a, b) is $b - a$. Note here we are assuming that $a, b \in \mathbb{R}$. Another positive feature is the fact that the length of an interval is *translation invariant*. If (a, b) is an interval and we slide it along the real line in the positive or negative direction a distance c, we don't damage the interval structure and we preserve the length. For example, if we consider $(1, 2)$ and slide it a distance 3 units to the right, we obtain the interval $(4, 5)$ and this new interval also has length 1. One last feature that will be of interest to us is the *additive* property of length. If A and B are disjoint intervals of length a and b, respectively, and their union is an interval, then the length of $A \cup B$ is $a + b$, $(1, 2) \cup [2, 3) = (1, 3)$. Understanding these desirable qualities will be key to the developing theory.

To generalize length, we will need to remove it from the interval context. Technically, we can define the length of a union of *any* two disjoint intervals to be the sum of the individual intervals, but this is as far as we can currently go. Even this, however, is slightly counterintuitive since we naturally think of length along a continuum. For instance, does it make sense to assign a "length" to the set $(1, 2) \cup (3, 4)$? As we develop the next few sections, we will want to keep the features of length mentioned above at the forefront, while our goal is to extend these features to more general types of sets. Ideally we would hope to develop a generalization of length which:

- applies to every subset of $\mathbb{R}$;
- preserves the length of an interval;
- preserves the translation invariance of length;
- preserves the additive feature of length.

Along the way we will see that a strengthened version of the fourth item is possible and we will be able to construct an extension which allows countable unions of disjoint sets. However, in order to meet the last three requirements we will have to give up on the first. In the end this will not be a severe setback; the collection of subsets of $\mathbb{R}$ to which our generalization will apply will be extremely rich, containing all the intervals, countable unions and intersections of intervals, and more complicated sets. However, as we will see in Sect. 8.4, the structure of $\mathbb{R}$ is ever elusive and there are sets which will not afford being "measured."

Rings and Measures

The path that we are headed down is one that leads to many generalizations and, as is typical in mathematics, there are two approaches we could take—begin with the specific and move to the more abstract or vice versa. The approach that seems most beneficial is a combination. Here we introduce two abstract notions and then move toward more specific objects. In the last section of this chapter we will return to the abstract setting and provide a bit more enticement. Our first definition gives structure to collections of subsets of $\mathbb{R}$, and these collections will ultimately comprise the types of sets we will integrate over.

Definition 8.1.1. A nonempty collection $\mathscr{R}$ of subsets of $\mathbb{R}$ is called a *ring (of sets)* *in* $\mathbb{R}$ if for each pair $A, B \in \mathscr{R}$ it follows that

(a) $A \cup B \in \mathscr{R}$ and
(b) $A \setminus B \in \mathscr{R}$.

The notation $A \setminus B$ represents the operation of set difference and is defined by $A \cap B^c$, where the complement of B is that relative to $\mathbb{R}$. Before discussing some of the basic properties of rings, let's consider a few examples.

Example 8.1.2. Let $\mathscr{R} = \{\emptyset, \mathbb{R}\}$. It should be easy to check that this set satisfies both of the hypotheses.

Example 8.1.3. Let $\mathscr{R} = \mathscr{P}(\mathbb{R})$, the power set of $\mathbb{R}$, defined to be the collection of all subsets of $\mathbb{R}$. Again, the two conditions should be easy to verify for this example.

Example 8.1.4. Let $\mathscr{R} = \{A \subseteq \mathbb{R} : A \text{ is finite}\}$. The fact that this collection is a ring in $\mathbb{R}$ follows since the union of two finite sets is still finite together with the fact that $A \setminus B \subseteq A$. As a technical point, if we consider $A \setminus A$ we have the empty set. Considering this as the set with zero elements, we can classify it as a finite subset of $\mathbb{R}$.

Example 8.1.5. Let $\mathscr{R}$ be the collection of all subsets of $\mathbb{R}$ which are finite unions of intervals in $\mathbb{R}$. Again, verifying that the two ring axioms hold is straightforward and the reader should keep in mind that we are only considering intervals of finite length. Also, the empty set is an element of $\mathscr{R}$ since we can represent it as the degenerate interval (a, a).

A natural question to ask is what else is in a given ring of sets. It is a simple exercise to check that the empty set is always present since we can represent it as $A \setminus A$, as seen in the last two examples. Also, if $A, B \in \mathscr{R}$, it's a quick exercise to show that $A \cap B \in \mathscr{R}$. What about the complement of a set in $\mathscr{R}$? Unfortunately it is not the case that all rings are closed under complementation. Consider Example 8.1.4—the complement of any finite set (having n elements) is the union of $n + 1$ intervals, a set which is uncountable, and hence not in $\mathscr{R}$. However, in Example 8.1.3, the complement of any set is included in the ring in question. Perhaps this is because the ring consists of all possible subsets and is an outlier of sorts. Example 8.1.2 also shares this property, in a trivial sense, and the key to both of these is not the sheer number of sets included or excluded but the fact that one particular set, namely $\mathbb{R}$, is an element of the ring. Exercise 8.1.2 will ask you to prove that this is always the case. As a matter of terminology, in the special case that $\mathbb{R} \in \mathscr{R}$ we call $\mathscr{R}$ an *algebra of sets in* $\mathbb{R}$.

As mentioned in the previous section, we will have the need to consider countable unions of intervals and the following definition extends the definition of a ring to accommodate this.

Definition 8.1.6. A ring $\mathscr{R}$ is called a *σ-ring* if for each countable collection of sets $\{A_n\}_{n=1}$ in $\mathscr{R}$, it follows that $\bigcup_{n=1}^{\infty} A_n$ is also in $\mathscr{R}$. If a σ-ring is also an algebra of sets, we call it a *σ-algebra*.

Of the previous examples, the first two satisfy this condition. Why is this the case? Why do the third and fourth fail?

Example 8.1.7. Let $\mathscr{R} = \{A \subseteq \mathbb{R} : A \text{ is finite or countable}\}$. To verify that $\mathscr{R}$ satisfies the σ-ring condition, we recall that a countable union of countable sets is countable.

Next we present the general notion which will allow us to realize our concrete extension of length.

Definition 8.1.8. Let $\mathscr{R}$ be a σ-ring in $\mathbb{R}$. A *positive measure* on $\mathscr{R}$ is a function $\mu : \mathscr{R} \to [0, \infty]$ which satisfies

(a) $\mu(\emptyset) = 0$;
(b) μ is *countably additive*: If $\{A_n\}_{n=1}$ in $\mathscr{R}$ is a pairwise disjoint collection of sets, then

$$\mu\left(\bigcup_{n=1}^{\infty} A_n\right) = \sum_{n=1}^{\infty} \mu(A_n). \tag{8.1}$$

Before moving on we should make a few observations. If a measure μ is defined on a ring (not a σ-ring), then we only require additivity—If $A, B \in \mathscr{R}$ with $A \cap B = \emptyset$, then $\mu(A \cup B) = \mu(A) + \mu(B)$. By induction this can be extended to any finite union of pairwise disjoint sets. As far as terminology, a positive measure is sometimes referred to as a *nonnegative, countably additive set function* and the triple $(\mathbb{R}, \mathscr{R}, \mu)$ is called a *measure space*. Just as a "vector space" consists of a collection of objects, called vectors, a field of scalars, and two operations, addition and scalar multiplication, a measure space is a mathematical structure consisting of a set X, a collection of subsets of X, call it $\mathscr{M}$ for the moment, and a set function that assigns a numerical value to each set in the collection $\mathscr{M}$. We'll discuss general measure spaces in Sect. 8.5. As a final remark, note that a measure is allowed to assign an infinite value to sets in the ring. This will often require us to be cautious when manipulating algebraic expressions since infinity does not follow the usual rules of arithmetic. If you recall from basic calculus, similar situations are encountered when dealing with limits. For instance, is $0 \cdot \infty = 0$? Or is it ∞? Along these same lines, is $\infty - \infty = 0$? These types of questions can only be answered in the context of a specific situation and we will rarely make general conclusions when infinity is lurking.

Example 8.1.9. Let $\mathscr{R} = \mathscr{P}(\mathbb{R})$ and let $x_0 \in \mathbb{R}$. Define $\mu : \mathscr{R} \to [0, \infty]$ by

$$\mu(A) = \begin{cases} 1, & x_0 \in A; \\ 0, & x_0 \notin A. \end{cases}$$

It should be clear that $\mu(\emptyset) = 0$. If $\{A_n\}_{n=1}$ is a pairwise disjoint collection of sets in $\mathscr{P}(\mathbb{R})$, then either x_0 is in exactly one of the sets or in none of the sets. If x_0 is in exactly one of the A_n's, then both sides of Eq. (8.1) equal one. If x_0 is in none of the sets in the union, the both sides of Eq. (8.1) are zero. A measure of this variety is called a *point mass measure* since the measure is placing all its "mass" on the point x_0.

Example 8.1.10. Let $\mathscr{R} = \mathscr{P}(\mathbb{R})$. Define $\mu : \mathscr{R} \to [0, \infty]$ by

$$\mu(A) = \begin{cases} |A|, & A \text{ is finite;} \\ \infty, & A \text{ is infinite.} \end{cases}$$

The symbol $|A|$ in this set context denotes the number of elements in A. Verify that this example, referred to as a *counting measure*, is countably additive.

The following theorem gives a few of the most relevant properties of measures.

Theorem 8.1.11 (Properties of Measures). *Let $\mathscr{R}$ be a σ-ring and let μ be a positive measure on $\mathscr{R}$.*

(a) Monotonicity—If $A, B \in \mathscr{R}$ with $A \subseteq B$, then $\mu(A) \leq \mu(B)$.
(b) Continuity for increasing unions—Suppose $\{A_n\}_{n=1}$ is in $\mathscr{R}$ with $A_1 \subseteq A_2 \subseteq A_3 \subseteq \ldots$ and $A = \bigcup_{n=1}^{\infty} A_n$. Then

$$\mu(A) = \lim_{n \to \infty} \mu(A_n).$$

(c) Continuity for decreasing intersections—Suppose $\{A_n\}_{n=1}$ is in $\mathscr{R}$ with $A_1 \supseteq A_2 \supseteq A_3 \supseteq \ldots$ and $A = \bigcap_{n=1}^{\infty} A_n$. If $\mu(A_1) < \infty$, then

$$\mu(A) = \lim_{n \to \infty} \mu(A_n).$$

(d) Countably subadditive—If $\{A_n\}_{n=1}$ is in $\mathscr{R}$, then $\mu\left(\bigcup_{n=1}^{\infty} A_n\right) \leq \sum_{n=1}^{\infty} \mu(A_n).$

Proof. For (a) observe that we can decompose $B = A \cup (B \setminus A)$, a disjoint union of sets in $\mathscr{R}$. Thus by the additivity of μ we have

$$\mu(B) = \mu(A \cup (B \setminus A)) = \mu(A) + \mu(B \setminus A) \geq \mu(A).$$

For (b) our goal is to relate A to a disjoint union of sets in $\mathscr{R}$ and then, as above, appeal to the additivity of μ. Define a collection of sets $\{B_j\}_{j=1}$ recursively by $B_1 = A_1$ and $B_j = A_j \setminus A_{j-1}$ for $j = 2, 3, 4, \ldots$; by construction notice that $B_j \in \mathscr{R}$ for $j = 1, 2, 3, \ldots$, $B_i \cap B_j = \emptyset$ when $i \neq j$, $A = \bigcup_{j=1}^{\infty} B_j$, and $A_n = \bigcup_{j=1}^{n} B_j$. Applying the additivity of μ to the last two observations yields $\mu(A) = \sum_{j=1}^{\infty} \mu(B_j)$ and $\mu(A_n) = \sum_{j=1}^{n} \mu(B_j)$. Combining these two statements we have

$$\mu(A) = \sum_{j=1}^{\infty} \mu(B_j) = \lim_{n \to \infty} \sum_{j=1}^{n} \mu(B_j) = \lim_{n \to \infty} \mu(A_n)$$

as desired.

For (c) a similar tactic is used. Define a collection of sets $\{C_n\}_{n=1}$ by $C_n = A_1 \setminus A_n$. Notice that $C_1 \subseteq C_2 \subseteq C_3 \subseteq \ldots$ and thus we aim to utilize part (b). As a second observation, it is clear that $\mu(C_n) = \mu(A_1 \setminus A_n) = \mu(A_1) - \mu(A_n)$ since $A_1 = A_n \cup (A_1 \setminus A_n)$ and $\mu(A_1) < \infty$. Moreover, $A_1 \setminus A = \bigcup_{n=1}^{\infty} C_n$, a fact that you are asked to prove in the exercises, and we see that

$$\mu(A_1 \setminus A) = \mu \left(\bigcup_{n=1}^{\infty} C_n \right) = \lim_{n \to \infty} \mu(C_n)$$

$$= \lim_{n \to \infty} \left(\mu(A_1) - \mu(A_n) \right)$$

$$= \mu(A_1) - \lim_{n \to \infty} \mu(A_n),$$

where we have used (b) in the second equality. As a last observation, $\mu(A_1 \setminus A) = \mu(A_1) - \mu(A)$, and, piecing these last two statements together, it follows that $\mu(A_1) - \mu(A) = \mu(A_1 \setminus A) = \mu(A_1) - \lim_{n \to \infty} \mu(A_n)$. The hypothesis $\mu(A_1) < \infty$ allows us to conclude $\mu(A) = \lim_{n \to \infty} \mu(A_n)$.

The proof of (d) is left as Exercise 8.1.9 $\qquad\qquad\qquad\qquad\qquad\qquad\square$

In the proof of (c) the hypothesis that $\mu(A_1) < \infty$ is crucial and Exercise 8.1.8 asks you to show that the theorem fails without it. The following example explores one particular measure in depth and will play a prominent role in the following sections.

Example 8.1.12. Let $\mathscr{E}$ be the collection of subsets of $\mathbb{R}$ which are finite unions of disjoint intervals (with any endpoint configuration); we shall refer to these as *elementary subsets* of $\mathbb{R}$. Again, keep in mind that we are only considering intervals of finite length. Since a finite union or difference of intervals can always be represented as a disjoint union of intervals (with some tinkering), we are certainly dealing with a ring. Is this set a σ-ring? Unfortunately it is not. For instance, the union of the intervals $(n, n+1], n = 1, 2, 3, \ldots$, is not an interval of finite length and therefore cannot be written as a finite union of disjoint intervals each of finite length. For a measure, define $m : \mathscr{E} \to [0, \infty)$ by

1. $m(\emptyset) = 0$;
2. if $A = \bigcup_{n=1}^{N} I_n \in \mathscr{E}$, a disjoint union of intervals, we set

$$m \left(\bigcup_{n=1}^{N} I_n \right) = \sum_{n=1}^{N} \ell(I_n),$$

where $\ell(I_n)$ represents the length of I_n.

This set function is additive (in fact countably additive, though this is not necessary to show that m is a measure on $\mathscr{E}$). The next theorem describes a subadditive property possessed by m, but first a covering lemma.

Lemma 8.1.13. *Let A be an elementary subset of $\mathbb{R}$. For each $\varepsilon > 0$ there is an open elementary set O with $A \subseteq O$ and $m(O) \leq m(A) + \varepsilon$.*

Proof. For the proof, we'll only consider elementary sets which are unions of closed intervals. The other endpoint configurations will follow from this case and will be explored in the exercises. To this end, suppose A is an elementary set which is a disjoint union of the form

$$A = \bigcup_{n=1}^{N} [a_n, b_n]$$

and suppose the endpoints are ordered so that

$$a_1 \leq b_1 < a_2 \leq b_2 < a_3 \leq b_3 < \ldots < a_N \leq b_N.$$

Let $\varepsilon > 0$ and choose

$$M = \min\{|a_n - b_{n-1}| : n = 2, 3, 4, \cdots, N\};$$

notice that M is the smallest distance between consecutive intervals in the union. At this point set $c = \min\{\varepsilon, M/4\}$ and define

$$O = \bigcup_{n=1}^{N} \left(a_n - \frac{c}{2N}, b_n + \frac{c}{2N}\right).$$

It follows that O is a disjoint union (by our choice of M) of open intervals, and is therefore an open elementary set. It is also clear that $A \subseteq O$ by construction; see Fig. 8.1. It follows that

$$m(O) = \sum_{n=1}^{N} \ell\left(a_n - \frac{c}{2N}, b_n + \frac{c}{2N}\right) = \sum_{n=1}^{N}\left(b_n - a_n + \frac{c}{N}\right)$$

$$= \sum_{n=1}^{N}(b_n - a_n) + c = m(A) + c \leq m(A) + \varepsilon$$

as desired. $\qquad\square$

Theorem 8.1.14. *Let A be an elementary set and let $\{A_n\}_{n=1}$ be a collection of elementary sets. If $A \subseteq \bigcup_{n=1}^{\infty} A_n$, then*

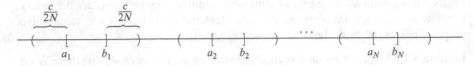

Fig. 8.1 Construction of the open set O

$$m(A) \le \sum_{n=1}^{\infty} m(A_n).$$

Proof. Here we only consider the case when A is an interval. The more general case follows in a similar manner and is requested in Exercise 8.1.11. Let A be an interval and let $\varepsilon > 0$. Our goal is to show that $m(A) \le \sum_{n=1}^{\infty} m(A_n) + \varepsilon$ from which the conclusion follows immediately. First, if $A = \emptyset$ or $A = [a, a]$, then there is nothing to prove since $m(A) = 0$ in both of these cases. Furthermore, if the series $\sum_{n=1}^{\infty} m(A_n)$ diverges, i.e. takes on an infinite value, then the inequality also holds. So we may assume that $m(A) = \ell(A) > 0$ and $\sum_{n=1}^{\infty} m(A_n) < \infty$. Without loss of generality we may also assume $0 < \varepsilon < m(A)$. Since A is an interval, we can find a closed interval B with $B \subseteq A$ and $m(A) \le m(B) + \varepsilon/2$. To clarify, if $A = [a, b]$, we could choose $B = [a + \varepsilon/4, b - \varepsilon/4]$.

By the lemma, for each $n \in \mathbb{N}$ we may choose an open elementary set B_n (a union of disjoint open intervals) with $A_n \subseteq B_n$ and $m(B_n) \le m(A_n) + \varepsilon/2^{n+1}$. Since B is a compact set with $B \subseteq A \subseteq \bigcup_{n=1}^{\infty} B_n$, applying the open cover characterization of compactness, there is a finite subcollection $\{B_{n_j}\}_{j=1}^{N}$ with $B \subseteq \bigcup_{j=1}^{N} B_{n_j}$. Further, since m is a measure on $\mathscr{E}$, Theorem 8.1.11(a) and Exercise 8.1.9(a) guarantee that $m(B) \le m\left(\bigcup_{j=1}^{N} B_{n_j}\right) \le \sum_{j=1}^{N} m(B_{n_j})$. Thus we may estimate $m(A)$ by

$$m(A) \le m(B) + \frac{\varepsilon}{2} \le \sum_{j=1}^{N} m(B_{n_j}) + \frac{\varepsilon}{2} \le \sum_{j=1}^{N} \left(m(A_{n_j}) + \frac{\varepsilon}{2^{n_j+1}} \right) + \frac{\varepsilon}{2}$$

$$\le \sum_{n=1}^{\infty} \left(m(A_n) + \frac{\varepsilon}{2^{n+1}} \right) + \frac{\varepsilon}{2} = \sum_{n=1}^{\infty} m(A_n) + \varepsilon.$$

$\square$

Exercises

Exercise 8.1.1. Suppose $\mathscr{R}$ is a ring of sets. Show that the intersection of any two sets in $\mathscr{R}$ is also in $\mathscr{R}$.

Exercise 8.1.2. Let $\mathscr{R}$ be an algebra of sets in $\mathbb{R}$. Show that $A \in \mathscr{R}$ if and only if $A^c \in \mathscr{R}$.

Exercise 8.1.3. Verify that the set function given in Example 8.1.10 is in fact a measure on $\mathscr{P}(\mathbb{R})$.

Exercise 8.1.4. Let $\mathscr{R} = \{A \subseteq \mathbb{R} : A \text{ or } \mathbb{R} \setminus A \text{ is countable}\}$. Show that $\mathscr{R}$ is a σ-algebra. Also, define μ by

$$\mu(A) = \begin{cases} 0, & A \text{ is countable}; \\ 1, & X \setminus A \text{ is countable}. \end{cases}$$

Show that μ is a measure on $\mathscr{R}$.

Exercise 8.1.5. Let $\mathscr{R}$ be a σ-ring of sets and suppose $\{A_n\}_{n=1}$ is a collection of sets in $\mathscr{R}$.

(a) Show that $\bigcap_{n=1}^{\infty} A_n = A_1 \setminus \left(\bigcup_{n=1}^{\infty} (A_1 \setminus A_n) \right)$.

(b) Show that $\bigcap_{n=1}^{\infty} A_n \in \mathscr{R}$.

Exercise 8.1.6. Let $\mathscr{R}$ be a ring of sets and let μ be a positive measure on $\mathscr{R}$. If $A, B \in \mathscr{R}$ with $A \cap B = \emptyset$ and $\mu(A \cup B) = \mu(A)$, does it follow that $\mu(B) = 0$? Explain.

Exercise 8.1.7. For A, A_n, and C_n as defined in the proof of Theorem 8.1.11(c), show that $A_1 \setminus A = \bigcup_{n=1}^{\infty} C_n$.

Exercise 8.1.8. Provide an example which shows that Theorem 8.1.11(c) fails if the hypothesis "$\mu(A_1) < \infty$" is removed.

Exercise 8.1.9. (a) Let $\mathscr{R}$ be a ring of sets and let μ be a positive measure on $\mathscr{R}$. Show that μ is subadditive—If $A, B \in \mathscr{R}$, not necessarily disjoint, then $\mu(A \cup B) \leq \mu(A) + \mu(B)$.

(b) Let $\mathscr{R}$ be a σ-ring of sets and let μ be a positive measure on $\mathscr{R}$. Show that μ is countably subadditive—If $\{A_n\}_{n=1}$ is a collection of sets in $\mathscr{R}$, then $\mu\left(\bigcup_{n=1}^{\infty} A_n\right) \leq \sum_{n=1}^{\infty} \mu(A_n)$.

Exercise 8.1.10 (Regularity of m).

(a) Let A be an elementary set and $\varepsilon > 0$. Show that there is a closed elementary set F with $F \subseteq A$ and $m(F) \geq m(A) - \varepsilon$.

(b) Complete the proof of Lemma 8.1.13 by considering the case where A is an arbitrary elementary set. To do this consider the closure of such a set.

Exercise 8.1.11. Complete the proof of Theorem 8.1.14 by showing that the theorem holds when A in an arbitrary elementary set.

8.2   Outer Measure on $\mathbb{R}$

In this section we focus on more specific objects, in particular the ring and measure defined in Example 8.1.12, and our goal is to find an extension of m which applies to all subsets of $\mathbb{R}$. However, in doing this we will sacrifice the additivity possessed by m.

Definition 8.2.1. Let $A \subseteq \mathbb{R}$ and define the *outer measure* of A by

$$m^*(A) = \inf \left\{ \sum_{n=1}^{\infty} \ell(I_n) : I_n \text{ is an open interval and } A \subseteq \bigcup_{n=1}^{\infty} I_n \right\}.$$

Outer measure on $\mathbb{R}$ is designed to preserve the length of an interval by identifying the most efficient means of covering a set with open intervals. As a first example let's consider the outer measure of the open interval $(0, 1)$. There are many ways to cover this set with intervals, for example, $(0, 1) \subseteq \bigcup_{n=2}^{\infty} \left(\frac{1}{n}, 1 \right)$ and

$$\sum_{n=2}^{\infty} \ell \left(\left(\frac{1}{n}, 1 \right) \right) = \sum_{n=1}^{\infty} \frac{n}{n+1} = \infty.$$

This is not a very efficient covering, however, due to the overlap in the sets $\left(\frac{1}{n}, 1 \right)$. On the other hand, for every $\varepsilon > 0$, $(0, 1) \subseteq (-\varepsilon, 1+\varepsilon)$ and $\ell((-\varepsilon, 1+\varepsilon)) = 1+2\varepsilon$. Thus, taking the infimum into account, the outer measure of $(0, 1)$ is less than or equal to 1. In terms of efficiency, the optimal covering for $(0, 1)$ is $(0, 1)$ itself and we can conclude that $m^*((0, 1)) = 1$. In actuality there are a few details to check here and Exercise 8.2.2 will ask you to show that outer measure preserves the length of any interval and extends the measure m defined in Example 8.1.12.

Equally important is the fact that outer measure is defined for every subset of $\mathbb{R}$, though this value may be infinite; as defined in the previous section, m only takes finite values. As a more extreme example let's consider the outer measure of the rationals, a countable set which is dense in $\mathbb{R}$. Countable sets are relatively small when it comes to infinite sets; however, in order to be dense, the rationals must be well dispersed throughout the real line. We've studied this structure quite a bit, but the point here is that outer measure is analyzing the set structure differently than ways we've encountered before and it may be hard to intuitively grasp what the answer should be. Consider the following: Let $\{r_1, r_2, r_3, \ldots\}$ be an enumeration of $\mathbb{Q}$ and, for $\varepsilon > 0$, choose

$$I_n = \left(r_n - \frac{\varepsilon}{2^{n+1}}, r_n + \frac{\varepsilon}{2^{n+1}} \right).$$

It follows that $\ell(I_n) = \varepsilon/2^n$ and thus $\sum_{n=1}^{\infty} \ell(I_n) = \varepsilon$. Ergo no matter how small we choose ε, it must be the case that $m^*(\mathbb{Q}) \leq \varepsilon$ and thus $m^*(\mathbb{Q}) = 0$.

This last example highlights a naive way of thinking which can be dangerous. Intuitively, the above calculation makes sense because the rationals contain no intervals, outer measure is an extension of length, and length is obtained via intervals. However, outer measure is more sensitive than this. If we were to continue in his manner of reasoning, we would also conclude that the irrationals have outer measure zero, but this is not the case. The next theorem will show that $m^*(\mathbb{R}) \le m^*(\mathbb{Q}) + m^*(\mathbb{I})$ from which it follows that $m^*(\mathbb{I}) = \infty$ since $m^*(\mathbb{R}) = \infty$ (Exercise 8.2.3). Outer measure is a generalization of the length of an interval, but is not tied to intervals in the intimate manner which binds length to the interval setting.

The following theorem summarizes the most relevant properties of outer measure.

Theorem 8.2.2 (Properties of Outer Measure on $\mathbb{R}$). *Let A and B be subsets of $\mathbb{R}$ and let $\mathbb{R}$.*

(a) $m^*(A)$ is defined, though the value may be infinite.
(b) $m^*(\emptyset) = 0$ and $m^*(\{a\}) = 0$.
(c) m^* preserves m—If $A \in \mathcal{E}$, then $m(A) = m^*(A)$.
(d) m^* is monotonic—If $A \subseteq B$, then $m^*(A) \le m^*(B)$.
(e) m^* is countably subadditive—If $\{A_n\}_{n=1}^{\infty}$ is in $\mathcal{P}(\mathbb{R})$, then

$$m^*\left(\bigcup_{n=1}^{\infty} A_n\right) \le \sum_{n=1}^{\infty} m^*(A_n).$$

Proof. The proof of properties (a)–(d) are straightforward and are requested in the exercises. For (e), let's consider two cases. If $\sum_{n=1}^{\infty} m^*(A_n) = \infty$, then the inequality is trivial. For the more substantial case, suppose $\sum_{n=1}^{\infty} m^*(A_n) < \infty$ and let $\varepsilon > 0$. By the definition of outer measure we may, for each n, find a collection of open intervals $\{I_j^n\}_{j=1}^{\infty}$ which cover A_n and satisfies

$$\sum_{j=1}^{\infty} \ell(I_j^n) \le m^*(A_n) + \frac{\varepsilon}{2^n}.$$

The doubly indexed set $\{I_j^n\}_{n,j=1}^{\infty}$ is a countable collection and covers $\bigcup_{n=1}^{\infty} A_n$. Using the definition of outer measure again and the Algebraic Limit Theorem for series, we have

$$m^*\left(\bigcup_{n=1}^{\infty} A_n\right) \le \sum_{n,j=1}^{\infty} \ell(I_j^n) = \sum_{n=1}^{\infty}\sum_{j=1}^{\infty} \ell(I_j^n)$$

$$\le \sum_{n=1}^{\infty}\left[m^*(A_n) + \frac{\varepsilon}{2^n}\right]$$

$$= \sum_{n=1}^{\infty} m^*(A_n) + \varepsilon$$

from which the conclusion follows. $\square$

Corollary 8.2.3. *If $A \subseteq \mathbb{R}$ is a countable set, then $m^*(A) = 0$.*

Proof. The conclusion follows immediately from properties (b) and (e).   □

A question that should have come to mind immediately is whether or not m^* is countably additive; this could be rephrased by asking whether or not m^* is a measure on $\mathscr{P}(\mathbb{R})$. The fact that m^* is an extension of m may lead one to think that m^* is countably additive, but we will see in Sect. 8.4 that m^* falls short in this regard. This construction will provide yet another example of the fact that subsets of $\mathbb{R}$ range from the simple to the diabolical.

Looking back at the list of properties of length that we discussed in Sect. 8.1, it is also natural to ask whether or not m^* is translation invariant. To be precise we first need a definition.

Definition 8.2.4. Let $A \subseteq \mathbb{R}$. For $t \in \mathbb{R}$, we define the *translation*, denoted $A + t$, of the set A by

$$A + t = \{a + t : a \in A\}.$$

Theorem 8.2.5. *Let $A \subseteq \mathbb{R}$ and $t \in \mathbb{R}$. Then $m^*(A + t) = m^*(A)$.*

Proof. Let $\{J_n\}_{n=1}$ be an arbitrary collection of open intervals which cover A. Using the translation invariance of ℓ we know that $\ell(J_n) = \ell(J_n + t)$. Moreover, the collection $\{J_n + t\}_{n=1}$ is an open cover for $A + t$. Together these two statements show that

$$m^*(A + t) \leq \sum_{n=1}^{\infty} \ell(J_n + t) = \sum_{n=1}^{\infty} \ell(J_n).$$

Furthermore, the fact that $\{J_n\}_{n=1}$ is an arbitrary open cover of A assures us that

$$m^*(A + t) \leq \inf \left\{ \sum_{n=1}^{\infty} \ell(J_n) : J_n \text{ is an interval and } A \subseteq \bigcup_{n=1}^{\infty} J_n \right\} = m^*(A).$$

The reverse inequality is analogous and we omit the details, though the reader is encouraged to complete the proof.   □

For a moment let's consider our position. We have m defined as a measure on $\mathscr{E}$ and m^* defined on $\mathscr{P}(\mathbb{R})$, though it lacks one of the measure properties we desire. Moreover, $\mathscr{E} \subseteq \mathscr{P}(\mathbb{R})$ and m^* agrees with m on $\mathscr{E}$. In short, we overshot our destination and our goal is now to restrict ourselves to a collection of sets $\mathscr{L}$ with $\mathscr{E} \subseteq \mathscr{L} \subseteq \mathscr{P}(\mathbb{R})$ such that m^* restricted to $\mathscr{L}$ is a measure. In other words, we want to exclude the subsets of $\mathbb{R}$ which are problematic with respect to the additivity of m^*.

To give an explicit construction for the collection $\mathscr{L}$, we need one more intermediary idea. From Sect. 1.1, recall the symmetric difference of two sets.

Combining this set operation with outer measure, we can define a semi-metric $d : \mathscr{P}(\mathbb{R}) \times \mathscr{P}(\mathbb{R}) \to [0, \infty]$ by the rule $d(A, B) = m^*(A \triangle B)$; the term semi-metric is discussed in the exercises. The following properties are immediate.

Proposition 8.2.6. *Let* $A, B, C, D \subseteq \mathbb{R}$.

(a) $d(A, \emptyset) = m^*(A)$
(b) $d(A, C) \leq d(A, B) + d(B, C)$
(c) $d(A \cup B, C \cup D) \leq d(A, C) + d(B, D)$
(d) $d(A \setminus B, C \setminus D) \leq d(A, C) + d(B, D)$
(e) *If* $m^*(B) < \infty$, *then* $|m^*(A) - m^*(B)| \leq d(A, B)$.

The proof of each of these is a straightforward application of the definition of d, though it may be useful to first review Exercise 1.1.5. Notice that part (b) is the triangle inequality for this semi-metric while part (e) resembles a continuity property.

Definition 8.2.7. Let $\overline{\mathscr{E}}$ denote the collection of subsets $A \subseteq \mathbb{R}$ for which there exists a sequence (A_n) in $\mathscr{E}$ with $d(A_n, A) \to 0$ as $n \to \infty$.

Another way of stating this definition would be to let $\overline{\mathscr{E}}$ be the collection of subsets of $\mathbb{R}$ which can be approximated by sets in $\mathscr{E}$ with respect to the semi-metric d. This also explains the use of the "overline" in the notation; the set $\overline{\mathscr{E}}$ is, in a manner of speaking, the closure of $\mathscr{E}$ with respect to the semi-metric d.

Theorem 8.2.8. *The collection* $\overline{\mathscr{E}}$ *is a ring of sets in* $\mathbb{R}$ *and* m^* *is a measure on* $\overline{\mathscr{E}}$.

Proof. The proof that $\overline{\mathscr{E}}$ is a ring follows directly from Proposition 8.2.6 and is left as Exercise 8.2.7. Our work here will be to show that m^* is additive on $\overline{\mathscr{E}}$. To this end, let $A, B \in \overline{\mathscr{E}}$ with $A \cap B = \emptyset$. Our goal is to show that we can make the value of $|m^*(A \cup B) - (m^*(A) + m^*(B))|$ arbitrarily small from which the equality we seek follows. According to the definition of $\overline{\mathscr{E}}$ there are sequences (A_n) and (B_n) in $\mathscr{E}$ with $d(A_n, A) \to 0$ and $d(B_n, B) \to 0$ as $n \to \infty$. As a consequence of Proposition 8.2.6(c), the quantity $d(A_n \cup B_n, A \cup B)$ also converges to 0. At this point recall that m^* agrees with m on $\mathscr{E}$, and is additive there. The sets $A_n, B_n \in \mathscr{E}$ we have obtained via the definition of $\overline{\mathscr{E}}$ are not necessarily disjoint, however, without loss of generality we may assume $A_n \cap B_n = \emptyset$. (We'll identify where we use this hypothesis and you will explore its removal in the exercises.) Using the additivity of m, we have

$$m^*(A_n \cup B_n) = m(A_n \cup B_n) = m(A_n) + m(B_n) = m^*(A_n) + m^*(B_n). \quad (8.2)$$

Also, the fact that the sets A_n and B_n are in $\mathscr{E}$ guarantees us that all the quantities in the previous line exist as finite values. At this point we're ready to estimate. The triangle inequality together with Eq. (8.2) confirms that $|m^*(A \cup B) - (m^*(A) + m^*(B))|$ is bounded above by

$$|m^*(A \cup B) - m^*(A_n \cup B_n)| + |m^*(A_n) - m^*(A)| + |m^*(B_n) - m^*(B)|$$

which is in turn bounded above by

$$d(A_n \cup B_n, A \cup B) + d(A_n, A) + d(B_n, B)$$

by Proposition 8.2.6(e). Since each quantity in the previous line converges to 0, it follows that the value of $|m^*(A \cup B) - (m^*(A) + m^*(B))|$ can be made as small as desired and we conclude that $m^*(A \cup B) = m^*(A) + m^*(B)$. $\qquad\square$

Let's return for a moment to the point in the proof where we made the additional assumption that $A_n \cap B_n = \emptyset$. We certainly used this assumption in Eq. (8.2), but where did we use the fact that the initial sets A and B are disjoint? The careful reader will notice that we never explicitly used this hypothesis, but it is hidden in the additional assumption we made that A_n and B_n are disjoint. However, the fact that A and B are disjoint is *not* sufficient to guarantee that A_n and B_n are disjoint, but it does allow us to construct disjoint subsets of A_n and B_n which will alleviate this issue. Exercise 8.2.9 will guide the way in completing this proof.

Moving forward again, we should observe that $\overline{\mathscr{E}}$ is not a σ-ring. To see this consider intervals of the form $(-n, n)$ where $n \in \mathbb{N}$. Each of these sets is in $\mathscr{E}$ and hence in $\overline{\mathscr{E}}$. However, $\mathbb{R} = \bigcup_{n=1}^{\infty}(-n, n)$ is not in $\overline{\mathscr{E}}$, and thus the collection is not closed with respect to countable unions. To show that $\mathbb{R}$ is not in $\overline{\mathscr{E}}$, we must prove that it is impossible to approximate $\mathbb{R}$ with sets in $\mathscr{E}$. For this step we begin by calculating $d(A, \mathbb{R})$ for an arbitrary $A \in \mathscr{E}$. First notice that $m^*(A) < \infty$ since $A \in \mathscr{E}$. Furthermore, $m^*(\mathbb{R}) \le m^*(A) + m^*(A^c)$ by the subadditivity of m^* and thus it must be the case that $m^*(A^c) = \infty$ since $m^*(\mathbb{R}) = \infty$ and $m^*(A) < \infty$. Piecing these two ideas together shows us that

$$d(A, \mathbb{R}) = m^*(A \Delta \mathbb{R}) = m^*(A^c) = \infty.$$

From this it is clear that no matter how a sequence of sets $(A_n) \subseteq \mathscr{E}$ is chosen, the quantity $d(A_n, \mathbb{R}) = \infty$ for all n and thus cannot possibly converge to 0.

As another example, let $\{r_1, r_2, r_3, \ldots\}$ be an enumeration of the rational numbers and define $A_n = \{r_k\}_{k=1}^{n}$. It's clear that $\mathbb{Q} \Delta A_n = \{r_k\}_{k=n+1}^{\infty}$ and $m^*(\mathbb{Q} \Delta A_n) = 0$ since this set is countable. Furthermore, each $A_n \in \mathscr{E}$ and we conclude that $\mathbb{Q} \in \overline{\mathscr{E}}$.

Though these two examples provide little insight as to what sets are or are not in $\overline{\mathscr{E}}$, Proposition 8.3.2 will give us a very tangible means of identifying sets in $\overline{\mathscr{E}}$.

As a final comment on outer measure, the following lemma shows that while $\overline{\mathscr{E}}$ is not a σ-ring, m^* is countably additive on $\overline{\mathscr{E}}$.

Lemma 8.2.9. *If $\{A_n\}_{n=1}^{\infty}$ in $\overline{\mathscr{E}}$ is a pairwise disjoint collection of sets and $A = \bigcup_{n=1}^{\infty} A_n$, then*

$$m^*(A) = \sum_{n=1}^{\infty} m^*(A_n).$$

Proof. The fact that outer measure is countably subadditive on $\mathscr{P}(\mathbb{R})$ guarantees us that

$$m^*(A) = m^* \left(\bigcup_{n=1}^{\infty} A_n \right) \leq \sum_{n=1}^{\infty} m^*(A_n)$$

and all that remains is to show the reverse inequality also holds. To see this notice that $\bigcup_{n=1}^{N} A_n \subseteq A$ for each $N \in \mathbb{N}$; furthermore, since m^* is additive on the ring $\overline{\mathscr{E}}$ and monotonic in general,

$$\sum_{n=1}^{N} m^*(A_n) = m^* \left(\bigcup_{n=1}^{N} A_n \right) \leq m^*(A).$$

Theorem 3.1.7 indicates that $\sum_{n=1}^{\infty} m^*(A_n) \leq m^*(A)$ whether the value of $m^*(A)$ is finite or infinite. $\qquad \square$

Exercises

Exercise 8.2.1. Let A and B be subsets of $\mathbb{R}$ with $A \subseteq B$. Show that $m^*(A) \leq m^*(B)$.

Exercise 8.2.2. Use Lemma 8.1.13 and Theorem 8.1.14 to show that $m(A) = m^*(A)$ for every $A \in \mathscr{E}$. It may help to begin by considering the simpler situation when A is an open interval of finite length.

Exercise 8.2.3. (a) Let $A \subseteq \mathbb{R}$ be a finite or countable set. Show that $m^*(A) = 0$ by using the ideas employed to show that $m^*(\mathbb{Q}) = 0$.
(b) Show that the outer measure of the Cantor set is 0.
(c) Show that $m^*(\mathbb{R}) = \infty$.

Exercise 8.2.4. (a) Show that any bounded subset of $\mathbb{R}$ has finite outer measure.
(b) Show that the outer measure of any nonempty open set is strictly greater than zero.
(c) Give an example of an unbounded, open set with finite outer measure.
(d) Give an example of an unbounded, closed set with outer measure zero.

Exercise 8.2.5. Explain why we use the terminology "semi-metric" in reference to the map $d : \mathscr{P}(\mathbb{R}) \times \mathscr{P}(\mathbb{R}) \to \mathbb{R}$ given by the rule $d(A, B) = m^*(A \Delta B)$. In other words, why is this not a true metric?

Exercise 8.2.6. Supply a proof for Proposition 8.2.6. **Hint:** For part (e) use parts (a) and (b).

Exercise 8.2.7. Show that $\overline{\mathscr{E}}$ is a ring of sets in $\mathbb{R}$.

Exercise 8.2.8. Let $A, B \subseteq \mathbb{R}$ and let $t \in \mathbb{R}$. Show each of the following statements:

(a) $(A + t)^c = A^c + t$;
(b) $(A + t) \cup (B + t) = (A \cup B) + t$;
(c) $(A + t) \cap (B + t) = (A \cap B) + t$;
(d) $(A + t) \setminus (B + t) = (A \setminus B) + t$;
(e) $(A + t) \Delta (B + t) = (A \Delta B) + t$;
(f) $A \in \overline{\mathscr{E}}$ if and only if $A + t \in \overline{\mathscr{E}}$.

Exercise 8.2.9. The goal of this exercise is to complete the verification that m^* is additive on $\overline{\mathscr{E}}$. At this point, reread the proof of Theorem 8.2.8 up through Eq. (8.2). The issue we need to address is the fact that the sets A_n and B_n obtained from the definition of $\overline{\mathscr{E}}$ need not be disjoint simply because A and B are disjoint, without which Eq. (8.2) is actually invalid. The exercises below will justify the use of the phrase "without loss of generality."

(a) First, define new sequences $C_n = A_n \setminus B_n$ and $D_n = B_n \setminus A_n$. Show that these sequences are in $\mathscr{E}$ and that $C_n \cap D_n = \emptyset$ for all $n \in \mathbb{N}$.
(b) Show that $d(C_n, A) \to 0$ and $d(D_n, B) \to 0$ as $n \to \infty$. As a consequence, $d(C_n \cup D_n, A \cup B) \to 0$ also.
(c) Finally, complete the proof by replacing A_n with C_n and B_n with D_n in Eq. (8.2).
(d) Where did you use the fact that $A \cap B = \emptyset$?

Exercise 8.2.10 (Inner measure on $\mathbb{R}$). If A is a subset of $\mathbb{R}$, we define the *inner measure* of A by

$$m_*(A) = \sup\{m^*(K) : K \subseteq A \text{ and } K \text{ is a compact set}\}.$$

(a) If K is a compact set, show that $m^*(K) = m_*(K)$.
(b) If $A \subseteq B$, show that $m_*(A) \le m_*(B)$.
(c) Show that $m_*(A) \le m^*(A)$.

8.3   Lebesgue Measure on $\mathbb{R}$

Definition 8.3.1. Let $\mathscr{L}$ denote the collection of subsets of $\mathbb{R}$ which can be written as a countable union of sets in $\overline{\mathscr{E}}$.

The purpose of this definition should be completely transparent. The ring $\overline{\mathscr{E}}$ generalized intervals of finite length but it is not a σ-ring. This new collection does form a σ-ring and will allow us to consider larger subsets of $\mathbb{R}$. Momentarily we will show that $\mathscr{L}$ is in fact a σ-algebra (a σ-ring which contains $\mathbb{R}$) and is well behaved with respect to outer measure, but first we develop a means to identify sets in $\mathscr{L}$ which are in the subcollection $\overline{\mathscr{E}}$.

Proposition 8.3.2. *Let $A \in \mathscr{L}$. Then $A \in \overline{\mathscr{E}}$ if and only if $m^*(A) < \infty$.*

Proof. First suppose $A \in \overline{\mathscr{E}}$ and choose a sequence (A_n) in $\mathscr{E}$ with $d(A, A_n) \to 0$ as $n \to \infty$. Choose $N \in \mathbb{N}$ with $d(A, A_N) < 1$. Also recall that $m^*(A_N) = m(A_N) < \infty$ since $A_N \in \mathscr{E}$. These two facts together with the triangle inequality for d produce the desired estimate,

$$m^*(A) = m^*(A \Delta \emptyset) = d(A, \emptyset) \leq d(A, A_N) + d(A_N, \emptyset) < 1 + m^*(A_N) < \infty.$$

For the converse, suppose $A \in \mathscr{L}$ with $m^*(A) < \infty$. From the definition of $\mathscr{L}$, there is a collection of sets $\{A_n\}_{n=1}$ in $\overline{\mathscr{E}}$ with $A = \bigcup_{n=1}^{\infty} A_n$. Our goal is now to find a sequence of sets in $\mathscr{E}$ which approximate A and to do this we will be forced to appeal to Lemma 8.2.9. To this end we first rewrite A as a countable union of pairwise disjoint sets by defining sets

$$B_1 = A_1, \quad B_2 = A_2 \setminus A_1, \quad B_3 = A_3 \setminus (A_1 \cup A_2),$$

and, in general,

$$B_n = A_n \setminus (A_1 \cup A_2 \cup \ldots \cup A_{n-1}).$$

Each of these sets is in $\overline{\mathscr{E}}$ since it is a ring; moreover, the B_n's are pairwise disjoint with $A = \bigcup_{n=1}^{\infty} B_n$, and thus $\sum_{n=1}^{\infty} m^*(B_n) = m^*(A) < \infty$ by Lemma 8.2.9.

At this point, consider the sequence $C_n = B_1 \cup \ldots \cup B_n \in \overline{\mathscr{E}}$. These sets certainly approximate A, but we need a sequence of sets in $\mathscr{E}$. The idea then is to approximate each of the C_n's with a set in $\mathscr{E}$ after which we will be able to approximate A in a transitive fashion. For each $m \in \mathbb{N}$, choose a set $D_m \in \mathscr{E}$ so that $d(C_m, D_m) < 1/m$ which is permissible since $C_m \in \overline{\mathscr{E}}$. To conclude that $d(A, D_m) \to 0$ as $m \to \infty$, let $\varepsilon > 0$ and choose M_1 so that $1/M_1 < \varepsilon/2$. Also, choose M_2 so that

$$\sum_{n=M_2}^{\infty} m^*(B_n) < \varepsilon/2$$

which is possible since the series converges. If $m > M = \max\{M_1, M_2\}$, the triangle inequality and the choice of M yield

$$d(A, D_m) \leq d(A, C_m) + d(C_m, D_m) = d\left(A, \bigcup_{n=1}^{m} B_n\right) + d(C_m, D_m)$$

$$= m^*\left(\bigcup_{n=m+1}^{\infty} B_n\right) + d(C_m, D_m) = \left(\sum_{n=m+1}^{\infty} m^*(B_n)\right) + d(C_m, D_m)$$

$$\leq \left(\sum_{n=M_2}^{\infty} m^*(B_n)\right) + 1/m$$

$$< \varepsilon/2 + \varepsilon/2 = \varepsilon,$$

completing the proof. $\qquad\qquad\qquad\qquad\qquad\qquad\qquad\qquad\qquad\qquad\qquad\qquad\square$

Theorem 8.3.3. *The collection $\mathscr{L}$ is a σ-algebra of sets in $\mathbb{R}$ and m^* is countably additive on $\mathscr{L}$.*

Proof. The proof that $\mathscr{L}$ is a σ-algebra is left as Exercise 8.3.1. To show that m^* is countably additive on $\mathscr{L}$, suppose that $\{A_n\}$ is a countable collection of pairwise disjoint sets in $\mathscr{L}$ and set $A = \bigcup_{n=1}^{\infty} A_n$. If there exists an n such that $m^*(A_n) = \infty$, then Eq. (8.1) holds trivially since the monotonicity of m^* implies that $m^*(A_n) \leq m^*(A)$. If $m^*(A_n) < \infty$ for all n, then Lemma 8.2.9 together with Proposition 8.3.2 provides the desired conclusion. $\square$

Definition 8.3.4. The collection $\mathscr{L}$ will henceforth be referred to as the *Lebesgue measurable subsets of* $\mathbb{R}$. Furthermore, the measure m defined on $\mathscr{L}$ by $m(A) = m^*(A)$ is called the *Lebesgue measure on* $\mathbb{R}$.

All of the results from Sect. 8.1 hold for $\mathscr{L}$ and m. In particular, $\mathscr{L}$ is closed with respect to countable intersections from which it easily follows that the Cantor set and others which can be written as a countable intersection of closed intervals are in $\mathscr{L}$. A statement about sets which are intersections of closed sets follows with Exercise 8.3.3. The fact that $\mathscr{L}$ is a σ-algebra also ensures us that $\mathscr{L}$ is closed with respect to complements. With this we see that the irrational numbers are Lebesgue measurable since we have already shown that the rational numbers are in $\overline{\mathscr{E}} \subseteq \mathscr{L}$.

Let us also take a moment to further understand the containment statement $\mathscr{E} \subseteq \overline{\mathscr{E}} \subseteq \mathscr{L} \subseteq \mathscr{P}(\mathbb{R})$. Proposition 8.3.2 allows us to know precisely when a Lebesgue measurable set is in $\overline{\mathscr{E}}$ and shows that $\overline{\mathscr{E}}$ is a proper subset of $\mathscr{L}$. The uppermost containment in this chain is a bit more interesting and has the potential to cause confusion if not carefully understood. Up to this point, our analysis has been focused on intervals, open, closed, and compact sets, and unions, intersections, and complements of these types of sets. Moreover, $\mathscr{L}$ contains all these basic types of sets and is closed with respect to the usual operations on them. The question then becomes, is $\mathscr{L} = \mathscr{P}(\mathbb{R})$? As alluded to in Sect. 8.1, this is not the case. While we have constructed a class of sets which is rich and more than adequate for modern analysis, we have by no means exhausted the subsets of $\mathbb{R}$. On the other hand, constructing a set which is not in $\mathscr{L}$ is difficult, and will be explored in Sect. 8.4. This construction will also verify that outer measure fails to be countably additive. That the most basic example of a nonmeasurable set demonstrates this failure should convince the reader of the delicate nature of the proceeding work. One fact needed for this construction is that m is translation invariant on $\mathscr{L}$.

Theorem 8.3.5. *Let $A \in \mathscr{L}$ and $t \in \mathbb{R}$. Then $A + t \in \mathscr{L}$ and $m(A + t) = m(A)$.*

Proof. We first show that $A + t \in \mathscr{L}$. Letting $A = \bigcup_{n=1}^{\infty} A_n$, where $A_n \in \overline{\mathscr{E}}$, we can write $A + t = \bigcup_{n=1}^{\infty} (A_n + t)$. By Exercise 8.2.8, the set $A_n + t$ is in $\overline{\mathscr{E}}$ for every $n \in \mathbb{N}$, and hence $A + t \in \mathscr{L}$. The equality of measures then follows from Theorem 8.2.5. $\square$

On occasion we will want to discuss smaller collections of Lebesgue measurable sets. If A is a Lebesgue measurable subset of $\mathbb{R}$, we will use the notation

$$\mathscr{L}(A) = \{B \in \mathscr{L} : B \subseteq A\}.$$

It follows easily that m is also a measure on this collection of sets. Equivalently this set could be defined as

$$\{B \subseteq A : \text{ there is a Lebesgue measurable set } T \text{ with } B = A \cap T\}.$$

The most common instance of this will involve collections of the form $\mathscr{L}([a, b])$. One reason, among many, for considering these smaller collections is that every set in $\mathscr{L}(A)$ has finite measure provided $m(A) < \infty$. This allows a bit more flexibility by avoiding the problems that arise in algebraic manipulations involving infinite values. For instance, if $B \in \mathscr{L}([0, 2])$, then $m([0, 2]) = m(B) + m(B^c)$ and we are justified in writing $m(B^c) = 2 - m(B)$. Keep in mind that in this situation, the complement is taken with respect to $[0, 2]$ and to be more formal we could write B^c as $[0, 2] \setminus B$. To see a more explicit example of how this is useful, we point the reader to Exercise 8.3.7; part (b) requires some significant machinery to extend the regularity condition to unbounded measurable sets, while part (a) follows quickly from the definition of outer measure and the discussion above.

On a historical note, the construction we have presented is not the construction originally given by Lebesgue in his doctoral thesis, accepted in 1902. Lebesgue's construction of "measurable sets" focused on subsets of bounded intervals and a comparison between the inner (Exercise 8.2.10) and outer measure of such subsets. This work, along with Lebesgue's integral, was not immediately accepted and met with much criticism. Despite this initial backlash, Lebesgue's work was eventually seen for its genius and intrinsic value and is the foundation of much of modern analysis. In the years following Lebesgue's work, there were several reconstructions of his theory. The first was presented by Carathéodory in 1914 and states that a set $A \subseteq \mathbb{R}$ is in $\mathscr{L}$ if and only if $m^*(T) = m^*(T \cap A) + m^*(T \cap A^c)$ for every $T \subseteq \mathbb{R}$. One of the most striking features of Carathéodory's construction is that it removed the restriction which forced Lebesgue to consider bounded subsets of $\mathbb{R}$. While there have been other variations on the construction of $\mathscr{L}$, Carathéodory's seems to be the most insightful. As we discussed, the issue with outer measure is that it is not additive on $\mathscr{P}(\mathbb{R})$; the above formulation brings this to the forefront and says that a set A is measurable if and only if it always splits into pieces on which m^* is additive.

Exercises

Exercise 8.3.1. Show that $\mathscr{L}$ is a σ-algebra of sets in $\mathbb{R}$.

Exercise 8.3.2. Let $A, B \in \mathscr{L}$. Show that $m(A) + m(B) = m(A \cup B) + m(A \cap B)$.

Exercise 8.3.3. (a) Show that every open subset of $\mathbb{R}$ is Lebesgue measurable.
(b) Show that every closed subset of $\mathbb{R}$ is Lebesgue measurable.

(c) Show that every compact subset of $\mathbb{R}$ is in $\overline{\mathscr{E}}$.

Exercise 8.3.4. Let $\mathscr{R}$ be the collection of all subsets of $\mathbb{R}$ with Lebesgue measure zero. Show that $\mathscr{R}$ is a σ-ring.

Exercise 8.3.5 (Completeness of $\mathscr{L}$).

(a) Let $A \subseteq \mathbb{R}$ with $m^*(A) = 0$. Show that $A \in \mathscr{L}$.
(b) Let $A, B \subseteq \mathbb{R}$ with $B \subseteq A$, $A \in \mathscr{L}$, and $m^*(A) = 0$. Show that $B \in \mathscr{L}$ with $m^*(B) = 0$.

Exercise 8.3.6. Let $A \subseteq \mathbb{R}$. Show that $A \in \mathscr{L}$ if and only if $A \cap T \in \overline{\mathscr{E}}$ for every $T \in \overline{\mathscr{E}}$.

Exercise 8.3.7 (Regularity of m).

(a) Let $A \in \mathscr{L}([a, b])$ and let $\varepsilon > 0$. Show that there is an open set O and a closed set F with $F \subseteq A \subseteq O$ such that $m(O) < m(A) + \varepsilon$ and $m(F) > m(A) - \varepsilon$.
(b) Let $A \in \mathscr{L}$ and $\varepsilon > 0$. Show that there is an open set O and a closed set F with $F \subseteq A \subseteq O$ such that $m(O \setminus F) < \varepsilon$. The three points below are meant to act as a guide.

   (i) To construct the open set, begin by considering sets of the form $A_j = A \cap [j, j + 1]$ and part (a) to show that there is an open set O_j with $A_j \subseteq O_j$ and $m(O_j \setminus A_j) < \varepsilon/2^{|j|+3}$.

   (ii) Letting $O = \bigcup_n^\infty O_j$, where the union is over all the integers, show that $A \subseteq O$ and $m(O \setminus A) < 3\varepsilon/8 < \varepsilon/2$.

   (iii) To construct the closed set F, consider (i) and (ii) applied to A^c. Finally, show that $m(O \setminus F) < \varepsilon$.

8.4 A Nonmeasurable Set

We now work to show that there are sets which are not Lebesgue measurable. The construction of such a set depends on the Axiom of Choice which is stated as follows.

Theorem 8.4.1 (Axiom of Choice). *Let $\mathscr{C}$ be any collection of nonempty sets. Then there is a function f defined on $\mathscr{C}$ so that $f(E) \in E$ for each $E \in \mathscr{C}$.*

In short, the statement above says that we can sort through the sets in $\mathscr{C}$ and choose a single element from each set. If $\mathscr{C}$ is a finite set, then of course we can do this. However, in the case when $\mathscr{C}$ is countable or even uncountable, the plausibility of such a process becomes unclear and the axiom allows us to rest assured that this is possible. Keep in mind that this statement is taken as an axiom and is not something that we are going to prove (or something that we can prove within the confines of our current system). The axiom has a long and tumultuous history and for a reference we encourage the reader to consider pp. 150–157 in [7]. In our particular application, we will take the collection $\mathscr{C}$ to be an uncountable collection of disjoint sets and then form a new set P which consists of a single element from each set in $\mathscr{C}$.

Theorem 8.4.2. *There is a set which is not Lebesgue measurable.*

Proof. To construct the set, we first define an equivalence relation on $(0, 1]$ by $x \sim y$ if $x - y \in \mathbb{Q}$. This relation partitions $(0, 1]$ into equivalence classes $\{E_\alpha\}$. Each class E_α contains at most countably many elements since $\mathbb{Q}$ is countable, and thus necessarily there are uncountably many such classes. By the Axiom of Choice we form a set P by choosing exactly one element from each of these equivalence classes.

To show P is not Lebesgue measurable, assume that it is. In order to generate a contradiction, one preliminary idea is necessary. Define an operation $\oplus$ on $(0, 1]$ by

$$x \oplus y = \begin{cases} x + y, & x + y \le 1; \\ x + y - 1, & x + y > 1. \end{cases}$$

Also, if $A \subseteq (0, 1]$ and $x \in (0, 1]$ we define the *wrap-around translation* $A \oplus x = \{a \oplus x : a \in A\}$. A modification of the proof of Theorem 8.3.5 shows that $A \oplus x$ is Lebesgue measurable and satisfies $m(A \oplus x) = m(A)$ whenever A is a Lebesgue measurable subset of $(0, 1]$. In particular, $m(P \oplus x) = m(P)$ for every $x \in (0, 1]$.

Now, let $\{r_1, r_2, r_3, \dots\}$ be an enumeration of $\mathbb{Q} \cap (0, 1]$ and define sets $P_n = P \oplus r_n$. The first observation here is that $P_n \cap P_m = \emptyset$ when $n \ne m$. To verify this we argue by contrapositive, i.e. suppose there is some x in this intersection and show that $n = m$. For x, we can find y and z in P with $x - y \oplus r_n = z \oplus r_m$ or equivalently $(y \oplus r_n) - (z \oplus r_m) = 0$. Considering the four cases that arise from the definition of $\sim$, it must be the case that $y - z$ is rational showing that $y \sim z$. However, since P was chosen to contain exactly one element from each class E_α, it follows that $y = z$. Resorting to cases again with the equation $y \oplus r_n = y \oplus r_m$ shows that $r_n = r_m$ from which it follows that $n = m$.

The second key observation is that $\cup_{n=1}^\infty P_n = (0, 1]$. We leave the proof of this fact as an exercise. For the contradiction, the countable additivity of m suggests that

$$1 = m((0, 1]) = m\left(\bigcup_{n=1}^{\infty} P_n\right) = \sum_{n=1}^{\infty} m(P_n) = \sum_{n=1}^{\infty} m(P),$$

but this situation is an absurdity. The only chance for a sum of this nature to be finite is for $m(P) = 0$ but the sum also has no hope of being equal to 1 unless $m(P) > 0$. Ergo, P is not Lebesgue measurable. $\square$

As promised in Sect. 8.3, the last equation of this proof also demonstrates that outer measure m^* fails to be countably additive on $\mathscr{P}(\mathbb{R})$.

Exercises

Exercise 8.4.1. Suppose A is a Lebesgue measurable subset of $(0, 1]$ and let $x \in (0, 1]$. Show that $A \oplus x$ is Lebesgue measurable and satisfies $m(A \oplus x) = m(A)$. For a hint, begin by rewriting A as a disjoint union of sets which depend on the translation parameter x.

Exercise 8.4.2. Using the definitions and notation of the proof of Theorem 8.4.2, show that $\cup_{n=1}^{\infty} P_n = (0, 1]$.

8.5 General Measure Theory

Measure theory plays a very significant role in many areas of mathematics including integration theory, probability, and functional analysis. The definitions and theorems of Sect. 8.1 can also be defined in greater generality. The statements of the definitions and theorems are essentially the same as those presented in the earlier sections but we also include several fundamental examples of measures.

Definition 8.5.1. Let X be a nonempty set. A collection $\mathcal{M}$ of subsets of X is called a *σ-ring of sets in X* if

(a) $A \setminus B \in \mathcal{M}$ whenever $A, B \in \mathcal{M}$;
(b) $\bigcup_{n=1}^{\infty} A_n \in \mathcal{M}$ whenever $\{A_n\}_{n=1}^{\infty}$ is in $\mathcal{M}$.

If $X \in \mathcal{M}$, we call $\mathcal{M}$ a *σ-algebra of sets in X*.

We have seen several examples of this when $X = \mathbb{R}$. The beauty is that now we can consider other base sets.

Example 8.5.2. Let X be any set. Then the power set of X, $\mathscr{P}(X)$, is a σ-algebra.

Example 8.5.3. Let $X = [a, b] \subseteq \mathbb{R}$ and define $\mathscr{L}([a, b])$ to be the collection of all subsets of $[a, b]$ which are Lebesgue measurable. These sets were discussed briefly in Sect. 8.3 and hold relevance to the discussion from the end of Sect. 7.5. You are asked to verify that this set forms a σ-algebra in the exercises. When doing this, keep in mind that we have already shown that $\mathscr{L}$ has this property.

Example 8.5.4. Let $X = \mathbb{R}$ and consider the collection of all open intervals. This set is not a σ-algebra but we can use it to construct one. Let $\mathscr{B}$ denote the collection of all subsets of $\mathbb{R}$ which can be formed from (repeatedly) taking countable unions and intersections, and complements of open intervals. By definition $\mathscr{B}$ is a σ-algebra and is referred to as the *Borel σ-algebra of* $\mathbb{R}$. Another way to phrase this is that $\mathscr{B}$ is the smallest σ-algebra which contains all the open subsets of $\mathbb{R}$. The existence of such a "smallest" collection is a statement which needs proving, but we will take it on faith.

As a point of caution, the set contains every open subset of $\mathbb{R}$ and every closed set, but there are many other sets in this collection. For a simple example, the interval $(1, 2]$ is in $\mathscr{B}$ since it can be written as a countable intersection of open sets,

$$(1, 2] = \bigcap_{n=1}^{\infty} \left(1, 2 + \frac{1}{n} \right).$$

For another example, the Cantor set is a Borel set since it can be written as a countable intersection of closed intervals. Further, $\mathscr{L}$ also contains all these sets and it may seem plausible that they are in fact the same σ-algebra. The work of the previous sections would be irrelevant if this were the case and, though we don't present an example here, there are Lebesgue measurable sets which are not Borel sets. The containment $\mathscr{B} \subseteq \mathscr{L}$ is obvious based on the definition of $\mathscr{B}$ and our construction of $\mathscr{L}$. However, there is more to be said and Exercise 8.5.4 encapsulates the true relationship between these two collections.

Definition 8.5.5. Let $\mathscr{M}$ be a σ-algebra in X. A *positive measure* on $\mathscr{M}$ is a function $\mu : \mathscr{M} \to [0, \infty]$ which satisfies

(a) $\mu(\emptyset) = 0$;
(b) μ is *countably additive*: If $\{A_n\}_{n=1}$ in $\mathscr{M}$ is a pairwise disjoint collection of sets, then

$$\mu \left(\bigcup_{n=1}^{\infty} A_n \right) = \sum_{n=1}^{\infty} \mu(A_n). \tag{8.3}$$

The triple $(X, \mathscr{M}, \mu)$ is called a *measure space on* X.

While our most important example of a measure will be Lebesgue measure on the real line, there are also several other examples which will hold significant value for us.

Example 8.5.6. The triple $([a, b], \mathscr{L}([a, b]), m)$ is a measure space. For notational convenience we will simply denote this triple by $\mathscr{L}([a, b])$. The fact that m is a measure on this collection of sets is a direct consequence of the fact that m is a measure on $\mathscr{L}$.

Example 8.5.7. Let $X = \mathbb{N}$ and set $\mathscr{M} = \mathscr{P}(\mathbb{N})$. Define $\mu : \mathscr{P}(\mathbb{N}) \to [0, \infty]$ by

$$\mu(A) = \begin{cases} |A|, & A \text{ is finite}; \\ \infty, & A \text{ is infinite}. \end{cases}$$

Again, this type of measure is referred to as a *counting measure*. The proof that this set function is countably additive is identical to the proof given in Exercise 8.1.3 for the case when $X = \mathbb{R}$.

We take our final example from the realm of probability theory.

Example 8.5.8. Let $(X, \mathcal{M}, \mu)$ be a measure space with $\mu(X) = 1$. In this situation we refer to the triple $(X, \mathcal{M}, \mu)$ as a *probability space* and call μ a *probability measure*. To see why the terminology is used, let $X = \{1, 2, 3, 4, 5, 6\}$, $\mathcal{M} = \mathcal{P}(X)$, and define $\mu : \mathcal{P}(X) \to [0, \infty]$ such that

1. $\mu(\{x\}) = \frac{1}{6}$ for each $x \in X$;
2. $\mu(A \cup B) = \mu(A) + \mu(B)$ whenever $A, B \in \mathcal{M}$ with $A \cap B = \emptyset$.

This example simulates the rolling of a six-sided die (assuming the die is fair, i.e. each outcome is equally likely) using the numbers 1 through 6 to represent the distinct sides. The set X represents all the outcomes of a single roll and the collection $\mathcal{M}$ represents all combinations of possible outcomes. The requirement that $\mu(X) = 1$ indicates that every outcome of the experiment is represented in the set X. For example, find the probability of rolling a 1 and the probability of rolling an even number. To roll a one, the event $\{1\}$ must occur and by definition this has probability $\mu(\{1\}) = \frac{1}{6}$ as expected. Rolling an even number is represented by the event $\{2, 4, 6\}$ and has probability

$$\mu(\{2, 4, 6\}) = \mu(\{2\}) + \mu(\{4\}) + \mu(\{6\}) = \frac{1}{2}.$$

In the most abstract sense, any discrete probability setting can be modeled as follows. Let $X = \{x_1, x_2, x_3, \ldots\}$ be a finite or countable set and let $\mathcal{M} = \mathcal{P}(X)$. Also let (a_n) be a sequence in $[0, \infty)$ with $\sum_{n=1}^{\infty} a_n = 1$. Then define $\mu : \mathcal{P}(X) \to [0, \infty]$ by

$$\mu(A) = \sum_{x_n \in A} a_n.$$

This set function is clearly a measure on $\mathcal{M}$ and satisfies the additional hypothesis that $\mu(X) = 1$ by choice of the sequence (a_n).

The theory of continuous probability spaces is necessarily more complicated than the discrete setting, and one of the most remarkable features of measure theory and the forth coming theory of Lebesgue integration is the fact that it provides a common framework for both discrete and continuous probability theory.

All of the properties of positive measures for σ-rings in $\mathbb{R}$ hold in this more general context and are restated in the next theorem.

Theorem 8.5.9 (Properties of Measures). *Let $\mathcal{M}$ be a σ-ring on a set X and let μ be a positive measure on $\mathcal{M}$.*

(a) Monotonicity—If $A, B \in \mathcal{M}$ with $A \subseteq B$, then $\mu(A) \leq \mu(B)$.
(b) Continuity for increasing unions—Suppose $\{A_n\}_{n=1}^{\infty}$ is in $\mathcal{M}$ with $A_1 \subseteq A_2 \subseteq A_3 \subseteq \ldots$ and $A = \bigcup_{n=1}^{\infty} A_n$. Then

$$\mu(A) = \lim_{n \to \infty} \mu(A_n).$$

(c) *Continuity for decreasing intersections—Suppose $\{A_n\}_{n=1}$ is in $\mathcal{M}$ with $A_1 \supseteq A_2 \supseteq A_3 \supseteq \dots$ and $A = \bigcap_{n=1}^{\infty} A_n$. If $\mu(A_1) < \infty$, then*

$$\mu(A) = \lim_{n \to \infty} \mu(A_n).$$

(d) *Countably subadditive—If $\{A_n\}_{n=1}$ is in $\mathcal{M}$, then $\mu\left(\bigcup_{n=1}^{\infty} A_n\right) \le \sum_{n=1}^{\infty} \mu(A_n)$.*

Exercises

Exercise 8.5.1. Let X be a set and let $\mathcal{M}$ be a σ-algebra in X.

(a) Show that the emptyset is in $\mathcal{M}$.
(b) Show that $\mathcal{M}$ is closed under countable intersections.
(c) Show that $\mathcal{M}$ is closed under complementation.

Exercise 8.5.2. Show that $\mathcal{L}([a, b])$ is a σ-algebra.

Exercise 8.5.3. Let X be any uncountable set and let

$$\mathcal{M} = \{A \subseteq X : A \text{ or } X \setminus A \text{ is countable}\}.$$

Define $\mu : \mathcal{M} \to [0, \infty]$ by

$$\mu(A) = \begin{cases} 0, & A \text{ is countable;} \\ 1, & X \setminus A \text{ is countable.} \end{cases}$$

Show that $\mathcal{M}$ is a σ-algebra on X and that μ is a measure.

Exercise 8.5.4. (a) Show that every Lebesgue measurable set can be written as the union of a Borel set and a set of Lebesgue measure zero. You may want to consider Exercise 8.3.7.
(b) Let $a \in \mathbb{R}$. Show that the set $\{a\}$ is a Borel set.
(c) Show that $\mathbb{Q}$ is a Borel set.

Exercise 8.5.5. Let $(X, \mathcal{M}, \mu)$ be a measure space and let $\mathcal{R} \subseteq \mathcal{M}$ be the collection of all measurable subsets of X with measure zero. Show that $\mathcal{R}$ is a σ-ring.

Chapter 9
Lebesgue Integration

Here we continue on our journey toward defining the Lebesgue integral. The previous chapter developed the theory of positive measures and measurable sets which will play the roles filled by length and intervals in the Riemann theory. The next phase of this process is to determine the class of functions to which we will apply our new integral. The final step will then be to construct the Lebesgue integral. As previously mentioned, this new integral will be concerned more specifically with the range of a function rather than its domain and will therefore be defined for functions which map into $\mathbb{R}$, but whose domain may be some other set, be it of numbers or other objects. In Sect. 7.5 and the introduction to Chap. 8 we spent a considerable amount of time discussing some of the major deficiencies of the Riemann integral. Once we have our formulation for the Lebesgue integral, we will demonstrate that it is a vast improvement upon its predecessor.

9.1 Measurable Functions

This section is devoted to developing a class of functions which we will later attempt to integrate. To gain as much ground as possible, we will be using a general measure space $(X, \mathcal{M}, \mu)$, where $\mathcal{M}$ is a σ-algebra and, unless otherwise stated, functions f defined on X and taking values in $[-\infty, \infty]$. For specific examples we will often return to the measure space $(\mathbb{R}, \mathcal{L}, m)$. As a matter of terminology, when we say that a set $E \subseteq X$ is measurable, we simply mean that E is an element of the given σ-algebra.

Definition 9.1.1. Let $(X, \mathcal{M}, \mu)$ be a measure space and let $f : X \to [-\infty, \infty]$. We say that f is a *measurable function* (or *$\mathcal{M}$-measurable* for emphasis) if for every $a \in \mathbb{R}$ the set $\{x : f(x) > a\}$ is a measurable set. In other words, if $f^{-1}((a, \infty)) = \{x : f(x) > a\} \in \mathcal{M}$ for every $a \in \mathbb{R}$.

In mathematical language, the definition above requires that the inverse image of every unbounded, open interval be an element of $\mathcal{M}$. If you recall, when we

M.A. Pons, *Real Analysis for the Undergraduate: With an Invitation to Functional Analysis*, 351
DOI 10.1007/978-1-4614-9638-0_9, © Springer Science+Business Media New York 2014

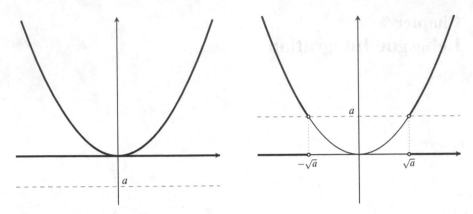

Fig. 9.1 Square function measurable

discussed continuity we came across a condition similar to this, equivalent to our original definition, which states that a function f is continuous on $\mathbb{R}$ if and only if the inverse image of every open set is open. The definition of measurable is relaxing that condition by only requiring that the inverse images be measurable. At the moment, this may not seem like a significant reduction, but it turns out to make tremendous difference, particularly when considering the pointwise limit of a sequence of measurable functions.

Example 9.1.2. Consider the measure space $(X, \mathcal{P}(X), \mu)$ where X is any set, $\mathcal{P}(X)$ is the power set of X, and μ is any measure on $\mathcal{P}(X)$. For specific examples we could consider those given in Examples 8.1.9 or 8.1.10. If f is any function on X which maps into $[-\infty, \infty]$, then f is measurable. This is due to the fact that *all* subsets of X are measurable. Though this may seem like a class of trivial examples, the two referenced will both play an interesting role when we define the Lebesgue integral.

Example 9.1.3. For a more concrete example, consider the function $f(x) = x^2$ and the measure space $(\mathbb{R}, \mathcal{L}, m)$. To show f is measurable we consider two cases: $a < 0$ and $a \geq 0$. If $a < 0$, then

$$\{x : f(x) > a\} = \{x : x^2 > a\} = \mathbb{R}$$

since $x^2 \geq 0 > a$ for every $x \in \mathbb{R}$. If $a \geq 0$,

$$\{x : f(x) > a\} = \{x : x < -\sqrt{a} \text{ or } x > \sqrt{a}\} = (-\infty, -\sqrt{a}) \cup (\sqrt{a}, \infty).$$

In either case, the set $\{x : f(x) > a\}$ is in $\mathcal{L}$ since $\mathcal{L}$ contains all the open sets and thus f is measurable (Fig. 9.1).

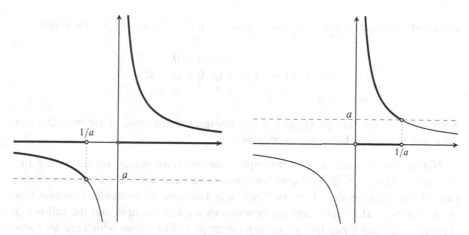

Fig. 9.2 Reciprocal function measurable

More generally, any continuous function f is measurable with respect to the measure space $(\mathbb{R}, \mathscr{L}, m)$ since the inverse image of any open set is open and $\mathscr{L}$ contains all open subsets of $\mathbb{R}$. This is essentially what was used in the previous example, though the fact that we were dealing with a specific function allowed us to identify the sets $\{x : f(x) > a\}$ explicitly.

Example 9.1.4. For a discontinuous example with respect to the same measure space, consider the function

$$f(x) = \begin{cases} \dfrac{1}{x}, & x \neq 0; \\ 0, & x = 0, \end{cases}$$

which has a discontinuity at zero. As before, the inverse image of (a, ∞) depends on the parameter a and we find

$$\{x : f(x) > a\} = \begin{cases} (-\infty, 1/a) \cup [0, \infty), & a < 0; \\ (0, \infty), & a = 0; \\ (0, 1/a), & a > 0. \end{cases}$$

Each of these sets is certainly in $\mathscr{L}$ and thus this discontinuous function is Lebesgue measurable. Notice also that the discontinuity is brought to light in the top row since the set there is not open (Fig. 9.2).

Example 9.1.5. Let f be the Dirichlet function, that is,

$$f(x) = \begin{cases} 1, & x \in \mathbb{Q}; \\ 0, & x \notin \mathbb{Q}, \end{cases}$$

and again consider the measure space $(\mathbb{R}, \mathcal{L}, m)$. Calculating the inverse images,

$$\{x : f(x) > a\} = \begin{cases} \mathbb{R}, \ a < 0; \\ \mathbb{Q}, \ 0 \le a < 1; \\ \emptyset, \ a \ge 1. \end{cases}$$

Since each of the sets $\mathbb{R}$, $\mathbb{Q}$, and $\emptyset$ is Lebesgue measurable, it follows that the Dirichlet function is Lebesgue measurable.

Notice that in each of these examples, the measure played no role, only the σ-algebra. Also, as with the Lebesgue measurable sets, and from these few examples, it may seem as though every function is Lebesgue measurable. However this is not the case. The next definition generalizes the last example and the following theorem will make it possible to construct examples of functions which are not measurable. As was also the case with $\mathcal{L}$, this is not a severe setback. The elementary functions developed in the first-year calculus sequence, even the discontinuous ones, are all Lebesgue measurable, and we shall see that much more is gained.

Definition 9.1.6. Let $(X, \mathcal{M}, \mu)$ be a measure space and let $E \subseteq X$. The *characteristic function of E*, denoted χ_E, is defined by

$$\chi_E(x) = \begin{cases} 1, \ x \in E; \\ 0, \ x \notin E. \end{cases}$$

More generally, a function s whose range consists of finitely many (finite) values is called a *simple function*. If $c_1, c_2, \ldots c_n$ are the distinct range values and $C_j = \{x : s(x) = c_j\}$, then $s(x) = \sum_{j=1}^{n} c_j \chi_{C_j}(x)$ is called the *canonical representation of s*.

With this terminology, the Dirichlet function is seen to be the characteristic function (or *indicator function*) for the rationals. These functions will provide the basic building blocks for defining our new integral and the theorem below identifies when they are measurable.

Theorem 9.1.7. *Let $(X, \mathcal{M}, \mu)$ be a measure space. The function χ_E is a measurable function if and only if E is a measurable set. A simple function s in canonical form is measurable if and only if C_j is measurable for $j = 1, 2, \ldots n$.*

Simple functions often have multiple representations and the canonical form is always ideal; other forms usually involve characteristic functions of sets which are not disjoint. For example, the functions $f(x) = \chi_{[0,1]}(x) + 2\chi_{(1,2]}(x)$ and $g(x) = \chi_{[0,2]}(x) + \chi_{(1,2]}(x)$ represent the same simple function, but the first is in canonical form while the second is not. Furthermore, the previous theorem concerning the measurability of a simple function does not apply to non-canonical forms and for this reason we will always assume that we are dealing with canonical representations.

Returning to the definition of a measurable function, it is natural to ask why the "greater than" symbol was used. Is this definition altered if equality is allowed or if we consider when $f(x) < a$? The following theorem demonstrates that this was an arbitrary choice since we are working within a σ-algebra. Exercise 9.1.1 will supply two more variations on this theme.

Theorem 9.1.8. *Let $(X, \mathcal{M}, \mu)$ be a measure space and let $f : X \to [-\infty, \infty]$. The following are equivalent:*

(a) $\{x : f(x) > a\}$ is a measurable set for every $a \in \mathbb{R}$;
(b) $\{x : f(x) \geq a\}$ is a measurable set for every $a \in \mathbb{R}$;
(c) $\{x : f(x) < a\}$ is a measurable set for every $a \in \mathbb{R}$;
(d) $\{x : f(x) \leq a\}$ is a measurable set for every $a \in \mathbb{R}$.

Proof. The equivalences $(a) \Leftrightarrow (d)$ and $(b) \Leftrightarrow (c)$ follow since the sets are complementary and $\mathcal{M}$ is a σ-algebra. The desired equivalence will then follow if we show $(a) \Leftrightarrow (b)$. To show $(a) \Rightarrow (b)$, let $a \in \mathbb{R}$ and observe the set equality

$$\{x : f(x) \geq a\} = \bigcap_{k=1}^{\infty} \left\{x : f(x) > a - \frac{1}{k}\right\}.$$

The fact that we are assuming (a) guarantees that the set $\{x : f(x) > a - 1/k\}$ is measurable for every $k \in \mathbb{N}$. From this it follows that the set on the right-hand side is measurable since $\mathcal{M}$ is a σ-algebra, and thus the set $\{x : f(x) \geq a\}$ is also measurable.

That $(b) \Rightarrow (a)$ follows similarly from the equality

$$\{x : f(x) > a\} = \bigcup_{k=1}^{\infty} \left\{x : f(x) \geq a + \frac{1}{k}\right\}$$

and the fact that $\mathcal{M}$ is σ-algebra. $\qquad\square$

The next theorem demonstrates that measurable functions respect the standard algebraic operations.

Proposition 9.1.9. *Let $(X, \mathcal{M}, \mu)$ be a measure space, let $f, g : X \to \mathbb{R}$ be measurable functions, and let $c \in \mathbb{R}$. Then $|f|$, cf, $f + g$ and fg are measurable.*

Proof. To show that $f + g$ is measurable, we'll show that the set $\{x : (f+g)(x) > a\}$ is measurable for each $a \in \mathbb{R}$. To this end, let a be a real number. First, we rewrite the inequality in question, $(f+g)(x) > a$, as $f(x) > a - g(x)$. Since this inequality is strict, we can find a rational number r with $f(x) > r > a - g(x)$. The set inclusion

$$\{x : (f + g)(x) > a\} \subseteq \bigcup_{r \in \mathbb{Q}} \left(\{x : f(x) > r\} \cap \{x : r > a - g(x)\} \right)$$

then follows immediately. The reverse inclusion clearly holds and we obtain the equality,

$$\{x : (f + g)(x) > a\} = \bigcup_{r \in \mathbb{Q}} \Big(\{x : f(x) > r\} \cap \{x : r > a - g(x)\} \Big)$$

$$= \bigcup_{r \in \mathbb{Q}} \Big(\{x : f(x) > r\} \cap \{x : g(x) > a - r\} \Big).$$

Applying our hypothesis that f and g are measurable, we conclude that the sets $\{x : f(x) > r\}$ and $\{x : g(x) > a - r\}$ are measurable for every rational r. Thus the intersection of these sets is measurable, as is the countable union over $\mathbb{Q}$, since $\mathcal{M}$ is a σ-algebra. Hence, for each $a \in \mathbb{R}$, $\{x : (f + g)(x) > a\}$ is a measurable set showing that the function $f + g$ is measurable.

Before showing the general statement for fg, let's consider the case when $f = g$ and show f^2 is measurable. If $a < 0$, it should be clear that $\{x : f(x)^2 > a\} = X$. If $a \geq 0$, then, mimicking Example 9.1.3, $\{x : f(x)^2 > a\} = \{x : f(x) < -\sqrt{a}\} \cup \{x : f(x) > \sqrt{a}\}$. In either case, we obtain measurable sets and it follows that f^2 is measurable on X.

For fg, we appeal to the identity

$$fg = \frac{1}{2}((f + g)^2 - f^2 - g^2).$$

Since f and g are measurable by hypothesis, each term inside the parenthesis on the right-hand side is measurable. Moreover, the entire term is measurable. That fg is measurable now follows from the fact that constant multiples of measurable functions are measurable; this last observation will be explored in the exercises. $\square$

A fact which we will exploit later on and which is implicitly shown by the preceding proposition is that the Lebesgue measurable functions form a real vector space, i.e. are closed with respect to addition and multiplication by real scalars and obey certain operational axioms with respect to these operations. The remainder of this section is devoted to an investigation of sequences of measurable functions. The next few statements will demonstrate the fact that measurability provides much more flexibility than continuity.

Definition 9.1.10. Let $(X, \mathcal{M}, \mu)$ be a measure space and let f be a measurable function. We define the *positive* and *negative parts of* f by

$$f_+(x) = \max\{f(x), 0\} \quad \text{and} \quad f_-(x) = \max\{-f(x), 0\},$$

respectively. If (f_n) is a sequence of measurable functions, we define the *supremum* and *infimum* functions as

$$(\sup f_n)(x) = \sup\{f_n(x) : n \in \mathbb{N}\} \quad \text{and} \quad (\inf f_n)(x) = \inf\{f_n(x) : n \in \mathbb{N}\}.$$

Two facts which will be of great importance later on are the decompositions

$$f = f_+ - f_- \quad \text{and} \quad |f| = f_+ + f_-.$$

It is also important to keep in mind that these definitions are pointwise. The following example aims to clarify this.

Example 9.1.11. Take $f_n(x) = x^{1/n}$ as a sequence of measurable functions with respect to the σ-algebra $\mathcal{L}([0, 1])$. To calculate sup f_n at the point x, we let $x \in [0, 1]$ and calculate $\sup\{f_n(x)\} = \sup\{x^{1/n}\}$. If $x = 0$, then this value is always zero and the corresponding value of (sup f_n)(0) is zero. If x is not zero, then the sequence $x^{1/n}$ is increasing and converges to one and hence (sup f_n)(x) = 1. Thus we find that, for this particular sequence,

$$(\sup f_n)(x) = \begin{cases} 0, & x = 0; \\ 1, & 0 < x \leq 1, \end{cases}$$

is a measurable but discontinuous function. In a similar fashion, we could calculate (inf f_n), which turns out to be simply the identity function $g(x) = x$. Thus the inf function is continuous and hence measurable. In both cases notice that the new function is measurable.

Proposition 9.1.12. *Let* $(X, \mathcal{M}, \mu)$ *be a measure space and let* (f_n) *be a sequence of measurable functions. Then* sup f_n *and* inf f_n *are measurable.*

Proof. To show that the function sup f_n is measurable we'll need to understand the inequality $\sup\{f_n(x)\} > a$. It should be clear that this inequality holds if there exists at least one n for which $f_n(x) > a$. With this in mind, we have

$$\{x : (\sup f_n)(x) > a\} = \{x : f_n(x) > a \text{ for some } n\} = \bigcup_{n=1}^{\infty} \{x : f_n(x) > a\}.$$

Notice that each set in the union on the right is measurable since the f_n's are measurable, from whence it follows that the union itself is measurable. By definition, we conclude that sup f_n is a measurable function. $\square$

Corollary 9.1.13. *Let* $(X, \mathcal{M}, \mu)$ *be a measure space and let* f *be a measurable function. Then* f_+ *and* f_- *are measurable.*

A similar result holds for the limit superior and inferior of a sequence of measurable functions. First we recall the definition of these two terms. Let (a_n) be a sequence in $[-\infty, \infty]$. Define a new sequence $b_n = \sup\{a_k : k \geq n\}$. By construction, (b_n) is a monotone non-increasing sequence and we set $\beta = \lim_{n \to \infty} b_n$. If (a_n) is bounded, then the Monotone Convergence Theorem for sequences guarantees that β is a real number. If (a_n) is unbounded, then $\beta = \infty$, but in either case $\beta \in [-\infty, \infty]$ and we write $\beta = \limsup_{n \to \infty} a_n$ to mean

$$\beta = \lim_{n \to \infty} b_n = \lim_{n \to \infty} (\sup\{a_k : k \geq n\}) = \inf_{n \geq 1} \{\sup\{a_k : k \geq n\}\}.$$

A similar definition holds for the limit inferior. Finally, recall that a sequence (a_n) of real numbers converges if and only if $\limsup a_n = \liminf a_n$; this common value must also be the limit of the sequence.

Definition 9.1.14. Let $(X, \mathcal{M}, \mu)$ be a measure space and let (f_n) be a sequence of measurable functions. We define the *limit superior* and *limit inferior* functions as

$$(\limsup f_n)(x) = \lim_{n \to \infty} (\sup\{f_k(x) : k \geq n\}) = \inf_{n \geq 1} (\sup\{f_k(x) : k \geq n\})$$

and

$$(\liminf f_n)(x) = \lim_{n \to \infty} (\inf\{f_k(x) : k \geq n\}) = \sup_{n \geq 1} (\inf\{f_k(x) : k \geq n\}).$$

Proposition 9.1.15. *Let $(X, \mathcal{M}, \mu)$ be a measure space and let (f_n) be a sequence of measurable functions. Then $\limsup f_n$ and $\liminf f_n$ are measurable.*

Proof. We'll show that the result holds for $\liminf f_n$. The $\limsup$ argument is completely analogous. First, define a new sequence $g_n(x) = \inf\{f_k(x) : k \geq n\}$. By Proposition 9.1.12, each function in this new sequence is measurable since the f_n's are measurable. Applying the proposition again, it follows that $g(x) = \sup_{n \geq 1}\{g_n(x)\}$ is also measurable. The conclusion now follows since,

$$g(x) = \sup_{n \geq 1}\{g_n(x)\} = \sup_{n \geq 1} (\inf\{f_k(x) : k \geq n\}) = (\liminf f_n)(x).$$

$\square$

To close this section we have a result which stands in contrast to the behavior of continuous, Riemann integrable, and differentiable sequences of functions with respect to the limit process.

Corollary 9.1.16. *Let $(X, \mathcal{M}, \mu)$ be a measure space and let (f_n) be a sequence of measurable functions. If the limit function $\lim_{n \to \infty} f_n(x) = f(x)$ exists for every $x \in X$, then f is measurable.*

Exercises

Exercise 9.1.1. Let $(X, \mathcal{M}, \mu)$ be a measure space and let $f : X \to [-\infty, \infty]$. For parts (a)–(c) it may be helpful to first rewrite the given set in set-builder notation, i.e. $f^{-1}((a, b)) = \{x : a < f(x) < b\}$.

(a) Show that f is a measurable function if and only if $f^{-1}((a,b))$ is a measurable set for all $a, b \in \mathbb{R}$ with $a \le b$.

(b) Show that f is a measurable function if and only if $f^{-1}(V)$ is a measurable set for every open set $V \subseteq \mathbb{R}$.

(c) Show that f is a measurable function if and only if $f^{-1}((r, \infty))$ is a measurable set for every $r \in \mathbb{Q}$.

(d) Suppose f is a measurable function. Show that the set $f^{-1}(\{a\})$ is a measurable set for every $a \in \mathbb{R}$.

Exercise 9.1.2. (a) Let $(X, \mathscr{M}, \mu)$ be a measure space and let $E \subseteq X$. For each $a \in \mathbb{R}$, identify the set $\{x : \chi_E(x) > a\}$ as a specific set in X.

(b) Supply a proof for the first statement in Theorem 9.1.7 using Definition 9.1.1.

(c) Give an example of a function which is not Lebesgue measurable.

(d) Supply a proof for the second statement in Theorem 9.1.7.

Exercise 9.1.3. Let $(X, \mathscr{M}, \mu)$ be a measure space.

(a) If f is measurable, show that $|f|$ and cf are measurable by applying Definition 9.1.1.

(b) If f is not measurable, show that cf is not measurable for every nonzero real number c.

(c) Give an example showing that $|f|$ may be measurable even if f is not.

(d) Give an example of two functions f and g which are not measurable but for which $f + g$ is measurable.

(e) In the statement of Proposition 9.1.9, why are we only considering functions which map into $\mathbb{R}$ and not functions which map into $[-\infty, \infty]$?

Exercise 9.1.4. Let $(X, \mathscr{M}, \mu)$ be a measure space.

(a) Let $f : X \to \mathbb{R}$ be a measurable function. Also suppose $g : \mathbb{R} \to \mathbb{R}$ is continuous. Show that $g \circ f$ is measurable.

(b) Use part (a) to supply another proof that $|f|$ and cf are measurable provided f is measurable.

(c) If f is measurable, show that $|f|^p$ is measurable for $1 < p < \infty$.

Exercise 9.1.5. Let $(X, \mathscr{M}, \mu)$ be a measure space and let $f : X \to [-\infty, \infty]$ be a measurable function.

(a) If $c \ge 0$, show that $(cf)_+ = cf_+$ and $(cf)_- = cf_-$.

(b) If $c < 0$, show that $(cf)_+ = -cf_-$ and $(cf)_- = -cf_+$.

Exercise 9.1.6. (a) Complete the proof of Proposition 9.1.12 by showing that $\inf f_n$ is a measurable function.

(b) Supply a proof for Corollary 9.1.13.

Exercise 9.1.7. (a) Complete the proof of Proposition 9.1.15 by showing that $\limsup f_n$ is a measurable function.

(b) Supply a proof for Corollary 9.1.16.

9.2 The Lebesgue Integral

In this section we define the Lebesgue integral. The first result shows that we can approximate measurable functions with simple functions, much like the Riemann integral depends on the fact that step functions can be used to approximate continuous functions.

Theorem 9.2.1 (Fundamental Approximation Theorem). *Let $(X, \mathscr{M}, \mu)$ be a measure space and let $f : X \to [0, \infty]$ be measurable. Then there exists a sequence of measurable, nonnegative, simple functions (s_n) with*

(a) $0 \le s_1(x) \le s_2(x) \le s_3(x) \le \cdots$ for every $x \in X$;
(b) $\lim_{n \to \infty} s_n(x) = f(x)$ for every $x \in X$.

Proof. We begin by constructing the sequence (s_n). Keep in mind that our focus here is on the range of the function f. For each $n \in \mathbb{N}$, partition the interval $[0, n)$ into $n2^n$ subintervals of length $1/2^n$ and set

$$s_n(x) = \begin{cases} n, & f(x) \ge n; \\ \dfrac{j-1}{2^n}, & \dfrac{j-1}{2^n} \le f(x) < \dfrac{j}{2^n}, \end{cases}$$

where $j = 1, 2, \ldots, n2^n$. It should be clear that these functions are nonnegative, simple functions taking on at most $n2^n + 1$ values. Also, each function is measurable since f is measurable. For a visual, Fig. 9.3 shows the graph of f with s_1 and s_2.

To show that the sequence is monotonic, let $x \in X$ and $n \in \mathbb{N}$. First consider the case that $f(x) \in [0, n)$. Then there exists a $j \in \{1, 2, \ldots, n2^n\}$ such that $f(x) \in \left[\frac{j-1}{2^n}, \frac{j}{2^n}\right)$. Let's call this interval I for the moment. In moving from n to $n + 1$, the interval I is split into two intervals,

$$I = \left[\frac{j-1}{2^n}, \frac{j-\frac{1}{2}}{2^n}\right) \cup \left[\frac{j-\frac{1}{2}}{2^n}, \frac{j}{2^n}\right) = \left[\frac{2j-2}{2^{n+1}}, \frac{2j-1}{2^{n+1}}\right) \cup \left[\frac{2j-1}{2^{n+1}}, \frac{2j}{2^{n+1}}\right)$$

and $f(x)$ is in exactly one of these. If $f(x)$ is in the left-hand subinterval, then $s_n(x) = (j-1)/2^n = (2j-2)/2^{n+1} = s_{n+1}(x)$. If $f(x)$ is in the right-hand subinterval, then $s_n(x) = (j-1)/2^n < (2j-1)/2^{n+1} = s_{n+1}(x)$. In either case, $s_n(x) \le s_{n+1}(x)$.

For the second case, suppose $f(x) \ge n$. It must be the case then that $s_n(x) = n$. Now if $f(x) \ge n+1$, then $s_{n+1}(x) = n+1$. On the other hand, if $f(x) \in [n, n+1)$, then $f(x) \in [(j-1)/2^{n+1}, j/2^{n+1})$ for some $j \in \{n2^{n+1} + 1, n2^{n+1} + 2, \ldots, (n+1)2^{n+1}\}$ in which case $s_{n+1}(x) = (j-1)/2^{n+1} \ge n$. Again, in either case, $s_n(x) \le s_{n+1}(x)$ and thus the sequence is monotonically increasing.

Finally, we show that the sequence s_n converges pointwise to f. Again we consider two cases. First, if $f(x) = \infty$, then $s_n(x) = n$ and the limit statement holds. On the other hand, suppose $f(x) < \infty$ and let $\varepsilon > 0$. We need to find N

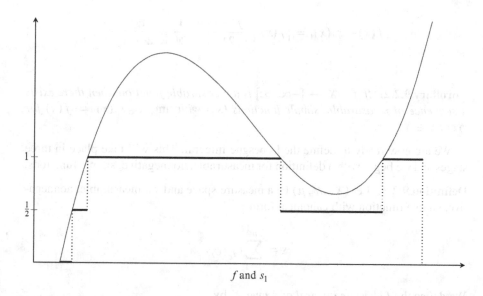

f and s_1

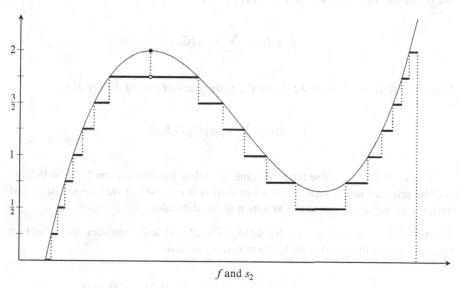

f and s_2

Fig. 9.3 The fundamental approximation theorem

so that $|f(x) - s_n(x)| < \varepsilon$ for all $n \geq N$. First, choose N_1 so that $1/2^{N_1} < \varepsilon$.
Also, choose N_2 so that $f(x) \in [0, N_2]$ and set $N = \max\{N_1, N_2\}$. Now, if $n \geq N$,
$f(x) \in [0, n]$ since $n \geq N_2$, and thus there exists a $j \in \{1, 2, \ldots, n2^n\}$ so that
$f(x) \in \left[\frac{j-1}{2^n}, \frac{j}{2^n}\right)$. The length of this interval is $1/2^n$ from which it follows that

$$|f(x) - s_n(x)| = \left| f(x) - \frac{j-1}{2^n} \right| \le \frac{1}{2^n} < \frac{1}{2^{N_1}} < \varepsilon.$$

$\square$

Corollary 9.2.2. *If* $f : X \to [-\infty, \infty]$ *is a measurable function, then there exists a sequence of measurable, simple functions* (s_n) *with* $\lim_{n \to \infty} s_n(x) = f(x)$ *for every* $x \in X$.

We are now ready to define the Lebesgue integral. This will take place in three stages and we begin with a definition for measurable, nonnegative, simple functions.

Definition 9.2.3. Let $(X, \mathscr{M}, \mu)$ be a measure space and s a measurable, nonnegative, simple function with canonical form

$$s = \sum_{j=1}^{n} c_j \chi_{C_j}.$$

We define the *Lebesgue integral of s over X* by

$$\int_X s \, d\mu = \sum_{j=1}^{n} c_j \mu(C_j).$$

Moreover, if $E \in \mathscr{M}$, then we define the *Lebesgue integral of s over E* by

$$\int_E s \, d\mu = \sum_{j=1}^{n} c_j \mu(C_j \cap E).$$

When considering these types of sums, we allow the convention $0 \cdot \infty = 0$. This will be relevant in the case when a function is 0 on a set of infinite measure and assigning a value of zero to this situation is our definition.

Example 9.2.4. We return to the setting $(\mathbb{R}, \mathscr{L}, m)$ and consider the Dirichlet function $\chi_{\mathbb{Q}}$ on the interval $[0, 1]$. To integrate, we have

$$\int_{[0,1]} \chi_{\mathbb{Q}} \, dm = 1 \cdot ([0, 1] \cap \mathbb{Q}) + 0 \cdot m([0, 1] \cap (\mathbb{I})) = 0.$$

It should be clear from this calculation that the integral would be zero no matter what set we integrate over due to the fact that the measure of the rationals is zero.

As noted previously, this function is not compatible with the Riemann integral, but, as demonstrated here, one of the simplest functions when considering the Lebesgue integral. One of the problems discussed concerning the Riemann integral is the fact that not enough functions are integrable, the Dirichlet function being one of the most basic examples of a function with this particular affliction. It is refreshing then that this simple version of the Lebesgue integral is designed to overcome this obstacle. Lebesgue shared this sentiment:

> A generalization made not for the vain pleasure of generalizing but in order to solve previously existing problems is always a fruitful generalization [17, p. 194].

Before moving on to the more general definition of the integral, we explore a means of constructing new measures via integration of simple functions.

Theorem 9.2.5. *Let* $(X, \mathcal{M}, \mu)$ *be a measure space and let* s *be a measurable, nonnegative, simple function. The set function* $\varphi : \mathcal{M} \to [0, \infty]$ *defined by*

$$\varphi(E) = \int_E s \, d\mu$$

is a positive measure on $\mathcal{M}$.

Proof. It should be clear that φ maps $\mathcal{M}$ into $[0, \infty]$ since s is nonnegative. Also, the fact that $\varphi(\emptyset) = 0$ is a consequence of the fact that μ is a measure together with the definition of the integral of a simple function. To show that φ is countably additive, let $\{A_k\}$ be a collection of pairwise disjoint sets in $\mathcal{M}$ and let A be their union. For the sake of notation, also suppose s has canonical form $\sum_{j=1}^{n} c_j \chi_{C_j}$. Some simple manipulations show that

$$\varphi(A) = \int_A s \, d\mu = \sum_{j=1}^{n} c_j \mu(C_j \cap A) = \sum_{j=1}^{n} c_j \mu \left(\bigcup_{k=1}^{\infty} (C_j \cap A_k) \right)$$

$$= \sum_{j=1}^{n} c_j \sum_{k=1}^{\infty} \mu(C_j \cap A_k) = \sum_{k=1}^{\infty} \sum_{j=1}^{n} c_j \mu(C_j \cap A_k)$$

$$= \sum_{k=1}^{\infty} \int_{A_k} s \, d\mu = \sum_{k=1}^{\infty} \varphi(A_k).$$

$\square$

For the second stage in our definition of the Lebesgue integral, we move to nonnegative, measurable functions.

Definition 9.2.6. Let $(X, \mathcal{M}, \mu)$ be a measure space. If $f : X \to [0, \infty]$ is measurable and $E \in \mathcal{M}$, we define the *Lebesgue integral of* f *over* E by

$$\int_E f \, d\mu = \sup \left\{ \int_E s \, d\mu : s \text{ is measurable and simple with } 0 \leq s \leq f \text{ on } E \right\}.$$

If this value is finite, we say that f is *Lebesgue integrable*.

By the Fundamental Approximation Theorem, the set over which the supremum is taken is not empty and thus the integral above is well defined, though the value may be infinite. Moreover, as mentioned previously, this integral partitions the range of the function rather than its domain. This partitioning is actually represented most clearly in the Fundamental Approximation Theorem. We leave it as an exercise to check that this new definition agrees with our previous definition in the case when f is a simple function.

Example 9.2.7. Let X be any set and fix an element $x_0 \in X$. Consider the measure space $(X, \mathscr{P}(X), \mu)$ where the measure μ is the point mass measure at x_0. As discussed in the previous section every function $f : X \to [-\infty, \infty]$ is measurable. If s is a measurable, nonnegative, simple function with canonical representation $s = \sum_{j=1}^{n} c_j \chi_{C_j}$, then x_0 is in exactly one of the sets in this decomposition, call it C_J, in which case $s(x_0) = c_J$ and

$$\int_X s \, d\mu = \sum_{j=1}^{n} c_j \mu(C_j) = c_J \mu(C_J) = c_J = s(x_0)$$

since μ assigns a value of zero to any set which does not contain x_0. Furthermore, if $f : X \to [0, \infty]$ it follows that

$$\int_X f \, d\mu = \sup \left\{ \int_X s \, d\mu : s \text{ is measurable and simple with } 0 \le s \le f \text{ on } X \right\}$$

$$= \sup \left\{ s(x_0) : s \text{ is measurable and simple with } 0 \le s \le f \text{ on } X \right\}$$

$$= f(x_0)$$

where the last equality follows from the Fundamental Approximation Theorem.

Example 9.2.8. Let μ be counting measure on the set $\mathscr{P}(\mathbb{N})$ as defined in Example 8.5.7. As above, since our σ-algebra is a power set, every function is measurable. If we take a function $f : \mathbb{N} \to [0, \infty)$, we can construct a sequence of simple functions, s_n, by

$$s_n(x) = \begin{cases} f(j), & x = j \in \{1, 2, \ldots, n\}; \\ 0, & x > n. \end{cases}$$

Applying Exercise 9.2.3 (because we don't know that the values $f(j)$ are distinct), we integrate s_n,

$$\int_{\mathbb{N}} s_n \, d\mu = \sum_{j=1}^{n} f(j) \mu(\{j\}) + 0 \cdot \mu(\{n+1, n+2, \ldots\}) = \sum_{j=1}^{n} f(j).$$

By definition of the integral and the Order Limit Theorem, we have

$$\sum_{j=1}^{\infty} f(j) \le \int_{\mathbb{N}} f \, d\mu.$$

At the moment, this is the best we can do, but equality does in fact hold in this situation. We'll continue this example in Sect. 9.3 where we will also see that we can extend it to functions $f : \mathbb{N} \to [0, \infty]$.

Definition 9.2.9. Let $(X, \mathcal{M}, \mu)$ be a measure space. If $f : X \to [-\infty, \infty]$ is measurable and $E \in \mathcal{M}$, then we define the *Lebesgue integral of* f *over* E by

$$\int_E f \, d\mu = \int_E f_+ \, d\mu - \int_E f_- \, d\mu$$

provided either $\int_E f_+ \, d\mu$ or $\int_E f_- \, d\mu$ is finite. If both these integrals are finite we say that f is *Lebesgue integrable*.

As a first observation, if $A \in \mathcal{M}$ and f is measurable, we can write

$$\int_A f \, d\mu = \int_X f\chi_A \, d\mu. \tag{9.1}$$

If f is a simple function, this follows from the fact that $\chi_A \chi_B = \chi_{A \cap B}$ and Exercise 9.2.3(b). The case for a general measurable function then follows quickly from the definition of the integral. This property can actually be taken to be the definition of the integral over subsets of X. It also allows for simplifications in proof by only considering integrals over the set X. The remainder of this section describes basic properties of our new integral. Many of them will bear resemblance to the key properties of the Riemann integral.

Theorem 9.2.10 (Monotonicity of the Lebesgue Integral). *Let* $(X, \mathcal{M}, \mu)$ *be a measure space.*

(a) If $A \in \mathcal{M}$ *and* f *and* g *are measurable with* $0 \le f(x) \le g(x)$ *on* A, *then*

$$\int_A f \, d\mu \le \int_A g \, d\mu.$$

(b) If $A, B \in \mathcal{M}$ *with* $A \subseteq B$ *and* f *is measurable with* $0 \le f(x)$ *on* B, *then*

$$\int_A f \, d\mu \le \int_B f \, d\mu.$$

Proof. For (a), consider any simple function which is measurable with $0 \le s(x) \le f(x)$ for all $x \in A$. The inequality relationship between f and g, then ensures us that $0 \le s(x) \le g(x)$ on A. Thus the set used to define the integral of f is a subset of that used to define the integral of g. The desired inequality is now a consequence of the familiar property of the supremum: sup $A \le$ sup B whenever $A \subseteq B$.

For (b), we use the fact that $f\chi_A \leq f\chi_B$ on X since $A \subseteq B$. Thus by (a) we have

$$\int_A f \, d\mu = \int_X f\chi_A \, d\mu \leq \int_X f\chi_B \, d\mu = \int_B f \, d\mu.$$

$\square$

Theorem 9.2.11 (Linearity of the Lebesgue Integral). *Let $(X, \mathscr{M}, \mu)$ be a measure space and let $A \in \mathscr{M}$. If f and g are Lebesgue integrable and $c \in \mathbb{R}$, then*

(a) $f + g$ is integrable and $\displaystyle\int_A f + g \, d\mu = \int_A f \, d\mu + \int_A g \, d\mu$;

(b) cf is integrable and $\displaystyle\int_A cf \, d\mu = c\int_A f \, d\mu$.

Proof. The proof of (b) is straightforward and requested of the reader. To get started on (a), let s and t be simple functions on X with canonical representation $\sum_{j=1}^{n} c_j \chi_{C_j}$ and $\sum_{k=1}^{m} d_k \chi_{D_k}$, respectively. Here it is important to notice that the sets $\{C_j \cap D_k\}_{j,k=1}^{n,m}$ are disjoint and their union, as j ranges from 1 to n and k ranges from 1 to m, is X. We can then write $s + t$ as

$$s + t = \sum_{j,k}(c_j + d_k)\chi_{C_j \cap D_k};$$

this may not be canonical since we cannot guarantee that the values of $c_j + d_k$ are distinct, but Exercise 9.2.3(a) gives us a means of integrating this function and we arrive at

$$\int_X s + t \, d\mu = \sum_{j,k}(c_j + d_k)\mu(C_j \cap D_k) = \sum_{j,k} c_j \mu(C_j \cap D_k) + d_k \mu(C_j \cap D_k).$$

Moreover, the same result allows us to write

$$\int_X s \, d\mu = \sum_{j,k} c_j \mu(C_j \cap D_k) \quad \text{and} \quad \int_X t \, d\mu = \sum_{j,k} d_k \mu(C_j \cap D_k).$$

The desired equality then follows immediately.

Section 9.3 will provide the machinery necessary to show that the statement holds when f and g are arbitrary, integrable functions. $\square$

The final result of this section provides an analog of the subdivision property of the Riemann integral.

Corollary 9.2.12. *Let $(X, \mathscr{M}, \mu)$ be a measure space and let $A, B, C \in \mathscr{M}$ with $A \cup B = C$ and $A \cap B = \emptyset$. If f is Lebesgue integrable, then*

$$\int_C f \, d\mu = \int_A f \, d\mu + \int_B f \, d\mu.$$

Proof. The proof follows immediately from the linearity of the integral paired with the identity $f\chi_C = f\chi_A + f\chi_B$. $\square$

Exercises

Exercise 9.2.1. Explain why the functions s_n constructed in Theorem 9.2.1 are measurable.

Exercise 9.2.2. Supply a proof for Corollary 9.2.2.

Exercise 9.2.3. (a) Let $(X, \mathcal{M}, \mu)$ be a measure space. Suppose s is a measurable, nonnegative, simple function with representation $\sum_{j=1}^{N} a_j \chi_{A_j}$ where the A_j's (in $\mathcal{M}$) are pairwise disjoint. Show that $\int_X s \, d\mu = \sum_{j=1}^{N} a_j \mu(A_j)$. Keep in mind that we are not assuming that this is a canonical representation here, i.e. the a_j's may not be distinct.
(b) Suppose s is a measurable, nonnegative, simple function with representation $\sum_{j=1}^{N} a_j \chi_{A_j}$, with $A_j \in \mathcal{M}$. Show that $\int_X s \, d\mu = \sum_{j=1}^{N} a_j \mu(A_j)$.

Exercise 9.2.4. Let $(X, \mathcal{M}, \mu)$ be a measure space. If s and t are measurable, nonnegative, simple functions with $s(x) \leq t(x)$ on X, show that $\int_X s \, d\mu \leq \int_X t \, d\mu$. Do this using the definition of the integral of a simple function and without referencing Theorem 9.2.10.

Exercise 9.2.5. Supply a proof for Theorem 9.2.11(b). You will need to consider several cases according to the development of the Lebesgue integral and whether $c \geq 0$ or $c < 0$.

Exercise 9.2.6. Let $(X, \mathcal{M}, \mu)$ be a measure space and suppose $f : X \to [0, \infty]$ is measurable. If $A \in \mathcal{M}$ with $\mu(A) > 0$ such that $f(x) = \infty$ for all $x \in A$, show that $\int_X f \, d\mu$ is infinite.

Exercise 9.2.7. Let $(X, \mathcal{M}, \mu)$ be a measure space.

(a) If $A \in \mathcal{M}$ and $f(x) = 0$ on A, show that $\int_A f \, d\mu = 0$.

(b) If $A \in \mathcal{M}$ with $\mu(A) = 0$ and f is measurable, show that $\int_A f \, d\mu = 0$.

(c) Suppose $A, B \in \mathcal{M}$ with $A \subseteq B$ and $\mu(B \setminus A) = 0$. If f is integrable on A, show that f is integrable on B with $\int_A f \, d\mu = \int_B f \, d\mu$.

(d) If $A \in \mathcal{M}$ and $f : X \to [0, \infty]$ is measurable with $\int_A f \, d\mu = 0$, show that
$\mu\left(\{x \in A : f(x) \neq 0\}\right) = 0$. To do this, consider sets $A_n = \{x \in A : f(x) > \frac{1}{n}\}$ and show $\mu(A_n) = 0$ for all $n \in \mathbb{N}$. How do these sets relate to the set in question?

Exercise 9.2.8. Consider the particular measure space $(\mathbb{R}, \mathcal{L}, m)$. If f and g are functions defined on $\mathbb{R}$ such that f is measurable and $m\left(\{x : f(x) \neq g(x)\}\right) = 0$, show that g is also measurable. It will help to begin by showing that the set on which the functions agree is a measurable set. Can you extend this result to a general measure space $(X, \mathcal{M}, \mu)$?

Exercise 9.2.9. Let $(X, \mathcal{M}, \mu)$ be a measure space.

(a) If f is measurable and $\mu\left(\{x : f(x) \neq 0\}\right) = 0$, show that f is integrable with
$$\int_X f \, d\mu = 0.$$

(b) If f and g are measurable and $\mu\left(\{x : f(x) \neq g(x)\}\right) = 0$, show that f is integrable if and only g is integrable with $\int_X f \, d\mu = \int_X g \, d\mu$.

Exercise 9.2.10. Let $(X, \mathcal{M}, \mu)$ be a measure space.

(a) If $A \in \mathcal{M}$ and f and g are integrable with $f(x) \leq g(x)$ on A, show that
$$\int_A f(x) \, d\mu \leq \int_A g \, d\mu.$$

(b) Give an example to show that Theorem 9.2.10(b) can fail without the nonnegative hypothesis, even if we further assume that f is integrable.

9.3 Limits and the Lebesgue Integral

Each of the main theorems in this section will be stated for an integral over an arbitrary measurable set, but, for the sake of simplicity, proofs will be given for the case when we are integrating over the entire set X. The general statement then follows from Eq. (9.1).

Theorem 9.3.1 (Lebesgue's Monotone Convergence Theorem). *Let $(X, \mathcal{M}, \mu)$ be a measure space and let $A \in \mathcal{M}$. If $f_n : X \to [0, \infty]$ is a sequence of measurable functions with $0 \leq f_1(x) \leq f_2(x) \leq \cdots$ for every $x \in A$, then the limit function $\lim_{n \to \infty} f_n(x) = f(x)$ is measurable with*

$$\lim_{n\to\infty} \int_A f_n \, d\mu = \int_A f \, d\mu.$$

Proof. The statement that the limit function f is measurable is a consequence of Theorem 9.1.16. The monotonicity of the integral now shows that

$$\int_X f_n \, d\mu \leq \int_X f_{n+1} \, d\mu \leq \int_X f \, d\mu$$

for every n. The first inequality along with the Monotone Convergence Theorem for sequences guarantees us that the limit of the integrals exists as a value in $[0, \infty]$, call it α. The second inequality then asserts that

$$\alpha \leq \int_X f \, d\mu$$

and we are left to show the reverse inequality. The matter is trivial if $\alpha = \infty$ and thus we further assume that $\alpha < \infty$. We first show that the inequality holds for any simple function s with $0 \leq s(x) \leq f(x)$ on X and then appeal to the definition of the integral to obtain our conclusion for f.

To this end, let s be a simple function with $0 \leq s(x) \leq f(x)$ on X and fix $0 < c < 1$. Also, for each n, define a set

$$E_n = \{x : f_n(x) \geq cs(x)\} = \{x : f_n(x) - cs(x) \geq 0\}.$$

Two observations are immediate. First, each E_n is measurable since f_n and s are measurable. Second, the sets are nested, $E_1 \subseteq E_2 \subseteq E_3 \subseteq \cdots$ since the functions f_n form a non-decreasing sequence. A third, not so obvious observation is the fact that $X = \bigcup_{n=1}^{\infty} E_n$. Recall from Theorem 9.2.5 that we can define positive measures via integration, i.e. the set function

$$\varphi(E) = \int_E s \, d\mu$$

is a measure and satisfies the properties of Theorem 8.1.11; in particular, φ satisfies property (b) and thus

$$\int_X s \, d\mu = \varphi(X) = \varphi\left(\bigcup_{n=1}^{\infty} E_n\right) = \lim_{n\to\infty} \varphi(E_n).$$

Furthermore, by definition of the set E_n, we also see that

$$\varphi(E_n) = \int_{E_n} s \, d\mu \leq \int_{E_n} \frac{1}{c} f_n \, d\mu \leq \frac{1}{c} \int_X f_n \, d\mu.$$

where we have used Theorems 9.2.10 and 9.2.11(b) to obtain the two inequalities. Applying a limit to this last statement reveals that

$$\int_X s\, d\mu = \lim_{n\to\infty} \varphi(E_n) \le \lim_{n\to\infty} \frac{1}{c}\int_X f_n\, d\mu = \frac{\alpha}{c}.$$

Since this last relationship holds for every $c \in (0, 1)$, it must be the case that

$$\int_X s\, d\mu \le \alpha.$$

Moreover, since s was taken to be an arbitrary simple function, Definition 9.2.6 and properties of the supremum guarantee us that

$$\int_X f\, d\mu \le \alpha.$$

□

Though this is the simplest convergence theorem, it has many applications and, in fact, is a catalyst for the other two main convergence theorems. It also plays a role in showing that the Lebesgue integral extends the Riemann integral. Before moving on to the other convergence theorems, we need to tie up a few loose ends from the previous section.

Proof (Proof of Theorem 9.2.11(a)). We've already shown that the statement holds for simple functions. Now assume f and g are measurable and nonnegative on X. By the Fundamental Approximation Theorem, we can find increasing sequences of measurable, nonnegative, simple functions (s_n) and (t_n) with s_n converging to f and t_n converging to g. By the Monotone Convergence Theorem, it follows that

$$\lim_{n\to\infty}\int_X s_n\, d\mu = \int_X f\, d\mu \quad\text{and}\quad \lim_{n\to\infty}\int_X t_n\, d\mu = \int_X g\, d\mu.$$

Adding these two equations together and using the linearity for simple functions, we see that

$$\lim_{n\to\infty}\int_X s_n + t_n\, d\mu = \int_X f\, d\mu + \int_X g\, d\mu.$$

Moreover, $s_n + t_n$ is an increasing sequence of nonnegative, measurable, simple functions with $s_n + t_n \to f + g$ and thus

$$\lim_{n\to\infty}\int_X s_n + t_n\, d\mu = \int_X f + g\, d\mu.$$

The conclusion now follows from the last two equations.

It is now left to show that the conclusion holds for two arbitrary integrable functions f and g. First, we must show that $f + g$ is integrable and then show equality of the integrals. If $x \in X$, it should be clear that

$$(f + g)_+(x) \le f_+(x) + g_+(x) \qquad \text{and} \qquad (f + g)_-(x) \le f_-(x) + g_-(x).$$

We investigate the first inequality and leave the second for the reader. Though it may seem elementary, we consider cases based on the sign of f and g at x. First, if both functions are nonnegative at x, then $(f + g)_+(x) = f_+(x) + g_+(x)$ since the positive part takes the larger of the function value and 0. Second, if both functions are negative at x, then $f + g$ is negative at x and all the terms in the inequality are zero, i.e. a case of equality. For the final case, suppose (without loss of generality) $f(x) \ge 0$ and $g(x) < 0$. Here it follows that

$$0 \le (f + g)_+(x) \le \max\{(f + g)(x), 0\} < f(x) = f_+(x) \le f_+(x) + g_+(x).$$

By the monotonicity of the integral and the case for nonnegative functions,

$$\int_X (f + g)_+ \, d\mu \le \int_X f_+ + g_+ \, d\mu = \int_X f_+ \, d\mu + \int_X g_+ \, d\mu < \infty$$

and

$$\int_X (f + g)_- \, d\mu \le \int_X f_- + g_- \, d\mu = \int_X f_- \, d\mu + \int_X g_- \, d\mu < \infty,$$

showing that $f + g$ is in fact Lebesgue integrable; of course, this depends on the fact that f and g are integrable as exhibited in the last inequality of the previous two statements.

Finally, we show that the integral equality holds. This hinges on the fact that we can represent $(f + g)$ as $(f + g)_+ - (f + g)_-$ and as $f_+ - f_- + g_+ - g_-$. Via transitivity and with some rearranging, we obtain

$$(f + g)_+ + f_- + g_- = (f + g)_- + f_+ + g_+.$$

Notice that each expression in this line is a nonnegative function. Thus we integrate both sides and apply the result for the case of nonnegative functions,

$$\int_X (f+g)_+ \, d\mu + \int_X f_- \, d\mu + \int_X g_- \, d\mu = \int_X (f+g)_- \, d\mu + \int_X f_+ \, d\mu + \int_X g_+ \, d\mu;$$

rearranging the sides once again we arrive at the desired conclusion

$$\int_X f + g \, d\mu = \int_X f \, d\mu + \int_X g \, d\mu.$$

$\square$

The following corollary asserts that the Lebesgue integral is designed as an absolutely convergent integral, a property not shared by the Riemann integral.

Corollary 9.3.2. *If $(X, \mathcal{M}, \mu)$ is a measure space, a measurable function f is Lebesgue integrable if and only if $|f|$ is integrable. Moreover, for every $A \in \mathcal{M}$ we have the inequality*

$$\left| \int_A f \, d\mu \right| \leq \int_A |f| \, d\mu.$$

As a final application we have an extension of Theorem 9.2.5. These two theorems provide us with a means of constructing new measures and gets at something broader. Given a set and a (σ)-ring or algebra, there are numerous measures to consider and often one seeks to understand the relationship between two distinct measures. The theorem below says that we can construct any number of new measures by integrating measurable, nonnegative functions and an immediate consequence of this is that the new measure has the same measure zero sets as the original (Exercise 9.2.7). The converse to this question then asks, given two measures μ_1 and μ_2, is it possible to represent μ_2 in terms of the μ_1 via integration against a measurable, nonnegative function. The answer to this is yes, provided μ_2 assigns a value of zero to every set for which μ_1 assigns a value of zero (there are also some additional hypothesis that must be placed on the measure space for this to hold). This result, known as the Radon–Nikodym Theorem, is well beyond the scope of what we are considering here, but has far reaching effects in the realm of analysis.

Theorem 9.3.3. *Let $(X, \mathcal{M}, \mu)$ be a measure space and let f be a measurable, nonnegative function. The set function $\varphi : \mathcal{M} \to [0, \infty]$ defined by*

$$\varphi(E) = \int_E f \, d\mu$$

is a positive measure on $\mathcal{M}$.

For the next convergence theorem you should recall the notion of $\liminf$.

Theorem 9.3.4 (Fatou's Lemma). *Let $(X, \mathcal{M}, \mu)$ be a measure space and let $A \in \mathcal{M}$. If $f_n : X \to [0, \infty]$ is a sequence of measurable functions, then*

$$\int_A \left(\liminf f_n \right) d\mu \leq \liminf \int_A f_n \, d\mu.$$

Proof. To begin, define a sequence of functions (g_n) by $g_n(x) = \inf_{k \geq n}\{f_k(x)\}$. By Proposition 9.1.12, the functions in this sequence are measurable, and monotonically non-decreasing by construction. By Proposition 9.1.15, the function $\liminf f_n$ is measurable and we can write $\liminf f_n = \lim_{n \to \infty} g_n$. Applying the Monotone Convergence Theorem,

$$\int_X \liminf f_n \, d\mu = \int_X \lim_{n \to \infty} g_n \, d\mu = \lim_{n \to \infty} \int_X g_n \, d\mu.$$

Using the properties of the infimum it follows that $g_n(x) \leq f_k(x)$ for every $x \in X$ and all $k \geq n$, from which the monotonicity of the integral implies

$$\int_X g_n \, d\mu \leq \inf_{k \geq n} \int_X f_k \, d\mu.$$

The conclusion is now apparent from the last two equations,

$$\int_X \liminf f_n \, d\mu \leq \lim_{n \to \infty} \left(\inf_{k \geq n} \int_X f_k \, d\mu \right) = \liminf \int_X f_n \, d\mu.$$

□

The fact that Fatou's Lemma involves the lim inf may seem distasteful at first, but this is actually its strength. For a given sequence of functions, it may be the case that there is no pointwise limit, or it may be that the limit of the sequence of integrals does not exist, or both problems may occur. The beauty of this theorem is then that the lim inf always exists, as a function, on the one hand, and as a value in $[0, \infty]$, on the other. One other drawback may be the fact that we are dealing with an inequality here instead of an equality. Unfortunately this cannot be resolved and Exercise 9.3.7 will demonstrate that strict inequality is possible. As a final note, as with the Monotone Convergence Theorem, this is only valid when dealing with nonnegative functions. The sequence of functions

$$f_n(x) = \begin{cases} -n, & \frac{1}{n} \leq x \leq \frac{2}{n}; \\ 0, & \text{else} \end{cases}$$

provides an example where the theorem fails for functions with negative values.

The final convergence theorem is the most general, applying to functions with negative values. The second conclusion of the version we present is not generally included in the statement of the theorem but will be relevant to Sect. 9.4.

Theorem 9.3.5 (Lebesgue's Dominated Convergence Theorem). *Let $(X, \mathcal{M}, \mu)$ be a measure space and let $A \in \mathcal{M}$. Suppose that $f_n : X \to [-\infty, \infty]$ is a sequence of measurable functions such that the limit function $\lim_{n \to \infty} f_n(x) = f(x)$ exists for every $x \in A$. Further suppose there exists a Lebesgue integrable function g such that $|f_n(x)| \leq g(x)$ for every $x \in A$. Then*

(a) f is Lebesgue integrable;

(b) $\displaystyle\lim_{n \to \infty} \int_A |f_n - f| \, d\mu = 0$;

(c) $\displaystyle\lim_{n \to \infty} \int_A f_n \, d\mu = \int_A f \, d\mu.$

Proof. For (a), we know that the function f is measurable from Theorem 9.1.16. Also, since (f_n) converges pointwise to f and $|f_n(x)| \leq g(x)$ on X, it must also be the case that $|f(x)| \leq g(x)$ on X. This inequality and the definition of the integral for nonnegative functions imply that

$$\int_X |f|\, d\mu \leq \int_X g\, d\mu$$

and Corollary 9.3.2 assures us that f is integrable.

For (b), we first observe that $|f_n - f| \leq 2g$ by the triangle inequality and the observation above. Rearranging this expression, we see that the sequence of functions $2g - |f_n - f|$ is a nonnegative sequence of measurable functions and we may thus apply Fatou's Lemma. Before doing so, notice that $\lim_{n\to\infty} |f_n - f| = \limsup |f_n - f| = \liminf |f_n - f| = 0$ on X from which it follows that $\liminf(2g - |f_n - f|) = 2g + \liminf(-|f_n - f|) = 2g - \limsup(|f_n - f|) = 2g$. Fatou's Lemma now yields

$$\int_X 2g\, d\mu = \int_X \liminf \left(2g - |f_n - f|\right) d\mu \leq \liminf \left(\int_X 2g - |f_n - f|\, d\mu \right).$$

Manipulating this last expression we find that

$$\int_X 2g\, d\mu \leq \int_X 2g\, d\mu + \liminf \left(-\int_X |f_n - f|\, d\mu \right)$$

$$= \int_X 2g\, d\mu - \limsup \int_X |f_n - f|\, d\mu$$

where we have used the fact that $\liminf(-a_n) = -\limsup(a_n)$ in the last equality. Rearranging this last inequality, it is clear that

$$\limsup \int_X |f_n - f|\, d\mu \leq 0.$$

However, $|f_n - f| \geq 0$ and thus it must be the case that

$$\liminf \int_X |f_n - f|\, d\mu \geq 0.$$

Stringing these two ideas together,

$$0 \leq \liminf \int_X |f_n - f|\, d\mu \leq \limsup \int_X |f_n - f|\, d\mu \leq 0$$

implies that the quantities in this last line are all zero and our conclusion is attained.

For (c), notice that the function $|f_n - f|$ is integrable, and Corollary 9.3.2 asserts that

$$0 \leq \left| \int_X f_n \, d\mu - \int_X f \, d\mu \right| = \left| \int_X f_n - f \, d\mu \right| \leq \int_X |f_n - f| \, d\mu.$$

An application of the Squeeze Theorem with the result of (b) produces the desired result. $\square$

The first conclusion of the Dominated Convergence Theorem is perhaps the most subtle and it is natural to ask if a similar statement can be made for a sequence of Riemann integrable functions. We have already seen that this isn't the case however, if we consider the sequence of functions

$$f_n(x) = \begin{cases} 1, & x \in \{r_1, r_2, \dots, r_n\}; \\ 0, & else, \end{cases}$$

where $\{r_1, r_2, r_3, \dots\}$ is an enumeration of $\mathbb{Q} \cap [0, 1]$. The functions f_n are all Riemann integrable on $[0, 1]$, bounded above by the function $g(x) = 1$ which is Riemann integrable on $[0, 1]$, but (f_n) converges pointwise to the Dirichlet function which is not Riemann integrable.

It is interesting to note, however, that a similar result does hold if we add the hypothesis that the limit function be Riemann integrable. This result first appeared in 1885 and is due to Arzelà. The article [9] provides an interesting proof as well as references to other proofs and the history of the problem. The theorem is stated below. As a matter of interest, the example above also shows that the Monotone Convergence Theorem fails in the Riemann setting. However an analog does hold if we again impose the additional assumption that the limit function f is Riemann integrable and we point the reader to [31].

Theorem 9.3.6. *Let $f_n : [a, b] \to \mathbb{R}$ be a sequence of Riemann integrable functions with pointwise limit $f : [a, b] \to \mathbb{R}$ which is also Riemann integrable. Suppose further that there is a constant $C > 0$ such that $|f_n(x)| < C$ for every $x \in [a, b]$ and $n \in \mathbb{N}$. Then*

$$\lim_{n \to \infty} \int_a^b f_n(x) \, dx = \int_0^1 f(x) \, dx.$$

The Monotone Convergence Theorem and the Dominated Convergence Theorem have not been stated in their most general form. The hypotheses of these theorems contain statements which are required to hold on some measurable set A. However, Exercises 9.2.7(b), (c) and 9.2.9 show that sets of measure zero have no bearing on the Lebesgue integral and thus we could generalize these statements to overlook subsets of A of measure zero. To be precise, let $(X, \mathcal{M}, \mu)$ be a measure space and let $A \in \mathcal{M}$. If a property P holds on A except on a set of measure zero, then we say that the property holds *almost everywhere* on A. For example, if $B \subseteq A$ with

$m(B) = 0$ and f and g are functions on A with $f(x) = g(x)$ for every x in $A \setminus B$, we say that $f = g$ almost everywhere on A. We often abbreviate this as $f = g$ a.e. For another example, Exercise 9.2.9(a) states that if a function $f = 0$ a.e., then the Lebesgue integral of f is zero.

Theorem 9.3.7. *Let $(X, \mathcal{M}, \mu)$ be a measure space and let $A \in \mathcal{M}$. Suppose that $f_n : X \to [-\infty, \infty]$ is a sequence of measurable functions such that the limit function $\lim_{n \to \infty} f_n(x) = f(x)$ exists a.e. on A. Further suppose there exists a Lebesgue integrable function g such that $|f_n(x)| \le g(x)$ for a.e. $x \in A$. Then*

(a) f is Lebesgue integrable;

(b) $\displaystyle \lim_{n \to \infty} \int_A |f_n - f| \, d\mu = 0$;

(c) $\displaystyle \lim_{n \to \infty} \int_A f_n \, d\mu = \int_A f \, d\mu$.

Proof. To see why this is true, we appeal to Exercise 8.5.5 which shows that the measure zero sets of a measure space always form a σ-ring. Each statement in the hypothesis is valid except on some measure zero subset of A, of which there are countably many. If we let B be the union of these, it follows from the mentioned exercise that the measure of B is zero. Each statement is now valid on $A \setminus B$ and the original version of the Dominated Convergence Theorem shows that the result holds on $A \setminus B$. The fact that f is integrable on A now follows from Exercise 9.2.7(c). The other two conclusions are also immediate consequences of this same exercise,

$$\lim_{n \to \infty} \int_A |f_n - f| \, d\mu = \lim_{n \to \infty} \int_{A \setminus B} |f_n - f| \, d\mu = 0$$

and

$$\lim_{n \to \infty} \int_A f_n \, d\mu = \lim_{n \to \infty} \int_{A \setminus B} f_n \, d\mu = \int_{A \setminus B} f \, d\mu = \int_A f \, d\mu.$$

$\square$

Comparing Riemann and Lebesgue

As discussed in earlier chapters, while the Riemann integral works beautifully for continuous functions, it is ill equipped to deal with general bounded functions, has no hope when considering unbounded functions, and fails abysmally when faced with a limit. As exhibited by Example 9.2.4, the Lebesgue integral doesn't waste any time in demonstrating that it can handle functions which are problematic for its predecessor. Moreover, the theorems of this section show that the Lebesgue integral is also much more well equipped with regard to the limit process. The question that should come to mind at this point concerns the relationship between these two integration theories. It would be a terrible loss if the Lebesgue theory was designed to handle these more complicated functions but somehow failed to apply to Riemann integrable functions, or if the Lebesgue integral returned a different value than the Riemann integral for elementary functions. The next theorem clarifies the situation by asserting that we have in fact generalized the Riemann integral. In other words, we haven't lost any ground, but we have gained tremendously.

Before continuing with the proof, it would be advisable to review the notation used for the Riemann integral.

Theorem 9.3.8. *If* $f : [a,b] \to \mathbb{R}$ *is Riemann integrable on* $[a,b]$, *then* f *is Lebesgue integrable on* $[a,b]$ *and*

$$\int_a^b f(x)\, dx = \int_{[a,b]} f\, dm.$$

Proof. Suppose f is Riemann integrable on the interval $[a,b]$. This allows us to find a sequence of partitions (P_n) of $[a,b]$ so that

$$\lim_{n\to\infty} U(f, P_n) - L(f, P_n) = 0$$

and

$$\lim_{n\to\infty} U(f, P_n) = \lim_{n\to\infty} L(f, P_n) = \int_a^b f\, dx; \tag{9.2}$$

we may also choose this sequence so that P_{n+1} is a refinement of P_n (Exercise 7.1.5). Our work is now to show that f is Lebesgue integrable and that the two integrals agree.

For notation, let $P_n = \{x_0, x_1, x_2, \ldots, x_N\}$ be a partition from the sequence above and set

$$L(f, P_n) = \sum_{k=1}^N m_k(x_k - x_{k-1}) \quad \text{and} \quad U(f, P_n) = \sum_{k=1}^N M_k(x_k - x_{k-1}).$$

In addition, define sequences (g_n) and (h_n) by

$$g_n(x) = \sum_{k=1}^N m_k \chi_{(x_{k-1}, x_k]} \quad \text{and} \quad h_n(x) = \sum_{k=1}^N M_k \chi_{(x_{k-1}, x_k]}.$$

It is apparent that g_n and h_n are measurable, simple functions with

$$\int_{[a,b]} g_n\, dm = L(f, P_n) \quad \text{and} \quad \int_{[a,b]} h_n\, dm = U(f, P_n). \tag{9.3}$$

Now, since (P_n) is a sequence of refinements, we have

$$g_1(x) \leq g_2(x) \leq \cdots \leq f(x) \leq \cdots \leq h_2(x) \leq h_1(x)$$

on $[a, b]$ and, by the Monotone Convergence Theorem for sequences, we may define limit functions g and h. By Theorem 9.1.16, these new functions are measurable and $g(x) \leq f(x) \leq h(x)$ for all $x \in [a, b]$. Moreover, since the function f is bounded (requirement for Riemann integral), the functions g_n are bounded and the Dominated Convergence Theorem provides the relationship

$$\int_{[a,b]} g \, dm = \lim_{n \to \infty} \int_{[a,b]} g_n \, dm. \tag{9.4}$$

We could make a similar conclusion for h_n but this isn't necessary. Continuing, observe that

$$h_1(x) - g_1(x) \geq h_2(x) - g_2(x) \geq h_3(x) - g_3(x) \geq \cdots \geq h(x) - g(x) \geq 0$$

and that $h_1 - g_1$ is integrable. Applying the Dominated Convergence Theorem, it follows that

$$0 \leq \int_{[a,b]} h - g \, dm = \lim_{n \to \infty} \int_{[a,b]} h_n - g_n \, dm = \lim_{n \to \infty} \left[U(f, P_n) - L(f, P_n) \right] = 0.$$

By Exercise 9.2.7(d) it must then be the case that $h(x) - g(x) = 0$ except possibly on a set of measure zero. Returning to the inequality relating f, g and h, we see that $g(x) = f(x)$ except possibly on a set of measure zero. Exercise 9.2.8 now implies that f is measurable and the fact that g is integrable along with Exercise 9.2.9(b) shows that

$$\int_{[a,b]} f \, dm = \int_{[a,b]} g \, dm.$$

The conclusion is now brought to light via this last equation and Eqs. (9.2)–(9.4),

$$\int_{[a,b]} f \, dm = \int_{[a,b]} g \, dm = \lim_{n \to \infty} \int_{[a,b]} g_n \, dm = \lim_{n \to \infty} L(f, P_n) = \int_a^b f(x) \, dx.$$

$$\square$$

With the above proof, we can prove the criterion established by Lebesgue which provides the clearest characterization of Riemann integrable functions.

Theorem 9.3.9 (Lebesgue's Criterion for Riemann Integrability). *Let* f : $[a, b] \to \mathbb{R}$ *be a bounded function. Then* f *is Riemann integrable if and only if the set of discontinuities of* f *has Lebesgue measure zero.*

Proof. As part of the proof we first obtain a characterization of continuity in terms of partitions of $[a, b]$. Let (P_n) be a sequence of partitions (each a refinement of the previous) of $[a, b]$ such that $\text{mesh}(P_n) \to 0$ as $n \to \infty$. Define two sequences of functions (g_n) and (h_n) as in the proof of the previous theorem. These sequences are monotonic,

$$g_1(x) \leq g_2(x) \leq \cdots \leq f(x) \leq \cdots \leq h_2(x) \leq h_1(x),$$

since we have a sequence of refinements. Furthermore, the limit functions g and h exist by the Monotone Convergence Theorem for sequences and satisfy $g(x) \leq f(x) \leq h(x)$. The claim is now that f is continuous at $c \in (a, b)$, where c is not a point of P_n for $n \in \mathbb{N}$, if and only if $g(c) = h(c)$.

First assume f is continuous at $c \in (a, b)$. Let $\varepsilon > 0$ and choose $\delta > 0$ such that $|x - c| < \delta$ implies $|f(x) - f(c)| < \varepsilon/4$. Also choose $N \in \mathbb{N}$ such that $\text{mesh}(P_N) < \delta$. Then $c \in (x_{k-1}, x_k]$ for some subinterval of P_N and thus $|x - c| < \delta$ for every $x \in (x_{k-1}, x_k]$. By the continuity assumption it must be the case that $|f(x) - f(c)| < \varepsilon/4$ for every $x \in (x_{k-1}, x_k]$. If we now consider $M_k = \sup\{f(x) : x \in (x_{k-1}, x_k]\}$ and $m_k = \inf\{f(x) : x \in (x_{k-1}, x_k]\}$, it is clear that $M_k \leq f(c) + \varepsilon/4$ while $m_k \geq f(c) - \varepsilon/4$. Taking these two inequalities in hand shows that $M_k - m_k \leq \varepsilon/2 < \varepsilon$ and, appealing to the definition of $h_N, g_N, h,$ and g, it follows that

$$h(c) - g(c) \leq h_N(c) - g_N(c) = M_k - m_k < \varepsilon$$

Hence $h(c) = g(c)$.

For the converse, assume $h(c) = g(c)$ for $c \in (a, b)$ where c is not a point in any of the chosen partitions P_n. For $\varepsilon > 0$, the definitions of h and g indicate that there is an $N \in \mathbb{N}$ such that $|h_N(c) - g_N(c)| < \varepsilon$. (An explicit choice for N is obtained by considering the fact that $|h_N(c) - g_N(c)| \leq |h_N(c) - h(c)| + |h(c) - g(c)| + |g(c) - g_N(c)|$ together with the fact that h and g are pointwise limit functions.) As before $c \in (x_{k-1}, x_k]$ for some subinterval in the partition P_N. We now set $\delta = \min\{c - x_{k-1}, x_k - c\} > 0$, which is positive since we are assuming that c is not an endpoint for this partition. Then if $|x - c| < \delta$, it follows that $x \in (x_{k-1}, x_k]$ and

$$|f(x) - f(c)| \leq M_k - m_k = h_N(c) - g_N(c) < \varepsilon$$

where we are again using the standard definitions of M_k and m_k; note the final equality follows from the definitions of h_N and g_N since $c \in (x_{k-1}, x_k]$. Our conclusion is now apparent though and f must be continuous at c.

With this characterization of continuity in hand we are now ready to show that the theorem holds. First suppose that f is Riemann integrable on $[a, b]$. By the proof of the previous theorem it follows that $g(x) = h(x)$ except possibly on a set of measure zero. Therefore f is continuous except possibly on a set of measure zero. On the other hand, suppose f is continuous on $a \subseteq [a, b]$ where

$m([a,b] \setminus A) = 0$. Then we know that $g(x) = h(x)$ for every $x \in A$, and, by construction, $g(x) = f(x) = h(x)$ for every $x \in A$. Now, the fact that f is bounded guarantees us that both g and h are Lebesgue integrable by the Dominated Convergence Theorem. Furthermore Eq. (9.3) holds, as does Eq. (9.4) and the corresponding integral expression concerning h. These two facts show us that

$$\lim_{n \to \infty} \left[U(f, P_n) - L(f, P_n) \right] = \lim_{n \to \infty} \int_{[a,b]} h_n - g_n \, dm = \int_{[a,b]} h - g \, dm$$

$$= \int_A h - g \, dm = 0$$

where we have also employed the fact that $m([a,b] \setminus A) = 0$ and that $g(x) = h(x)$ on A in the last two equalities. Therefore we conclude that f is Riemann integrable.

□

With this characterization it is easy to see that the product of two Riemann integrable functions is integrable and that a composition $h = f \circ g$ is integrable if f is integrable and g is continuous. We explored simple cases of these statements in Chap. 7 and proofs of these more general statements are requested in the exercises. Furthermore, in Exercise 7.1.9, we showed that Thomae's function integrates to zero, but we had assumed there that the function is integrable. As it turns out, Thomae's function is continuous at each irrational point of $[0, 1]$ and discontinuous at each rational point and is hence Riemann integrable by Lebesgue's criterion.

We have made the case that the Lebesgue integral is an improvement upon its predecessor but we are also obligated to point out that the integral also has its own deficiencies; we restrict ourselves now to the measure space $(\mathbb{R}, \mathscr{L}, m)$. One obvious drawback is the need to develop a substantial amount of measure theory before considering the Lebesgue integral, even in the simple case of real valued functions of a single real variable. A second issue arises from the fact that the Lebesgue integral is designed to be an absolutely convergent integral, i.e. f is Lebesgue integrable if and only if $|f|$ is Lebesgue integrable. Thus if $|f|$ is not Lebesgue integrable, then neither is f. An example of a function f where $|f|$ is not Lebesgue integrable is $f(x) = \sin(x)/x$ defined on the interval $(0, \infty)$. To understand this, notice that for a fixed $j \in \mathbb{N}$,

$$\int_{j\pi}^{(j+1)\pi} \left| \frac{\sin(x)}{x} \right| \, dm \geq \frac{1}{(j+1)\pi} \int_{j\pi}^{(j+1)\pi} |\sin(x)| \, dm = \frac{2}{(j+1)\pi}.$$

This then implies that

$$\int_0^{n\pi} \left| \frac{\sin(x)}{x} \right| \, dm = \sum_{j=0}^{n-1} \int_{j\pi}^{(j+1)\pi} \left| \frac{\sin(x)}{x} \right| \, dm \geq \sum_{j=0}^{n-1} \frac{2}{(j+1)\pi}$$

and it follows that

$$\int_0^\infty \left| \frac{\sin(x)}{x} \right| \, dm \geq \sum_{j=0}^{n-1} \frac{2}{(n+1)\pi}$$

for all $n \in \mathbb{N}$. However, the partial sums on the right of the previous inequality diverge to infinity and hence the integral also diverges to infinity. This is startling when we consider the fact that the improper Riemann integral (though we did not discuss these) of $\sin(x)/x$ over the interval $(0, \infty)$ does exist and has value 2. The drawback here is then that we cannot apply Lebesgue theory techniques to the study of improper Riemann integrals.

A third concern relates to the Fundamental Theorem of Calculus. Sadly, it is true that there exist functions F which are differentiable at every point in an interval $[a, b]$, but such that their derivatives are not Lebesgue integrable. Such functions provide a counterexample to the desired statement that

$$\int_{[a,b]} F' \, dm = F(b) - F(a)$$

as the left-hand side of this expression does not exist for these functions. There are also examples where the function F is differentiable almost everywhere on $[a, b]$, the derivative F' is Lebesgue integrable, but for which the above equality does not hold. Thus, as with the Riemann integral, an additional hypothesis is needed to guarantee that the fundamental theorem holds in this setting. For an example of such a function, see [10].

To close this section, we mention that there is another integration theory which overcomes these deficiencies. This integration technique was developed around 1960 and its construction is more akin to the Riemann integral; for this reason, it is referred to as the *generalized Riemann integral* or the *Henstock–Kurzweil integral* (for its founders). The most notable improvement of this new theory over the Lebesgue integral is the fact that a strong version of the fundamental theorem holds: *if $F : [a, b] \to \mathbb{R}$ is differentiable on $[a, b]$, then F' has a generalized Riemann integral and*

$$\int_a^b F' \, dx = F(b) - F(a).$$

The article [4] outlines the basic construction of this integral and requests that it replaces Lebesgue's as the central integration technique of modern mathematics.

Exercises

Exercise 9.3.1. Show that the sets E_n defined in the proof of the Monotone Convergence Theorem are measurable, nested, and that their union is X. To see the last property, for a point $x \in X$, consider two cases: $f(x) = 0$ and $f(x) > 0$.

Exercise 9.3.2. (a) Complete Example 9.2.8 by showing that equality does hold.
(b) Extend this example by showing equality holds for any measurable function $f : \mathbb{N} \to [0, \infty]$.

Exercise 9.3.3. Give an example to show that the Monotone Convergence Theorem does not hold for the Riemann integral, i.e. give an example of an increasing sequence of nonnegative functions which converges pointwise to a function f and for which the conclusion of the theorem does not hold for the Riemann integral.

Exercise 9.3.4. Supply a proof for Corollary 9.3.2.

Exercise 9.3.5. (a) Let $(X, \mathcal{M}, \mu)$ be a measure space and suppose $f_n : X \to [0, \infty]$ is a sequence of measurable functions. If $f(x) = \sum_{n=1}^{\infty} f_n(x)$, show that

$$\int_X f \, d\mu = \sum_{n=1}^{\infty} \int_X f_n \, d\mu.$$

(b) Supply a proof for Theorem 9.3.3. To show that the set function is countably additive the following observation will be useful: If $\{B_k\}$ is a pairwise disjoint collection of sets and B is their union, then $\chi_B = \sum_{k=1}^{\infty} \chi_{B_k}$.
(c) If φ is defined as in the statement of Theorem 9.3.3, and g is nonnegative and measurable, show that

$$\int_X g \, d\varphi = \int_X gf \, d\mu.$$

Exercise 9.3.6. (a) Let $(X, \mathcal{M}, \mu)$ be a measure space and let $A \in \mathcal{M}$. Suppose that $f_n : X \to [0, \infty]$ is a sequence of measurable functions with $f_1(x) \geq f_2(x) \geq f_3(x) \geq \cdots \geq 0$ for every $x \in A$ and define $f : X \to [0, \infty]$ by $f(x) = \lim_{n \to \infty} f_n(x)$. Further assume that f_1 is Lebesgue integrable. Show that

$$\lim_{n \to \infty} \int_A f_n \, d\mu = \int_A f \, d\mu.$$

(b) Give an example to show that this statement does not hold if the condition "f_1 is Lebesgue integrable" is removed.

Exercise 9.3.7. Let $(X, \mathcal{M}, \mu)$ be a measure space and let $E \in \mathcal{M}$. Set $f_n = \chi_E$ if n is odd and $f_n = 1 - \chi_E$ if n is even. Does the conclusion of Fatou's Lemma hold for this sequence of functions? Provide the relevant calculations.

Exercise 9.3.8. Let $(X, \mathcal{M}, \mu)$ be a measure space. If f and g are measurable functions which are equal a.e. on X, i.e. $\mu(\{x : f(x) \neq g(x)\}) = 0$, then we write $f \sim g$. Show that the relation defined above is an equivalence relation on the set of functions which are measurable with respect to the measure space $(X, \mathcal{M}, \mu)$.

Exercise 9.3.9. Let $(X, \mathcal{M}, \mu)$ be a measure space with $A \in \mathcal{M}$ and suppose f and g are measurable on A with $0 \leq f(x) \leq g(x)$ a.e. on A. Show that

$$\int_A f \, d\mu \leq \int_A g \, d\mu.$$

Exercise 9.3.10. State and prove a general version of the Monotone Convergence Theorem using an almost everywhere condition as in the statement of Theorem 9.3.7.

Exercise 9.3.11. (a) Suppose f and g are Riemann integrable on $[a, b]$. Show that fg is Riemann integrable.

(b) Suppose f and g are Riemann integrable on $[a, b]$ and that f is nonnegative. If f integrates to zero, show that fg integrates to zero.

(c) Suppose f is Riemann integrable on $[a, b]$ with $m \leq f(x) \leq M$ for all $x \in [a, b]$. Further suppose g is continuous on $[m, M]$. Show that $g \circ f$ is Riemann integrable on $[a, b]$.

(d) Suppose f and g are Riemann integrable on $[a, b]$. By considering the function $h = (c|f| - d|g|)^2$, where $c, d \in \mathbb{R}$, show that

$$\left| \int_a^b fg \, dx \right| \leq \left(\int_a^b f^2 \, dx \right)^{1/2} \left(\int_a^b g^2 \, dx \right)^{1/2}.$$

You will want to make particular choices for c and d, and you may also want to consider what happens if either integral on the right is zero.

9.4 The L^p Spaces

Let $(X, \mathcal{M}, \mu)$ be a measure space and let $1 \leq p < \infty$. Our goal in this section is to consider all measurable functions which satisfy

$$\int_X |f|^p \, d\mu < \infty.$$

As a collection, our hope is that this set forms a vector space and that we can identify
a norm which will result in a complete space. We investigated this in Sect. 7.5 for
the Riemann integral and encountered some resistance. In fact we identified two
particular problems there. The Lebesgue integral was designed to alleviate the first.
The following will eliminate the second.

Definition 9.4.1. Let $(X, \mathcal{M}, \mu)$ be a measure space and let $1 \leq p < \infty$. We define
$\tilde{L}^p(X, \mathcal{M}, \mu)$ to be the set of all measurable functions $f : X \to [-\infty, \infty]$ such that

$$\int_X |f|^p \, d\mu < \infty$$

and define a rule $\| \cdot \|_p : \tilde{L}^p \to [0, \infty)$ by

$$\|f\|_p = \left(\int_X |f|^p \, d\mu \right)^{1/p}.$$

It turns out that the collection $\tilde{L}^p$ is a vector space if we apply the usual rules of
function addition and scalar multiplication, but the rule $\| \cdot \|_p$ does not define a
norm. The problem here is the fact that there are nonzero functions which integrate
to zero; take a function which is zero except on a set of measure zero. The Riemann
integral applied to discontinuous functions suffers from a similar ailment. However,
Exercise 9.2.7(d) shows that such a function is equal to zero almost everywhere, if
we assume the function is measurable. Thus if we associate all these functions, we
will have a solution to our problem. To be more detailed, we have to work within
the structure of equivalence classes.

Referring to Exercise 9.3.8, equality almost everywhere with respect to the
measure μ defines an equivalence relation on the collection of all functions which
are measurable with respect to μ. This equivalence relation then partitions the set of
all measurable functions into equivalence classes. In other words, for a measurable
function $f : X \to [-\infty, \infty]$, the set of all functions which are a.e. equal to f can
be identified as

$$[f] = \{g : X \to [-\infty, \infty] : g \text{ is measurable with } g = f \text{ a.e.}\}.$$

For example, with respect to the measure space $(\mathbb{R}, \mathcal{L}, m)$, the Dirichlet function
is in the same class as the zero function. Moreover, if f and g are in the same
equivalence class and are integrable, then Exercise 9.2.9 guarantees that

$$\int_X f \, d\mu = \int_X g \, d\mu.$$

Hence for a given class of integrable functions, we can define

$$\int_X [f] \, d\mu = \int_X g \, d\mu \tag{9.5}$$

where g is any representative function in $[f]$.

Definition 9.4.2. Let $(X, \mathcal{M}, \mu)$ be a measure space and let $1 \leq p < \infty$. Define the set $L^p(X, \mathcal{M}, \mu)$ to be the collection of equivalence classes $[f]$ of measurable functions which satisfy

$$\int_X [|f|^p] \, d\mu = \int_X |g|^p \, d\mu < \infty$$

for some (and therefore all) $g \in [f]$. Also, define $\| \cdot \|_p : L^p \to [0, \infty)$ by

$$\|[f]\|_p = \left(\int_X |g|^p \, d\mu \right)^{1/p}$$

where g is any representative in the class $[f]$.

For the sake of convenience, we will often suppress the notation and write $L^p(\mu)$ or simply L^p for the space $L^p(X, \mathcal{M}, \mu)$. It should be apparent that $L^1(X, \mathcal{M}, \mu)$ is simply the collection of all functions which are Lebesgue integrable with respect to $(X, \mathcal{M}, \mu)$. Also, though it is important to maintain a distinction between the sets $\tilde{L}^p$ and L^p, we generally write $\|f\|_p$ for $\|[f]\|_p$ and think in terms of functions instead of equivalence classes. The reason for this is that when working with an equivalence class, it is often sufficient to work with a single element of the class, though there will be circumstances when this is inappropriate and in these cases we will use the more informative bracket notation.

To show that L^p is a vector space we need two operations. We use the standard definitions of function addition and scalar multiplication within the equivalence class structure and define

$$[f] + [g] = [f + g] \qquad \text{and} \qquad c[f] = [cf].$$

At this point it is easy to show that L^p is a vector space. Indeed, if $[f], [g] \in L^p$, the fact that $[f + g]$ is a vector in L^p follows from Exercise 9.4.1(a) combined with the monotonicity and linearity of the Lebesgue integral; the fact that L^p is closed with respect to scalar multiplication follows from the linearity of the integral. The other operational vector space axioms are easily verified. Keep in mind that the zero vector in these spaces is the equivalence class of the zero function.

For norm properties, it is clear that $\| \cdot \|_p$ maps L^p into $[0, \infty)$ by definition and that $\|c[f]\|_p = |c| \|[f]\|_p$. The fact that the norm outputs zero only for the zero vector is a consequence of the equivalence class structure we have imposed together with the result of Exercise 9.2.7(d). This leaves the triangle inequality. As it turns out, this is a nontrivial fact standing in contrast to the other normed linear space axioms. There are several steps in proving the triangle inequality in this setting and it is formally stated as Theorem 9.4.6.

Definition 9.4.3. If $1 < p < \infty$, we define the *conjugate exponent of p* to be the unique real number q such that

$$\frac{1}{p} + \frac{1}{q} = 1.$$

This can be rewritten as $q = p/(p-1)$ from which it is obvious that $1 < q < \infty$. If $p = 1$, we define $q = \infty$ to be the conjugate exponent and vice versa if $p = \infty$.

The following technical lemma provides an interesting relationship between conjugate exponents and is a crucial step along our path to proving the triangle inequality for L^p.

Lemma 9.4.4. *If $x, y \geq 0$ and p, q are conjugate exponents, then*

$$xy \leq \frac{x^p}{p} + \frac{y^q}{q}.$$

Proof. Fix $y \geq 0$. If $y = 0$, then the inequality clearly holds. For $y > 0$, the proof reduces to a simple analysis of the function $f(x) = xy - x^p/p$ on the interval $[0, \infty)$. We will show that $f(x) \leq y^q/q$ for every $x \in [0, \infty)$, i.e. f has an absolute maximum value at y^q/q. A review of basic techniques for identifying absolute extrema may be in order and we refer the reader to any calculus textbook. First observe that $f'(x) = y - x^{p-1}$ and that this function has exactly one zero in the interval $(0, \infty)$ occurring at $c = y^{1/(p-1)}$. Furthermore, if $x < y^{1/(p-1)}$, then $y - x^{p-1} > 0$ and if $x > y^{1/(p-1)}$, then $y - x^{p-1} < 0$. Analyzing this information in terms of the function f, we see that the function is increasing to the left of $c = y^{1/(p-1)}$ and decreasing on the right (on our given domain), and therefore c is the absolute maximum of f on $[0, \infty)$. Also, keeping in mind that p, q are conjugate exponents,

$$f(c) = yy^{1/(p-1)} - \frac{(y^{1/(p-1)})^p}{p} = y^{p/(p-1)} - \frac{y^{p/(p-1)}}{p} = y^{p/(p-1)}\left(1 - \frac{1}{p}\right) = \frac{y^q}{q}.$$

Combining these two observations shows that for any $x \in (0, \infty)$,

$$xy - \frac{x^p}{p} = f(x) \leq f(c) = \frac{y^q}{q}.$$

For our conclusion, the fact that y was an arbitrary nonnegative real number guarantees that

$$xy \leq \frac{x^p}{p} + \frac{y^q}{q}$$

for all $x, y \in [0, \infty)$. $\square$

The next two statements provide inequality relationships for functions in the spaces under consideration. The first is one of the most often employed tools of mathematicians working in such spaces and the second is the triangle inequality.

Theorem 9.4.5 (Hölder's Inequality). *Let $1 < p, q < \infty$ be conjugate exponents. If $f \in L^p$ and $g \in L^q$, then*

$$\left| \int_X fg \, d\mu \right| \leq \int_X |fg| \, d\mu \leq \|f\|_p \|g\|_q.$$

Proof. Suppose $[f] \in L^p$ and $[g] \in L^q$. We first show that the inequality on the right holds. Before doing so, observe that it is sufficient to work with representatives f and g from the classes $[f]$ and $[g]$, respectively. Now, if either $f \sim 0$ or $g \sim 0$, then the desired inequality holds by definition of $\| \cdot \|_p$ and $\| \cdot \|_q$. Therefore we may assume that $\|f\|_p$ and $\|g\|_q$ are both nonzero. Applying Lemma 9.4.4 with $x = |f|/\|f\|_p$ and $y = |g|/\|g\|_q$,

$$\left(\frac{|f|}{\|f\|_p} \right) \left(\frac{|g|}{\|g\|_q} \right) \leq \frac{1}{p} \left(\frac{|f|}{\|f\|_p} \right)^p + \frac{1}{q} \left(\frac{|g|}{\|g\|_q} \right)^q = \frac{|f|^p}{p\|f\|_p^p} + \frac{|g|^q}{q\|g\|_q^q}.$$

If we now integrate this inequality and appeal to the linearity of the integral, we have

$$\frac{1}{\|f\|_p\|g\|_q} \int_X |fg| \, d\mu \leq \frac{1}{p\|f\|_p^p} \int_X |f|^p \, d\mu + \frac{1}{q\|g\|_q^q} \int_X |g|^q \, d\mu = \frac{1}{p} + \frac{1}{q} = 1.$$

Multiplying on the left and right by $\|f\|_p\|g\|_q$ produces the desired inequality.

The left-hand inequality is now a consequence of Corollary 9.3.2 since we know that the function fg is Lebesgue integrable. $\quad\square$

Theorem 9.4.6 (Minkowski's Inequality). *If $1 \leq p < \infty$ and $f, g \in L^p$, then*

$$\|f + g\|_p \leq \|f\|_p + \|g\|_p.$$

Proof. As above we work with functions f and g which are representative functions from the classes $[f], [g] \in L^p$. As a first case, if $p = 1$, then the triangle inequality for absolute value and linearity of the integral produce the inequality we seek. Now suppose $1 < p < \infty$. If $\|f + g\|_p = 0$, then the inequality must hold by definition of $\| \cdot \|_p$. For the more significant case, suppose $\|f + g\|_p > 0$ and let q be the conjugate exponent of p. We may then write $p = q(p - 1)$ and $1/q = (p - 1)/p$, and thus

$$\|(f + g)^{p-1}\|_q = \left(\int_X |f + g|^{(p-1)q} \, d\mu \right)^{1/q}$$

$$= \left(\int_X |f + g|^p \, d\mu \right)^{(p-1)/p} = \|f + g\|_p^{p-1}.$$

This shows that $(f + g)^{p-1}$ is in L^q. To derive the inequality we seek, notice now that

$$\|f + g\|_p^p = \int_X |f + g|^p \, d\mu = \int_X |f + g||f + g|^{p-1} \, d\mu$$

$$\leq \int_X |f||f + g|^{p-1} \, d\mu + \int_X |g||f + g|^{p-1} \, d\mu$$

$$\leq \|f\|_p \|(f + g)^{p-1}\|_q + \|g\|_p \|(f + g)^{p-1}\|_q$$

$$= \|f\|_p \|(f + g)\|_p^{p-1} + \|g\|_p \|(f + g)\|_p^{p-1}$$

$$= \|f + g\|_p^{p-1} (\|f\|_p + \|g\|_p)$$

where the first inequality is a result of the triangle inequality in $\mathbb{R}$ and the second is a consequence of Hölder's inequality. Dividing the left- and right-hand side of this last inequality by the quantity $\|f + g\|_p^{p-1}$ shows that $\|f + g\|_p \leq \|f\|_p + \|g\|_p$.   □

Minkowski's Inequality provides the last piece of the puzzle in the verification that L^p is a normed linear space and we state the following theorem simply as a reference tool.

Theorem 9.4.7. *The space L^p equipped with the norm $\| \cdot \|_p$ is a normed linear space.*

Our next task is to show that we are working with a complete normed linear space. We have seen several proofs of completeness up to this point and it is a remarkable feat that the following proof holds for *any* positive measure μ. A drawback here is that the proof is quite abstract in the sense that no properties of the particular measure are employed. Our first example after the proof will provide a means of comparing the abstract proof with a proof for a specific measure.

Theorem 9.4.8 (Riesz–Fischer Theorem). *If $1 \leq p < \infty$ and $(X, \mathcal{M}, \mu)$ is a measure space, then the space $L^p(X, \mathcal{M}, \mu)$ is complete.*

Proof. Let (f_n) be a Cauchy sequence in L^p. By Exercise 2.5.5 it suffices to show that (f_n) has a convergent subsequence with limit f in L^p and hence our first task is to identify this subsequence. Appealing to the definition of Cauchy sequence, for each $k \in \mathbb{N}$, we can find $N_k \in \mathbb{N}$ so that $\|f_m - f_n\|_p < 1/2^k$ whenever $m, n \geq N_k$. In addition, construct an increasing sequence of natural numbers by choosing $n_1 = N_1$ and, for $k \geq 2$, set $n_k = \max\{n_{k-1}, N_k\} + 1$. At this point we have a subsequence (f_{n_k}) with the property that $\|f_{n_{k+1}} - f_{n_k}\|_p < 1/2^k$ for every $k \in \mathbb{N}$. We now work to show that (f_{n_k}) converges to a function $f \in L^p$. For convenience we will simply refer to f_{n_k} as f_k.

In order to show that (f_k) converges, we must find a limit candidate. To begin, we will find a function f such that (f_k) converges to f pointwise almost everywhere on X; one of the latter steps in the proof will be to show that this function is actually in L^p and that f_k converges to f in the L^p norm. Continuing, we proceed as is typical in verifying completeness of a space and aim to show that the sequence of values $(f_k(x))$ is Cauchy in $\mathbb{R}$. There are, however, some subtleties in this setting that require extra care and quite a bit more work. We begin by defining a sequence of functions (g_k) by $g_1(x) = |f_1(x)|$ and

$$g_k(x) = |f_1(x)| + |f_2(x) - f_1(x)| + |f_3(x) - f_2(x)| + \cdots + |f_k(x) - f_{k-1}(x)|$$

for $k \geq 2$. Notice that

$$0 \leq g_1(x) \leq g_2(x) \leq g_3(x) \leq \cdots$$

and

$$
\begin{aligned}
\|g_k\|_p &= \Big\| |f_1| + |f_2 - f_1| + |f_3 - f_2| + \cdots + |f_k - f_{k-1}| \Big\|_p \\
&\leq \|f_1\|_p + \|f_2 - f_1\|_p + \|f_3 - f_2\|_p + \cdots + \|f_k - f_{k-1}\|_p \\
&= \|f_1\|_p + \sum_{j=2}^{k} \|f_j - f_{j-1}\|_p \\
&< \|f_1\|_p + 1
\end{aligned}
\tag{9.6}
$$

where we have used the fact

$$\sum_{j=2}^{k} \|f_j - f_{j-1}\|_p < \sum_{j=1}^{\infty} \|f_j - f_{j-1}\|_p < \sum_{j=1}^{\infty} \frac{1}{2^j} = 1$$

in the last inequality. This shows that $g_k \in L^p$ for every $k \in \mathbb{N}$.

At this point, set $g(x) = \lim_{k \to \infty} g_k(x)$. It is immediately clear that $|g(x)| = g(x)$ and $g(x)^p = \lim_{k \to \infty} g_k(x)^p$ for every $x \in X$ (Exercise 9.4.9 will ask you to explain this equality). Moreover, the sequence $(g_k(x)^p)$ is increasing for each $x \in X$ and an application of the Monotone Convergence Theorem shows that

$$
\begin{aligned}
\|g\|_p &= \left(\int_X |g|^p \, d\mu \right)^{1/p} = \left(\int_X \lim_{k \to \infty} g_k^p \, d\mu \right)^{1/p} = \left(\lim_{k \to \infty} \int_X g_k^p \, d\mu \right)^{1/p} \\
&= \lim_{k \to \infty} \left(\int_X g_k^p \, d\mu \right)^{1/p} = \lim_{k \to \infty} \|g_k\|_p \leq \|f_1\|_p + 1
\end{aligned}
$$

where the last inequality is a result of Eq. (9.6). This shows that $g \in L^p$ since $\|f_1\|_p < \infty$ and thus it must be the case that g takes on a finite value almost everywhere on X. Indeed, if g assumed an infinite value on a set of positive measure, then the integral would also be infinite contradicting the fact that $g \in L^p$; we are also using the fact that g is nonnegative here.

Now we turn our attention back the numerical sequence $(f_k(x))$. Let $A \subseteq X$ be the set where g is finite. Using the fact that g is the pointwise limit of (g_k) shows that the numerical sequence $(g_k(x))_{k=1}^{\infty}$ is Cauchy in $\mathbb{R}$ for every $x \in A$. Furthermore, if $\varepsilon > 0$ and $x \in A$, we can find $N \in \mathbb{N}$ so that $|g_k(x) - g_m(x)| = g_k(x) - g_m(x) < \varepsilon$ whenever $k > m \geq N$ and thus

$$|f_k(x) - f_m(x)| = \left| \sum_{j=m+1}^{k} f_j(x) - f_{j-1}(x) \right| \leq \sum_{j=m+1}^{k} |f_j(x) - f_{j-1}(x)|$$

$$= g_k(x) - g_m(x) < \varepsilon$$

whenever $k > m \geq N$ and $x \in A$. Therefore $(f_k(x))_{k=1}^{\infty}$ is Cauchy in $\mathbb{R}$ for every $x \in A$. With this we can define a function $f : X \to [-\infty, \infty]$ such that $f(x) = \lim_{k \to \infty} f_k(x)$ for $x \in A$ and $f(x) = 0$ for $x \in X \setminus A$. This function is measurable by Exercise 9.2.8 and Corollary 9.1.16 since $\mu(X \setminus A) = 0$.

Now it remains to show that $f \in L^p$ and $\|f_k - f\|_p \to 0$ as $k \to \infty$. Let $\varepsilon > 0$ and choose $N \in \mathbb{N}$ so that $\|f_k - f_m\|_p < \varepsilon/2$ whenever $k, m \geq N$. Then, if $k \geq N$,

$$\|f_k - f\|_p^p = \int_X |f_k - f|^p \, d\mu = \int_X \liminf |f_k - f_m|^p \, d\mu$$

$$\leq \liminf \int_X |f_k - f_m|^p \, d\mu$$

$$= \liminf \|f_k - f_m\|_p^p$$

$$\leq (\varepsilon/2)^p$$

from which it follows that $\|f_k - f\|_p \leq \varepsilon/2 < \varepsilon$ whenever $k \geq N$. This shows two things. First, the function $f_k - f$ is in L^p for every $k \in \mathbb{N}$. Thus we may write $f = f_k - (f_k - f)$, a sum of functions in L^p. The fact that our space is closed with respect to addition allows us to conclude that $f \in L^p$. Second, this is exactly what it means to say that $\|f_k - f\|_p \to 0$ as $k \to \infty$ and the proof is complete (...just like our space!).   □

Example 9.4.9. For $1 \leq p < \infty$ and μ counting measure on $\mathbb{N}$, we can identify ℓ^p as the space $L^p(\mathbb{N}, \mathscr{P}(\mathbb{N}), \mu)$. As a preliminary, let $f : \mathbb{N} \to [-\infty, \infty]$. Utilizing our initial definition of a sequence as a function on $\mathbb{N}$, we may write $f(n) = a_n$. Thus every such function defines a sequence of real numbers (allowing infinity as a term in such a sequence) and vice versa. Furthermore, if $f : \mathbb{N} \to [0, \infty]$, by Exercise 9.3.2 we have

$$\int_{\mathbb{N}} f \, d\mu = \sum_{n=1}^{\infty} f(n) = \sum_{n=1}^{\infty} a_n.$$

Now, for our specific setting, if $f \in L^p(\mathbb{N}, \mathscr{P}(\mathbb{N}), \mu)$, then $|f(n)|^p = |a_n|^p$ and

$$\|f\|_p = \left(\int_{\mathbb{N}} |f|^p \, d\mu\right)^{1/p} = \left(\sum_{n=1}^{\infty} |a_n|^p\right)^{1/p} = \|(a_n)\|_p,$$

showing that $(a_n) \in \ell^p$. On the other hand, if we take a sequence $(a_n) \in \ell^p$, then we can define a function $f : \mathbb{N} \to (-\infty, \infty)$ by $f(n) = a_n$. Such a function is measurable since our σ-algebra is a power set and the above equality between integral and summation holds, showing that $f \in L^p(\mathbb{N}, \mathscr{P}(\mathbb{N}), \mu)$. This seems to confirm that these sets are the same with equal norm, but there is one subtle detail to consider. In our definition of ℓ^p, we took sequences of real numbers, so there could be some confusion from the initial discussion of functions $f : \mathbb{N} \to [-\infty, \infty]$. However, in the specific context of $L^p(\mathbb{N}, \mathscr{P}(\mathbb{N}), \mu)$, we are forced to only consider functions $f : \mathbb{N} \to (-\infty, \infty)$ since we are dealing with a counting measure, i.e. if f assumed an infinite value, then this would contradict the fact that $f \in L^p$. We this we may conclude that $L^p = \ell^p$.

Example 9.4.10. For another example, we can use a similar tactic to represent $\mathbb{R}^2$ with the Euclidean norm as an L^2 space. Take $X = \{1, 2\}$ (any set with two elements will suffice) and let $\mathscr{M} = \mathscr{P}(X)$. For our measure we take the counting measure. If (x_1, x_2) is any point in $\mathbb{R}^2$, then we can represent this as a function $f : X \to (-\infty, \infty)$ by $f(1) = x_1$ and $f(2) = x_2$ and vice versa if $f : X \to [-\infty, \infty]$, keeping in mind that infinity is an allowable value. Now, if we take a function in $f \in L^2(X, \mathscr{M}, \mu)$, then we can assume that f takes on only finite values and is therefore identified (in a bijective fashion) with a point in $\mathbb{R}^2$. Moreover, any such function is a simple function and thus

$$\|f\|_2 = \left(\int_X |f|^2 \, d\mu\right)^{1/2} = \sqrt{|x_1|^2 + |x_2|^2}.$$

Going one step further, in this setting Hölder's inequality reduces to the Cauchy–Schwarz inequality. Indeed, if $f \in L^2$ is associated with the point $x = (x_1, x_2)$ and $g \in L^2$ is associated with $y = (y_1, y_2)$, then by definition, fg represents the point $(x_1 y_1, x_2 y_2)$ and

$$|\langle x, y \rangle| = |x_1 y_1 + x_2 y_2| = \left|\int_X fg \, d\mu\right| \le \int_X |fg| \, d\mu$$

$$\le \left(\int_X |f|^2 \, d\mu\right)^{1/2} \left(\int_X |g|^2 \, d\mu\right)^{1/2} = \|x\|\|y\|$$

Notice here that we are using Hölder's inequality in the pleasant case when the conjugate indices are equal, namely $p = q = 2$. In the Exercises you will show that the spaces $L^2(X, \mathscr{M}, \mu)$ are all Hilbert spaces. It is typical in this setting to refer to Hölder's inequality as the Cauchy–Schwarz inequality; the computation above should therefore give you some help in identifying the appropriate inner product in any L^2 space.

Return now to the discussion of Sect. 7.5 where we showed that the normed linear space $(C([a,b]), \|\cdot\|_1)$ is not complete. After proving this, we attempted to enlarge our set of functions in order to find a complete space with respect to this norm. We then considered the set of all Riemann integrable functions, but in doing so we lost a valuable norm property. The question we now consider is if there is a collection of functions X with $C([a,b]) \subseteq X$ such that $(X, \|\cdot\|_1)$ is a complete normed linear space. With a few more specific details we would call X the *completion* of $(C([a,b]), \|\cdot\|_1)$.

As it turns out the space $L^1([a,b])$ is exactly the space we seek. Theorem 9.3.8 shows that the norm $\|\cdot\|_1$, when considered for continuous functions, returns the same value using either the Riemann integral or the Lebesgue integral. Thus to show that L^1 completes $C([a,b])$ we must show that given any Cauchy sequence in $(C([a,b]), \|\cdot\|_1)$, there is an L^1 function which acts as the limit function with respect to $\|\cdot\|_1$. The proof will not be given here as there are topological considerations which must be dealt with before proving the statement. As further enticement to seek out and understand these more complex ideas, we also recall that the space $(C([a,b]), \|\cdot\|_2)$ is not complete. Can you guess which of our new spaces will be its completion?

Example 9.4.11 (Volterra Operator). Of the topics covered in a basic analysis course, the concepts of continuity and integration seem to be the two which carry over most prevalently to functional analysis. Spaces of continuous functions provide exquisite examples of function spaces and, in the larger sense, the study of continuous (bounded) operators pervades the subject. The concept of integration also plays a prominent role as not only it allows us to construct norms on many interesting spaces, but it also appears in the operator context rather frequently. Here, in our final example, we explore one integral operator of great importance, the *Volterra operator*. This operator was explored in Exercise 7.5.9 but now we place it in the more general context of the Lebesgue spaces.

To begin, consider the measure space $L^2([0,1], \mathscr{L}([0,1]), m)$. We will eventually have two variables here and thus we will use dx and dt to represent the Lebesgue measure on the real line for these two variables rather than $dm(x)$ and $dm(t)$. For $x, t \in [0,1]$, define a function g by the rule

$$g(x,t) = \begin{cases} 1 & t \in [0,x]; \\ 0 & t \in (x,1]. \end{cases}$$

For a fixed $x \in [0,1]$, we then see that

$$\int_0^1 g(x,t)\,dt = \int_0^x 1\,dt = x.$$

This then implies that

$$\int_0^1 \left(\int_0^1 g(x,t)\,dt \right) dx = \int_0^1 x\,dx = \frac{1}{2}.$$

Notice also that $|g(x,t)|^2 = g(x,t)$ for all $x,t \in [0,1]$. Now, for $f \in L^2$, define the Volterra operator $V : L^2 \to L^2$ by

$$Vf(x) = \int_0^x f(t)\,dt;$$

such an object should look familiar from our study of the Fundamental Theorem of Calculus. If f is continuous, then this operator returns an antiderivative of f and we aim to show that this operator is bounded. Notice that to compute the norm of Vf in L^2, we will integrate with respect to x. The following computation will work with the square of the norms in L^2 to avoid fractional powers and we employ Corollary 9.3.2 and the Cauchy–Schwarz inequality:

$$\begin{aligned}
\|Vf\|_2^2 = \int_0^1 |Vf(x)|^2\,dx &= \int_0^1 \left| \int_0^x f(t)\,dt \right|^2 dx \\
&= \int_0^1 \left| \int_0^1 g(x,t)f(t)\,dt \right|^2 dx \\
&\le \int_0^1 \left(\int_0^1 |g(x,t)||f(t)|\,dt \right)^2 dx \\
&\le \int_0^1 \left(\int_0^1 |g(x,t)|^2\,dt \right) \left(\int_0^1 |f(t)|^2\,dt \right) dx \\
&= \|f\|_2^2 \int_0^1 \left(\int_0^1 |g(x,t)|^2\,dt \right) dx \\
&= \frac{1}{2} \|f\|_2^2.
\end{aligned}$$

With this we see that

$$\|Vf\|_2 \le \frac{1}{\sqrt{2}}\|f\|_2$$

and therefore conclude that V is bounded.

Exercises

Exercise 9.4.1. (a) Let $1 \leq p < \infty$ and let $x, y \in \mathbb{R}^+$. Show that $(x + y)^p \leq 2^p(x^p + y^p)$.

(b) Let $0 < p < 1$ and let $x, y \geq 0$. Show that $(x + y)^p \leq x^p + y^p$. To do this, begin by considering the function $f(t) = (1 + t)^p - 1 - t^p$ on the interval $[0, \infty)$ and the RaceTrack Principle (Theorem 5.4.5).

Exercise 9.4.2. Let $(X, \mathcal{M}, \mu)$ be a measure space and let $1 \leq p < \infty$. Show that $L^p(X, \mathcal{M}, \mu)$ is a vector space.

Exercise 9.4.3. (a) Let $(X, \mathcal{M}, \mu)$ be a measure space. Show that $L^2(X, \mathcal{M}, \mu)$ is a Hilbert space.

(b) Consider the particular measure space $([a, b], \mathscr{L}([a, b]), m)$. For $1 \leq p < \infty$, $p \neq 2$, show that the norm $\| \cdot \|_p$ does not arise as an inner product in the sense of Proposition 1.5.15.

Exercise 9.4.4. Let $(X, \mathcal{M}, \mu)$ be a measure space with $\mu(X) < \infty$.

(a) If $1 < q < \infty$, use Hölder's Inequality to show that $L^q \subseteq L^1$.

(b) If $1 \leq p < q < \infty$, use (a) to show that $L^q \subseteq L^p$.

Exercise 9.4.5. Let $(X, \mathcal{M}, \mu)$ be a measure space and let $1 \leq p < r < q < \infty$. Show that $L^p \cap L^q \subseteq L^r$. To do this, begin by finding $\alpha \in (0, 1)$ such that $r = (1 - \alpha)p + \alpha q$. Then use this to show the desired set inclusion.

Exercise 9.4.6. Let $(X, \mathcal{M}, \mu)$ be a measure space and let $p, q, r > 1$ with

$$\frac{1}{p} + \frac{1}{q} + \frac{1}{r} = 1.$$

If $f \in L^p$, $g \in L^q$, and $h \in L^r$, show that

$$\int_X |fgh|\, d\mu \leq \|f\|_p \|g\|_q \|h\|_r.$$

Exercise 9.4.7. Let $1 < p < q < \infty$ and consider the measure space $(\mathbb{R}, \mathscr{L}, m)$. Is there a containment statement that can be made for the spaces L^p and L^q in this setting? Explain in detail.

Exercise 9.4.8. Let $(X, \mathcal{M}, \mu)$ be a measure space and let $0 < p < 1$. Define $L^p(X, \mathcal{M}, \mu)$ as in Definition 9.4.2.

(a) Show that $L^p(X, \mathcal{M}, \mu)$ is a vector space.

(b) Show that $\| \cdot \|_p$ does not define a norm in this setting.

(c) Show that the rule $d(f, g) = \int_X |f - g|^p\, d\mu$ does define a metric on this space.

Exercise 9.4.9. Prove the assertion from the proof of Theorem 9.4.8 that $g(x)^p = \lim_{k \to \infty} g_k(x)^p$ for every $x \in X$. To do this, consider two cases, $g(x) < \infty$ and $g(x) = \infty$.

Exercise 9.4.10. Let $(X, \mathcal{M}, \mu)$ be a measure space and let $f : X \to [-\infty, \infty]$ be measurable. We say f is *essentially bounded* if there is an $M \in \mathbb{R}$ such that $0 < M < \infty$ and $|f(x)| \leq M$ for a.e. $x \in X$. In this case we say M is an *essential upper bound for* f.

(a) For the measure space $\mathcal{L}$ or $\mathcal{L}([a, b])$, show that any bounded measurable function is essentially bounded.
(b) Consider the measure space $\mathcal{L}([a, b])$ and let $f : [a, b] \to \mathbb{R}$ be continuous. Show that f is essentially bounded on $[a, b]$.
(c) For the measure space $\mathcal{L}$, find an example of a function that is essentially bounded but not bounded.

Exercise 9.4.11. Let $(X, \mathcal{M}, \mu)$ be a measure space and let $f : X \to [-\infty, \infty]$ be measurable. The *essential supremum of* f is defined to be the number $\|f\|_\infty$ satisfying

(a) $|f(x)| \leq \|f\|_\infty$ for a.e. $x \in X$;
(b) if $M < \|f\|_\infty$, then there is a set $A \in \mathcal{M}$ with $\mu(A) > 0$ and $|f(x)| > M$ for every $x \in A$.

If $\|f\|_\infty < \infty$, the above definition says that $\|f\|_\infty$ is an essential upper bound for f while no number less than $\|f\|_\infty$ has this property.

(a) Suppose that f is essentially bounded. Show that

$$\|f\|_\infty = \inf\{M : M \text{ is an essential upper bound for } f\}.$$

(b) Suppose that f is essentially bounded and let g be a measurable function with $f = g$ a.e. Show that $\|f\|_\infty = \|g\|_\infty$.

Exercise 9.4.12. Let $(X, \mathcal{M}, \mu)$ be a measure space. Define the set $L^\infty(X, \mathcal{M}, \mu)$ to be the collection of equivalence classes $[f]$ of essentially bounded functions, that is $\|f\|_\infty < \infty$ for some (and therefore all) $f \in [f]$.

(a) Show that L^∞ is a vector space.
(b) Define $\|[f]\|_\infty = \|g\|_\infty$ where g is any representative in $[f]$. Show that $\| \cdot \|_\infty$ defines a norm on L^∞.
(c) Prove a version of Hölder's inequality for $p = 1$ and $q = \infty$.
(d) Follow the outline below to show that L^∞ is a complete space. Let (f_n) be a Cauchy sequence in L^∞.

(i) Define sets

$$A_k = \{x : |f_k(x)| > \|f_k\|_\infty\}, \quad B_{m,n} = \{x : |f_n(x) - f_m(x)| > \|f_n - f_m\|_\infty\},$$

and

$$F = \left(\bigcup_{k=1}^{\infty} A_k \right) \cup \left(\bigcup_{m,n=1}^{\infty} B_{m,n} \right).$$

Argue that F is measurable with $\mu(F) = 0$.

(ii) For $x \in X \setminus F$, show that $(f_n(x))$ is a Cauchy sequence in $\mathbb{R}$ with $|f_n(x)| \le \|f_n\|_\infty \le M$ for all $n \in \mathbb{N}$. Be sure to explain the existence of such an M.

(iii) Define f to be the pointwise limit of (f_n) on $X \setminus F$ and zero on F. Explain why part (b) guarantees the existence of f. Also show that f is measurable with $\|f\|_\infty \le M$ to conclude that $f \in L^\infty$.

(iv) For $\varepsilon > 0$, show that there exists $N \in \mathbb{N}$ such that $|f_n(x) - f(x)| < \varepsilon$ for all $n \ge N$ and all $x \in X \setminus F$. Use this information to conclude that $\|f_n - f\|_\infty \to 0$ as $n \to \infty$.

Exercise 9.4.13. Suppose μ is counting measure on $\mathbb{N}$. Extend Example 9.4.9 to the case $p = \infty$.

Exercise 9.4.14. Show that $L^\infty([0, 1], m) \subseteq L^p([0, 1], m)$ for $1 \le p < \infty$.

Exercise 9.4.15. Let $(X, \mathcal{M}, \mu)$ be a measure space with $\mu(X) < \infty$. If $f \in L^\infty$, show that

$$\lim_{p \to \infty} \|f\|_p = \|f\|_\infty.$$

Exercise 9.4.16. Let $(X, \mathcal{M}, \mu)$ be a measure space and let $g \in L^\infty(\mu)$. Show that the multiplication operator $M_g : L^2 \to L^2$ maps into L^2 and is bounded.

Exercise 9.4.17. Consider the measure space $([0, 1], \mathcal{L}([0, 1]), m)$ and let V be the Volterra operator from Exercise 9.4.11. For $c \in [0, 1]$, show that the set $K_c = \{f : f = 0 \text{ a.e. on } [0, c]\}$ is an invariant subspace for V.

Epilogue

The core of this book is devoted to a study of functions whose domain and codomain are subsets of $\mathbb{R}$. Historically, the development of these topics was formalized by (roughly) 1880; however, this development followed a much more meandering path than we have presented here; the questions answered by the theory of infinite series, and the differential and integral calculus had been in circulation since the ancient world. Moreover, much of the progress made on these problems was done without the firm foundation we rely on today. This may seem absurd after journeying through the structured course we have followed, but this is not atypical in mathematics. Indeed, the rigorous definitions for the real number system, functions, and limits were developed in the mid-nineteenth century, long after many of the fundamental results of analysis were believed to be true. It was this period of restructuring in the 1800s, led by Cauchy, which transformed what had been known as "calculus" into what we now call analysis. The two wonderful texts [8] and [13] take a historical approach to the development of analysis and present the ideas we have considered in a more historically accurate manner.

So what has happened since the 1880s? The simplest answer is, a lot! In analysis, the end of the nineteenth century was marked by the introduction of the Lebesgue integral which, as we have seen, provided a more solid relationship between the integral and the limit. However, at this same time, the focus of analysis was shifting. This change came about as mathematicians became interested in studying not individual functions, but, rather, collections of functions; we have seen examples of this in the ℓ^p spaces, continuous function spaces, and L^p spaces. This change in perspective led to the birth what we now call functional analysis (and point-set topology). As it evolved, this new branch of analysis centered around infinite dimensional vector spaces whose elements are functions and linear operators on those spaces. This evolution produced new spaces in which to ask many of the same questions as those asked previously and also provided examples where many of the familiar properties of the real number system do not hold. With this shift, the scope of inquiry changed incredibly. The function theory of real-variable maps is tremendously rich, but the real number system, and its finite dimensional cousins, is almost too ideal a place when working with linear maps. We can blame this on

M.A. Pons, *Real Analysis for the Undergraduate: With an Invitation to Functional Analysis*, 397
DOI 10.1007/978-1-4614-9638-0, © Springer Science+Business Media New York 2014

the fact that $\mathbb{R}$ is one dimensional if you like, i.e. $\mathbb{R}$ is rather boring in the larger collection of all possible vector spaces, whereas this new infinite dimensional setting provided a chance for analysis to push its limits, pardon the word play, in ways that are not possible in finite dimensional spaces. We provide a few examples to convince you of the more diabolical, and hence more delightful, possibilities of infinite dimensional vector spaces.

- Every linear operator acting on a finite dimensional space is bounded and hence continuous. There are, however, examples of linear operators on infinite dimensional spaces which are not bounded; we saw an example of an unbounded operator in Sect. 4.5.
- A general definition for compact sets involves open covers as presented in our version of the Heine–Borel Theorem. In any finite dimensional normed linear space, this is equivalent to the set being closed and bounded. This is not the case in infinite dimensional spaces, and it is therefore interesting to seek out conditions (such as the set being closed and bounded) which are equivalent to compactness. An example of such a condition is provided by the Arzela–Ascoli Theorem and states that a set $K \subseteq C([a, b])$ is compact if and only if K is closed, bounded, and equicontinuous.
- If X is a vector space of dimension n and $T : X \to X$ is a linear operator, then T has exactly n eigenvalues (counting multiplicities). For X an infinite dimensional space, we have seen examples where the operator has no eigenvalues and where every possible scalar is an eigenvalue. Moreover, there is the more general notion of spectrum which arises as a result of the fact that a linear operator that is one-to-one need not be onto in the infinite dimensional setting.

The material in Sects. 1.5, 2.5, 3.4, 4.5, 5.6, 6.5, 7.5, 8.5, and 9.4 has been provided to give you a brief glimpse into the mathematics that has occurred in the last 130 years, and, collectively, these sections provide an introduction to the most basic concepts and examples of functional analysis. The inclusion of these topics is partly a personal preference as this author is a devotee. A second reason (and somewhat more satisfying for the reader) is the fact that the material is relevant to anyone who hopes to study analysis of any sort as spaces of functions and operators on them pervade almost every sub-area of modern analysis! This is true even of the applied subfields such as mathematical physics and the study of differential and partial differential equations. The third reason is an aesthetic one. It is my opinion that the most beautiful results in all of mathematics are those that reach into the other subfields of the subject to provide insight, and functional analysis is a prime example of such a collaborative effort. Its beauty hinges on the fact that it sits at the intersection of several branches of mathematics: analysis, algebra, topology, and geometry, and uses this positioning to influence and be influenced by its mathematical cousins.

The field of functional analysis has grown tremendously since its beginning in the early twentieth century and it would be impossible to give a brief yet satisfying overview in just a few pages. We will therefore not attempt a lengthy discussion here, but rather we will provide a brief road map through a few of the most basic concepts. For texts, we recommend [28] for the undergraduate wishing to jump right in. For the student willing to commit to some prerequisite material, courses in abstract algebra, point-set topology, and complex variables, we recommend the text [18].

In beginning a proper study of functions of any sort, one must first understand the structure of the underlying domain and codomain spaces. To understand linear operators, it is therefore necessary to understand the basic structures of Banach and Hilbert spaces; completeness is such a desirable trait that nearly all of the results in functional analysis are for these types of spaces though there are a few which pertain to the simpler class of normed linear spaces. The Hahn–Banach Theorem is an example of such a statement. It is also important to keep in mind that while Banach and Hilbert spaces have many similarities, the presence of an inner product in a Hilbert space provides a geometric flavor not present in the Banach space setting. This distinction in behavior (though not necessarily the geometric flavor) is seen most readily in the study of dual spaces. A dual space is the collection of all linear functionals on a given Banach/Hilbert space. The Riesz Representation Theorem states that any Hilbert space is its own dual (we also use the term self-dual), but this behavior is not common among Banach spaces. As a simple example, for $1 \leq p < \infty$ and q the conjugate exponent of p, the dual space of ℓ^p is isometrically isomorphic to the space ℓ^q.

With an understanding of complete normed linear spaces, a proper investigation of linear operators follows. The spectrum of an operator was introduced in Sect. 5.6. While the concept itself seems simple enough, a full understanding of the spectrum of an operator requires a thorough knowledge of Banach algebra techniques that we will not delve into here. There are also other point sets associated with a given operator. The set of eigenvalues of an operator is called the point spectrum of the operator. Then, there is the approximate point spectrum, the compression spectrum, the residual spectrum, and the continuous spectrum; these sets are all subsets of the spectrum. There is also the numerical range of an operator. Identifying this set for a given operator seems to be harder than identifying the spectrum, but the numerical range of an operator has the interesting property that it is always a convex set.

The adjoint of an operator acting on a Hilbert space is another elementary principle. If $(X, \langle \cdot, \cdot \rangle)$ is a Hilbert space and $T : X \to X$ is a bounded linear operator, then the *adjoint* of T is the unique linear operator T^* which satisfies

$$\langle Tx, y \rangle = \langle x, T^*y \rangle$$

for every $x, y \in X$. For a matrix operator acting on $\mathbb{R}^n$ with the Euclidean inner product, it is easy to see that the matrix of the adjoint is the transpose of the original matrix. Upon establishing the existence of such an operator for a general Hilbert space operator one can show that T^* is bounded, provided T is bounded. Recalling the Volterra operator from Example 9.4.11, as an operator on $L^2([0, 1])$, its adjoint is given by

$$(V^* f)(x) = \int_x^1 f(t)\, dt.$$

An equally interesting example is the multiplication operator M_g from Exercise 9.4.16. Can you determine its adjoint?

Finally, as with functions on $\mathbb{R}$, we can change perspective and consider not only individual operators and their properties, but also the properties of a collective of operators. Let X and Y be normed linear spaces and let $T : X \to Y$ be bounded. Also, let $\mathscr{B}(X, Y)$ denote the set of all bounded linear operators between X and Y. This set forms a vector space with the pointwise operations of addition and scalar multiplication. For a norm, we set

$$\|T\| = \sup\{\|Tx\|_Y : x \in X \text{ with } \|x\|_X \le 1\}.$$

The fact that this quantity exists follows from the fact that the operator T is bounded, after which it is easy to check that the norm properties hold. There is an equivalent formulation for the norm which appears to depend more readily on the boundedness of the operator,

$$\|T\| = \inf\{M > 0 : \|Tx\|_Y \le M\|x\|_X \text{ for all } x \in X\}.$$

With this norm, the space $\mathscr{B}(X, Y)$ is not necessarily complete, but it is complete provided that Y is complete. Thus if X and Y are both Banach or Hilbert spaces, then $\mathscr{B}(X, Y)$ forms another Banach space. Of particular interest is the collection $\mathscr{B}(X)$ which is the set of all bounded linear operators from a space to itself.

One important point here is the distinction between showing that an operator is bounded and calculating the norm of an operator. For an example where computing the norm is extremely simple, take the forward shift on ℓ^2. For this operator we saw that

$$\|T(a_n)\|_2 = \|(a_n)\|_2$$

for every $(a_n) \in \ell^2$ from which it follows that $\|T\| = 1$. For an example where computing the norm is more involved, but the value is not unexpected, take the multiplication operator $M_g : L^2 \to L^2$. In Exercise 9.4.16 you showed that $\|M_g\| \le \|g\|_\infty$. It turns out that this is an equality and verifying this fact is a straightforward application of properties of the Lebesgue integral. For a more complicated example consider again the Volterra operator. We showed that

$$\|Vf\|_2 \le \frac{1}{\sqrt{2}}\|f\|_2$$

for every f in L^2 which implies that $V : L^2 \to L^2$ is bounded and that $\|V\| \le 1/\sqrt{2}$. Computing the exact norm of this operator (and most others) is more challenging though one can show that

$$\|V\| = \frac{2}{\pi},$$

which is not at all obvious from our previous calculation.

As with most of the topics we have discussed, it is imperative to understand how seemingly different concepts fit together. The three notions just discussed above engage with each other in a beautiful interplay which represents the fact that functional analysis lies at the intersection of several areas of mathematics. The connection between these topics involves several mathematical ideas: the algebraic properties of adjoint and spectrum, the analytic notion of completeness and norm, and the topological concept of convergence. To see how these ideas come together, we look again to the Volterra operator. To compute the norm, one uses the fact that $\|T^*T\| = \|T\|^2$ for any operator. This reduces the question to computing the norm of the operator V^*V. Properties specific to this operator allow one to conclude that the norm of V^*V is the absolute value of its largest eigenvalue, which follows as a result of the Spectral Theorem for Compact Self-adjoint Operators. This demonstrates two things. One, the adjoint and spectrum of an operator can provide information about the norm of an operator indicating the strong tie between these seemingly different ideas. Furthermore, these relationships can be exploited in the reverse direction which further demonstrates the interplay mentioned above.

Second and finally, while the Volterra operator is easily understood by the beginner (it is no more complicated in its definition than the multiplication operator or the forward shift mentioned above except for the fact that the integral is used, but still this is elementary), its features are more delicate. The machinery mentioned above to compute the norm of this operator is substantial as is often the case with seemingly simple things. In this particular instance, the difference in behavior as compared to the multiplication operator is due to the fact that integration affects a function in a much more intricate manner than the operation of multiplication.

In these closing remarks, it is my hope that this brief discussion has not overwhelmed you, but that it has provided you with an inviting glimpse into the wider world of abstract mathematics. Indeed, in these last few pages I have purposefully included a variety of concepts and statements with little to no explanation so as to entice you to investigate these topics. And, while it may *seem* overwhelming (and we have only just begun), do not be daunted. On the other hand, do not expect to understand everything instantaneously. These ideas are advanced in nature and often take years to sink in fully, even for the most capable mathematician. So, while the task set before you is not a simple one, keep in mind that nothing truly valuable is ever easily acquired.

References

1. Abbott, S.: Understanding Analysis. Springer, New York (2001)
2. Axler, S.: Linear Algebra Done Right. Springer, New York (1996)
3. Axler, S.: Down with determinants! Am. Math. Mon. **102**, 139–154 (1995)
4. Bartle, R.G.: Return to the Riemann integral. Am. Math. Mon. **103**, 625–632 (1996)
5. Beanland, K., Roberts, J.W., Stevenson, C.: Modifications of Thomae's function and differentiability. Am. Math. Mon. **116**(6), 531–535 (2009)
6. Beauzamy, B.: Introduction to Operator Theory and Invariant Subspaces. North-Holland, Amsterdam (1988)
7. Bressoud, D.: A Radical Approach to Lebesgue's Theory of Integration. MAA Textbooks. Cambridge University Press, Cambridge (2008)
8. Bressoud, D.: A Radical Approach to Real Analysis, 2nd edn. Mathematical Association of America, Washington (2007)
9. de Silva, N.: A concise, elementary proof of Arzelà's bounded convergence theorem. Am. Math. Mon. **117**(10), 918–920 (2010)
10. Darst, R.B.: Some Cantor sets and Cantor functions. Math. Mag. **45**, 2–7 (1972)
11. Drago, G., Lamberti, P.D., Toni, P.: A "bouquet" of discontinuous functions for beginners of mathematical analysis. Am. Math. Mon. **118**(9), 799–811 (2011)
12. Halmos, P.: A Hilbert Space Problem Book. Springer, New York (1982)
13. Harier, E., Wanner, G.: Analysis by Its History. Springer, New York (1996)
14. Kirkwood, J.R.: An Introduction to Analysis. Waveland, Prospect Heights (1995)
15. Knopp, K.: Theory and Application of Infinite Series. Blackie, London (1951)
16. Kubrusly, C.: Elements of Operator Theory. Birkhäuser, Boston (2001)
17. Lebesgue, H.: Measure and the Integral. Holden-Day, San Francisco (1966)
18. MacCluer, B.D.: Elementary Functional Analysis. Springer, New York (2009)
19. Martinez-Avendano, R., Rosenthal, P.: An Introduction to Operators on the Hardy-Hilbert Space. Springer, New York (2006)
20. Mihaila, I.: The rationals of the Cantor set. Coll. Math. J. **35**(4), 251–255 (2004)
21. Nadler, S.B., Jr.: A proof of Darboux's theorem. Am. Math. Mon. **117**(2), 174–175 (2010)
22. Niven, I.: Numbers: Rational and Irrational. New Mathematical Library, vol. 1. Random House, New York (1961)
23. Niven, I.: Irrational Numbers. The Carus Mathematical Monographs, vol. 11. Mathematical Association of America, New York (1956)
24. Olmsted, J.M.H.: Real Variables. Appleton-Century-Crofts, New York (1959)
25. Patty, C.W.: Foundations of Topology. Waveland, Prospect Heights (1997)
26. Pérez, D., Quintana, Y.: A survey on the Weierstrass approximation theorem (English, Spanish summary). Divulg. Mat. **16**(1), 231–247 (2008)
27. Radjavi, H., Rosenthal, P.: The invariant subspace problem. Math. Intell. **4**(1), 33–37 (1982)

M.A. Pons, *Real Analysis for the Undergraduate: With an Invitation to Functional Analysis*, 403
DOI 10.1007/978-1-4614-9638-0, © Springer Science+Business Media New York 2014

28. Rynne, B.P., Youngson, M.A.: Linear Functional Analysis. Springer, London (2008)
29. Saxe, K.: Beginning Functional Analysis. Springer, New York (2002)
30. Strichartz, R.S.: The Way of Analysis. Jones and Bartlett, Boston (1995)
31. Thompson, B.: Monotone convergence theorem for the Riemann integral. Am. Math. Mon. **117**(6), 547–550 (2010)
32. Tucker, T.: Rethinking rigor in calculus: the role of the mean value theorem. Am. Math. Mon. **104**(3), 231–240 (1997)

Index

M.A. Pons, *Real Analysis for the Undergraduate: With an Invitation to Functional Analysis*, 405
DOI 10.1007/978-1-4614-9638-0, © Springer Science+Business Media New York 2014

Printed in the United States
By Bookmasters